PHYLOGENETICS

Photograph courtesy of Dr. George W. Byers, University of Kansas

Willi Hennig

PHYLOGENETICS

The Theory and Practice
of Phylogenetic Systematics

E. O. WILEY

University of Kansas, Lawrence

A WILEY-INTERSCIENCE PUBLICATION

JOHN WILEY & SONS
New York • Chichester • Brisbane • Toronto • Singapore

Library of Congress Cataloging in Publication Data

Wiley, E. O.
   Phylogenetics: the theory and practice of phylogenetic
systematics.

   "A Wiley-Interscience publication."
   Bibliography: p.
   Includes index.
   1. Phylogeny. 2. Biology—Classification.
I. Title.
QH83.W52   574′.012   81-5080
ISBN 0-471-05975-7   AACR2

Printed in the United States of America

10 9 8 7 6 5

To Karen

Preface

This is a book about systematics and how the results of systematic research can be applied to studying the pattern and processes of evolution. The past twenty or so years have seen tremendous changes in biological systematics. Although some of these changes have occurred because of the discovery of previously unobservable characters, the most profound changes have taken place on the methodological and philosophical levels. Systematists have become more critical about the methods they employ and the biological and philosophical bases for these methods.

The first half of this century saw evolutionary theory march ahead of systematics, but in a rather curious manner. Evolutionary theorists became disinterested in the pattern of organic descent and concentrated on various processes purported to occur on the populational level of analysis. This resulted in the generally accepted theory known as the Synthetic Theory of Evolution, or neo-Darwinism. In itself the neo-Darwinian theory is an admirable accomplishment. However, it is not enough. What is needed now is a better understanding of the origin of species, and, as Waddington (1957) says, why there are tigers and elephants and other such things. To approach such an understanding we must first have something to understand. This something is a phylogenetic tree, a pattern of organismic descent.

Phylogenetic systematics, or simply "phylogenetics," is not just another approach to systematics. It is an approach to systematics designed to estimate the pattern of phylogenetic descent that is needed to deduce the processes of evolution concerned with the origin of species. The classifications that result from phylogenetic analysis are critical tools for evolutionary studies. Phylogenetics is also more than the handmaiden of evolution, however, for its underlying philosophy provides a way of viewing nature, asking questions and solving problems associated with the evolution of organisms.

I wrote this book to outline what I perceive as the philosophy and methodology of phylogenetics as a systematic discipline. As such, it is not restricted simply to the methods for reconstructing phylogenetic relationships and presenting these relationships in the form of a classification. Rather, it is also directed toward an understanding of the

evolution of species and the biological entities that comprise the history
of descent with modification. Further, the phylogeneticist must also be
a taxonomic scholar familiar with methods for dealing with specimens
and characters, ways of assessing taxonomic literature, and various rules
of nomenclature. These subjects are also dealt with.

Phylogenetic systematics is an approach to systematics that accomplishes an ordering of organic diversity in such a way that our ideas
concerning the inferred evolutionary relationships among organisms can
be scientifically discussed and evaluated. Much reaction has been directed toward this approach from its critics. I believe that most of this
reaction stems from a lack of understanding of phylogenetics. My major
purpose in writing this book was to clearly and simply present phylogenetic systematics (to the best of my ability) in the hope that others will
understand its goals and methods. Only through understanding can prof-
itable criticism and subsequent improvement follow.

E. O. WILEY

Lawrence, Kansas
May 1981

Acknowledgments

To my friends and colleagues at the University of Kansas, my thanks for their tolerance and many hours of argument and discussion. My thanks to the students in my course on phylogenetics for their critical attitudes and for demanding fuller explanations of phylogenetic concepts. Steven P. Churchill (University of Kansas) provided invaluable help with various botanical literature and was good enough to discuss various botanical studies with me. Those who read and provided valuable comments on various chapters include Donn Rosen (American Museum of Natural History), James S. Farris (State University of New York at Stony Brook), James Dale Smith (University of California at Fullerton), Colin Patterson and Peter Forey (British Museum, Natural History), Darrell Frost and Greg Glass (University of Kansas). Kåre Bremer (University of Stockholm) and Norman I. Platnick (American Museum of Natural History) were kind enough to review my analyses of their works. Four reviewers provided invaluable comments on specific chapters covering subjects of special expertise. Leslie Marcus and Nancy Neff (City University of New York and the American Museum of Natural History) painstakingly reviewed the sections on statistical analysis and multivariate techniques in Chapter 10. John Endler (University of Utah) reviewed the chapter on species (Chapter 2) and provided many valuable criticisms. Peter Stevens (Harvard University) reviewed Chapter 11 and provided many valuable criticisms, especially in the area of botanical literature and the botanical code. Don Buth (University of California at Los Angeles) reviewed parts of Chapter 10 on biochemical systematics. My special thanks to Bill and Sara Fink (Harvard University), colleagues who have spent many hours discussing phylogenetics over the past few years and who provided comments and criticisms on the book as a whole. My special thanks also to David Hull (University of Wisconsin-Milwaukee) for providing extensive and excellent reviews of most of the manuscript and whose efforts greatly improved the final draft. I would like to thank Donn Rosen and Gareth Nelson (American Museum of Natural History) for taking what must have seemed an infinite number of hours discussing Hennig's system to me during the tenure of my graduate training. Review, comment, and encouragement do not imply acceptance of my particular point of view and any mistakes in logic or fact remain my responsibility. I also thank Joan McCabe Moore for typing the text, Tiki Grantham for typing the bibliography, Jan Elder and Jane Rawlings for typing the index, and Sharon Hagen for reproducing previously published illustrations. Their skills are greatly appreciated. I would also like to thank several people at John Wiley & Sons for their help and support during the writing of this book. Mary M. Conway oversaw my efforts from start to finish and I greatly appreciate her skill and help. Theresa Danielson's skill in copyediting the manuscript resulted in many improvements in style and organization. I also thank Linda Dugan for seeing the book through the production phase. Finally, I thank Karen Wiley who provided extensive help in making the original drawings and indexing, and without whose support this project would have been impossible.

E. O. W.

Contents

Chapter 1

Introduction

In 1950 the German entomologist Willi Hennig published *Grundzüge einer Theorie der Phylogenetischen Systematik*. Hennig's book contained five basic ideas which began a major revolution in systematics:

1. The relationships lending to the cohesion of living and extinct organisms are genealogical ("blood") relationships.
2. Such relationships exist for individuals within populations, between populations, and between species.
3. All other types of relationship (i.e., phenotypic and genetic) are phenomena correlated with genealogical descent and thus are best understood within the context of genealogical descent with modification (quite literally "evolution").
4. The genealogical relationships among populations and species may be recovered (discovered) by searching for particular characters which document these relationships.
5. The best general classification of organisms is one that exactly reflects the genealogical relationships among these organisms.

Hennig's ideas were first discussed in an American journal by Kiriakoff (1959). Wide discussion of phylogenetic methods in English-language journals, however, came after the publication of Hennig's revised book (1966) and Brundin's (1966) work on chironomid midges. Earlier English-language applications of Hennig's methods include Koponen (1968: mosses) and Nelson (1969: fishes). More recent discussions and applications of Hennig's methods applied to plants include Bremer (1976a,b, 1978a,b), Bremer and Wanntorp (1978), 1979), and Humphries (1979). More philosophical discussions, including criticisms of other methods include Schlee (1968, 1971), Nelson (1972a, 1972b, 1974), Kavanaugh (1972), Cracraft (1974a), Wiley (1975, 1976), Farris (1977, 1980), and Mickevich (1978). An excellent short summary of Hennig's own views is Hennig (1965). A historical analysis of Hennig's ideas is given by

1

Dupuis (1978). Hennig (1950, 1965, 1966, 1969, 1975) had many ideas other than the five basic points listed above. Some of these ideas are still considered basic to the discipline (e.g., monophyly) while others have been discarded (e.g., hierarchial ranks determined by absolute age). However modified the discipline might seem today, these five basic ideas continue to provide the major theoretical bases for what I understand to be "phylogenetic systematics."

The major purposes of this book are to introduce phylogenetics as I perceive it and to survey general practices in taxonomy. I do not claim, nor should the reader infer, that all phylogeneticists agree with my perception of phylogenetics. The discipline is not monothetic (see Hull, 1980). In many respects the system described here differs considerably from some of Hennig's ideas. But in many respects what may seem novel can be found in one form or another in Hennig's writings. Nevertheless, phylogenetics has progressed considerably in the past 14 years. To give you an abstracted summary of the phylogenetic system today, I present the following list of statements:

1.   Organismic diversity has been produced by speciation and character modification. Speciation encompasses a number of modes of lineage splitting as well as hybridization and probably symbiosis. Character modification may accompany speciation, cause speciation, or proceed independently of speciation.

2.   The historical course of evolution is both a continuum and a discontinuum. On the species level, each and every species shares the continuum of historical descent through the genealogical links of their shared common ancestors. This is true even of the most distantly related species, given that life itself has a single source. Discontinuua at the level of both species and higher "natural" taxa are the products of speciation.

3.   A phylogenetic tree is a representation of the historical course of speciation. In the phylogenetic system this is true even for trees of supraspecific taxa because each is hypothesized to have originated as a single species. Therefore, a phylogenetic tree of species is both necessary and sufficient to follow the history of evolution on both the specific and supraspecific levels of biological organization.

4.   It is the business of phylogenetic systematists to attempt to recover this history of speciation and to order it into a system of words that makes this history comprehensible.

5.   The history of speciation may be recovered when speciation is coupled with character modification or when the rate of speciation does not proceed faster than the rate of character evolution. This process is known as reconstructing phylogenetic (genealogical) relationships. Such a reconstruction is considered a *hypothesis* subject to further rigorous testing.

6. Reconstructing phylogenetic relationships begins by posing hypotheses about the species dealt with. Consider a simple array of three species, *Fundulus nottii*, *F. escambiae*, and *F. lineolatus* (killifishes of the family Cyprinodontidae). Pose the hypothesis "*F. nottii* is more closely related to *F. escambiae* than to *F. lineolatus*; that is, *F. nottii* and *F. escambiae* share a common ancestral species that is not shared by *F. lineolatus*." The only characters that support this particular hypothesis are homologues shared by *F. nottii* and *F. escambiae* that can be inferred to have originated only in their common ancestor *after* the origin of *F. lineolatus*. Conversely, we might falsify this hypothesis with a shared homology between *F. lineolatus* and *F. escambiae* which can be inferred to have originated only in their common ancestor and not in the common ancestor of all three species. Such homologies are termed *synapomorphies* and the testing of homologues to determine which is synapomorphic is the essence of phylogenetic analyses.

7. Because natural supraspecific taxa arise as species and can be fully comprehended by a tree of their included species, it follows logically that the procedures used for relating species can also be used to relate natural supraspecific taxa. Such natural supraspecific taxa are termed *monophyletic groups* or *clades*.

8. There is a natural phylogenetic tree of species existing as a historical by-product of evolution. A classification of these species in a way that exactly reflects their history is also natural. While we might very well comprehend the historical course of species evolution by inspecting the tree, a system of names defining natural groups of species and their relationships eases communication about these species. The major purpose of a phylogenetic classification is to condense and summarize the inferred history of speciation as reflected by our best hypotheses of that history.

TOPICS COVERED

The remaining part of this chapter is concerned with definitions of basic terms, the relationship between phylogenetic systematics and other areas of comparative biology, and an introduction to the philosophy of science.

A major part of this book deals with the nature of species (Chapter 2) and higher taxa (i.e., supraspecific taxa; Chapter 3). My purpose in discussing species extensively is to develop a biologically relevant species concept that can be applied fruitfully to phylogenetics and to explore the implications of this concept to speciation theory, biogeography, and the practical aspects of species recognition. Supraspecific taxa are dealt with as the natural result of classifying the history of speciation. The emphasis in Chapter 3 is on the development of the concept of the "natural supraspecific taxon" and that such a taxon is monophyletic sensu Hennig (i.e., it is a taxon which includes an ancestor and all of that ancestor's descendants).

After developing concepts concerning the entities which are related in phylogenetic trees, I turn to a consideration of the trees themselves (Chapter 4). The emphasis in Chapter 4 is on the hierarchy of organismic relationships and the arrays of phylogenetic trees that we may infer from observing nature. It is important to understand how our concepts of evolutionary entities affect the patterns (trees) we may perceive in nature. Chapter 5 deals with characters and how characters may be used to test an array of competing phylogenetic trees for the same taxa. Of particular concern are: (1) the establishment of a concept of homology and (2) the methods that can be used to determine which of two homologues is in an apomorphic (derived) condition and which is in a plesiomorphic (primitive) condition. The chapter ends with three examples of fully developed phylogenetic analyses and a consideration of certain computer algorithms as tools for phylogenetic argumentation.

Chapter 6 concerns phylogenetic classifications. The various ways of constructing a phylogenetic classification are discussed. A system based on the Linnaean hierarchy is detailed. One of the major problems with phylogenetics has been classification. The problem has been both practical and theoretical. The practical problem has been the inability of phylogeneticists to convince some parts of the systematic community that rigorously phylogenetic classifications are useful constructs. The emphasis in Chapter 6 is placed on a system of conventions that results in simple and convenient phylogenetic classification and on demonstration of the utility of such classifications.

Chapter 7 contains critiques of two of the three alternate approaches to the phylogenetic system: evolutionary taxonomy and phenetics. I do not pretend that this book is a balanced account of either of these approaches, as one might expect in a text which purports to survey dispassionately the entire field of systematics. Thus, the student should not expect to learn how to do phenetics or evolutionary taxonomy. Rather, the chapter contains the reasons I and others reject these approaches.

Chapter 8 is an introduction to historical biogeography. Biogeography is not systematics. Rather, systematics furnishes one array of basic data for the biogeographer. I believe it appropriate to discuss the various approaches to historical biogeography and to outline briefly one approach that is closely linked to phylogenetic systematics, vicariance biogeography.

Chapters 9 and 10 are directed toward various practical matters in systematics. Chapter 9 begins with a consideration of specimens. Topics considered include specimen selection, field collecting, and curating. In Chapter 10, various classes of characters (e.g., morphological, biochemical, etc.) are discussed. I emphasize nonmorphological characters because many of the previous examples (especially in Chapter 5) have concerned morphology. Examples that show the phylogenetic application of these various classes of characters to the phylogenetic system are discussed. Chapter 10 ends with a consideration of quantitative character analysis as a tool for basic description and data summarization and as a tool for taxon discrimination and character interrelationships (various multivariate techniques). I assume a basic background in statistics.

Chapter 11 is directed toward systematic publication and the three major rules of nomenclature. Two of the more important aspects of systematics are a knowledge of previous literature and the clarity of a systematic publication. Additionally, the mark of a good taxonomic scholar is a familiarity with the rules of nomenclature that govern various aspects of the nomenclatural history of the taxon or group of interest. Because this book is about systematics in general, all three major codes (bacteriological, botanical, and zoological) are discussed and contrasted.

TERMS AND CONCEPTS

Before going further it is necessary to define explicitly or at least characterize some of the terms the reader will encounter throughout the book. Some of these terms will be familiar, but others will not. Some will be familiar but the manner in which I use them may seem novel. For example, my use of the word "homologue" is very restrictive and this must be understood from the start. I have organized the terms and concepts dealt with under the general topics to which they pertain. Certain synonyms are also noted.

Terms for Disciplines

1.   **General and comparative biology.**   Nelson (1970) has pointed out that biology can be divided into two basic areas defined by the research of biologists. General biology is concerned with investigating biological processes. The organisms utilized are those best suited for studying the particular process of interest. A cell physiologist interested in the energetics of mitochondria or the geneticist interested in sex-linked genes picks the organism best suited for studying the phenomenon. Comparative biology is concerned with the study of diversity and the explanation of that diversity. The ethologist as general biologist may be interested in studying a stimulus-response reaction of a particular kind. The ethologist as comparative biologist is interested in how common that stimulus-response reaction is among different species, how this phenomenon may have originated, and why it was retained in the organisms now displaying it. Systematics is, of course, one part of comparative biology.

The phylogeneticist is interested in estimating the pattern of organismic diversity from which may be inferred the historical course of evolution. Any and all comparative data are potentially useful. Thus, phylogenetics is the potential repository of all comparative information.

2. **Systematics.**   Systematics is the study of organismic diversity as that diversity is relevant to some specified kind of relationship thought to exist among populations, species, or higher taxa. This definition is somewhat narrower than the concepts of systematics discussed by such workers as Mayr (1969) and Nelson (1970) who equate systematics with what Nelson would term comparative biology. There is no doubt that all comparative data are potential systematic data. However, not all comparative biologists are systematics. One can, for example, compare facets of the ecology of several species without practicing what I would term "systematics."

3. **Taxonomy.**   Taxonomy comprises the theory and practice of describing the diversity of organisms and ordering this diversity into a system of words that conveys information concerning the kind of relationship between organisms that the investigator thinks is relevant. In other words, taxonomy comprises description and classification and the theory and history of classification.

4. **Phylogenetic systematics.**   One approach to systematics that (1) attempts to recover the phylogenetic (genealogical) relationships among groups of organisms and (2) produces classifications that exactly reflect those genealogical relationships. We shall see that this genealogical history at the level of species and higher taxa is a history of speciation. The theory and practice of phylogenetic systematics is herein termed **phylogenetics.** Those who practice phylogenetics are termed **phylogeneticists.** Mayr (1969) has coined the terms "cladistics" and "cladists." Without spending too much energy, I will simply say that I do not prefer "cladism" for two reasons. (1) It implies a preoccupation with branching pattern and a de-emphasis on characters. While it is true that phylogeneticists classify by inferred branching pattern, we are equally concerned with character evolution. Indeed, *character evolution permits the reconstruction of phylogenies.* (2) There seems to be a real controversy over who is and who is not a cladist (see Hull, 1980, and references therein).

Definitions of Groupings of Organisms

1. **Taxon.**   A grouping of organisms given a proper name, or, a grouping that could be given a proper name but is not named as a matter of convention.

2. **Natural taxon.**   As discussed in Chapter 3, a species or a group of species (= supraspecific taxon) that exists in nature as a result of a unique history of descent with modification (i.e., evolution). Angios-

permae is hypothesized a natural taxon because all of its member species share a unique common history not shared by any species that is not an angiosperm. Invertebrata is a taxon, but it is not a natural taxon because some of its members (echinoderms for example) share a unique history of descent with noninvertebrates (specifically with Vertebrata) that is not shared by other invertebrates.

3. **Species.** As discussed in Chapter 2, species-as-taxa are lineages that are independent of other lineages in the sense that they may evolve independently of other such lineages. Species comprise the highest level of taxonomic organization on which the processes of evolution may work.

4. **Monophyletic group.** A natural taxon composed of two or more species comprises a monophyletic group. Another term for a monophyletic group is a clade. (Species, insofar as they are natural, are *a priori* monophyletic. In other words, the terms "monophyletic," "paraphyletic," or "polyphyletic" do not really apply to natural species.)

5. **Paraphyletic and polyphyletic groups.** In the phylogenetic system paraphyletic and polyphyletic groups are nonnatural taxa. More formal definitions and the controversy surrounding these terms are discussed in Chapters 4 and 7.

6. **Sister group.** A sister group is a species or a higher monophyletic taxon that is hypothesized to be the closest genealogical relative of a given taxon exclusive of the ancestral species of both taxa. Thus, when we say that two taxa are sister groups, we mean that they are hypothesized to share an ancestral species not shared by any other taxon.

7. **Out-group.** An out-group is a species or higher monophyletic taxon that is examined in the course of a phylogenetic study to determine which of two homologous characters may be inferred to be apomorphic. One or several out-groups may be examined for each decision. The most critical out-group comparisons involve the sister group of the taxon studied. Out-group comparisons are discussed in Chapters 4 and 5.

Definitions Concerning Phylogenetic History and Evolution

1. **Genealogy and genealogical descent.** A genealogy is a graphic representation of the descent of one or more descendants from an ancestor. Family trees may be defined as genealogies on the level of individual organisms. Phylogenetic trees may be defined as genealogies on the level of populations, species, and higher taxa. Both kinds of trees may be *divergent* (as in the case of asexual clones and most metazoan speciation) or *reticulate* (as in genealogies involving individuals belonging to sexually reproducing species or as in speciation via hybridization). Because the relationships between (for example) two descendent species and their ancestor are genealogical and because a phylogenetic tree is composed of hypothetical or actual ancestral species, I shall term the relationships between such species **phylogenetic relationships.**

2. Cladogenesis. Cladogenesis is branching divergent evolution. The term is usually applied to phylogenetic patterns exhibited by species, but populations within a species may also show a branching pattern of interrelationships. When applied to species, cladogenesis may result from any one of an array of speciation mechanisms which result in two (or more) species where only one species existed before the speciation event. When applied to populations within a species, cladogenesis is the result of geographic subdivision (vicariance) leading to different local populations not thought to represent independent evolutionary lineages. Cladogenesis differs from reticulate speciation in that reticulate speciation results in the origin of one new species from two previously existing species whereas cladogenesis results in two or more species (of which at least one species is new) from a single previously existing species.

3. Anagenesis. "Anagenesis" is a synonym of "phyletic evolution" and these terms are used interchangeably in the text. Anagenesis/phyletic evolution is the process by which a genetic or phenotypic character changes within a species whether that change is random or nonrandom, slow or rapid.

4. Speciation. Speciation refers to an array of processes that result in the origin of one or more new species.

5. Speciation event. A speciation event is the historical result of speciation. Whereas "speciation" refers to an array of processes, "speciation event" refers to a particular and historically unique speciation which has proceeded via one of the array of processes that has resulted in the origin of individual species. No particular time frame is associated with the term. Thus, a speciation event may be instantaneous or protracted.

6. Vicariance event. A vicariance event is a geographic separation of a continuous biota such that the biota becomes two or more geographic subunits. A vicariance event may result in (a) speciation, (b) intraspecific geographic variation, or (c) no apparent impact on the subdivided populations. The first result leads to a phylogenetic tree of species. The second result leads to a phylogenetic tree of populations within species. The third results in a phylogenetic tree of populations within species but is complicated by the fact that there are no characters that can be applied to resolving the cladistic structure of the relationships obtaining between these populations.

Definitions Relating to the Attributes of Organisms

1. Character. A character is a feature (attribute, observable part) of an organism. In practical applications, a character is a part or attribute of an organism that may be described, figured, measured, weighed, counted, scored, or otherwise communicated by one biologist to other biologists. If one cannot convince one's colleagues that a feature or attribute exists, then those colleagues are apt to doubt the reality of the character (of course, these colleagues may be wrong). In this context, the word "character" is synonymous with what many workers term "character state." In this context (following Hennig, 1966), "red flower" is a character of species A. In the other context, "red" is a character state of the character "flower color."

2. Evolutionary novelty. An evolutionary novelty is an inherited change of a previously existing character to a new character. The novelty is the homologue of the original character (as discussed later). Systematists who reconstruct phylogenies are usually interested in novelties that are fixed (frequency = 100%) in natural groups (although polymorphic conditions may be analyzed with appropriate techniques). Most evolutionary novelties are parts of a polymorphic phase of character evolution within a species during their origin and fixation. The exceptions to this generalization are cases of instant speciation where the novelty arises concurrently with the species in which it occurs. Also, the polymorphism may be carried through several speciation events and thus can potentially be fixed in two or more lineages independently.

3. Homologues. The concept of homology and the various types of homologous characters are discussed in greater depth in Chapter 5. For purposes of introduction, we may say that characters found in taxa are homologous if one of the following conditions is met:

a. Two characters are homologues if one is derived directly from the other. In such a case we have an evolutionary novelty and its preexisting homologue. Such a **pair of homologues** is termed an **evolutionary transformation series**. The original, preexisting character is the **plesiomorphic** member of the pair, whereas the evolutionary novelty is the **apomorphic** member of the pair. For example, in Fig. 1.1a we have two characters, "square" and "triangle." If "square" and "triangle" form a homologous pair and "square" evolved into "triangle," then "square" is plesiomorphic relative to the apomorphic "triangle." The pair form an evolutionary transformation series of "square" → "triangle."

b. Three or more characters are homologues if each is part of a character transformation series in which one character gave rise to the next in linear sequence. Each step in such a transformation series is characterized by the replacement of one preexisting homologue for a more apomorphic homologue in one or more lineages. For example, in Fig. 1.1b the characters "circle," "square," and "triangle" are part of a transformation series in which "circle" gave rise to "square" and "square" gave rise to "triangle." All three characters are homologues. Note that all three characters are found in different parts of the phylogeny in different terminal taxa. Replacement in a transformation series does not necessarily mean that the replaced characters are lost.

c. One character found in two or more taxa is homologous in all of

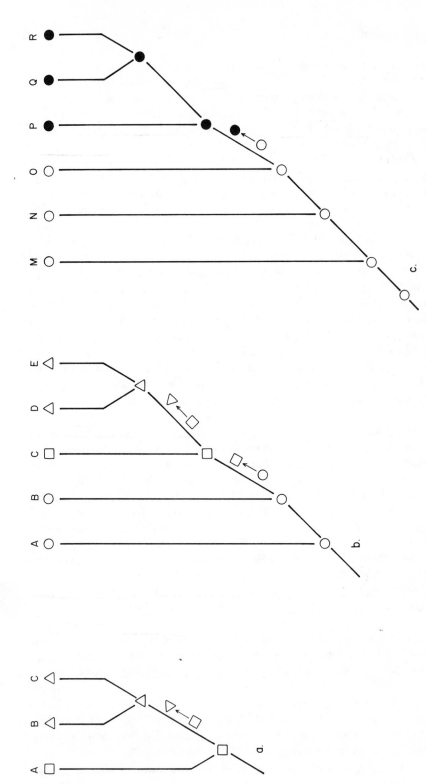

Figure 1.1 The concepts of homology, plesiomorphy, and apomorphy. Symbols are characters. Letters are taxa. Character evolution is indicated by arrows between characters. See text for discussion.

these taxa if the most recent common ancestor of these taxa also has (or can be inferred to have had) the character. Such a character is termed a **shared homologue**. If the shared homologue is found in (or inferred to occur in) the common ancestor of these taxa but is thought to have originated as an evolutionary novelty in an *earlier ancestor,* then the homologue is termed a **symplesiomorphy**. If the shared homologue is found in (or inferred to occur in) the common ancestor and if it is thought to have originated in that ancestor and not in an earlier ancestor, then the shared homologue is termed a **synapomorphy**. We may contrast these two types of homologues in reference to Fig. 1.1c. Taxa M, N, and O share the character "circle." This character is found in the common ancestor of all three taxa and thus can be taken as a homologue. "Circle," however, is not thought to have arisen as an evolutionary novelty in the common ancestor of M, N, and O, but rather, it is thought to have arisen in an earlier ancestor. Thus, "circle" is a symplesiomorphy of these three taxa. In contrast, taxa P, Q, and R share the character "dot." This character is thought to have arisen in the common ancestor of these three taxa and in no earlier ancestor. Thus, "dot" is a synapomorphy of P, Q, and R. Finally note that "circle" and "dot" form a transformation series, with "circle" being the plesiomorphic homologue and "dot" being the apomorphic homologue.

We shall see in Chapter 5 that the terms "apomorphic" and "plesiomorphic" are relative terms that contrast one character with its homologue(s) and that all characters of evolutionary significance arise as evolutionary novelties and as apomorphies but may become plesiomorphies because of subsequent character evolution. This may be illustrated briefly by reference to Fig 1.1b. The character "square" arose as an apomorphic homologue of the character "circle." But because "square" was transformed into "triangle" in one lineage, "square" is a plesiomorphic relative to "triangle."

4. **Homonomies and serial homologues.** The concept of homology developed above applies only to what many workers term "special homologues." *When I use the term "homologue" throughout the remainder of this book I refer only to homologue in the sense of "special homology."* The term "homonome" may be applied to those characters of two or more organisms which have the same development, are found on different parts of the organism, and whose developmental pathways have a single evolutionary origin (Riedl, 1978). As such, homonomy is a special case of special homology. In other words, homonomies may be accounted for by a single evolutionary event which produced their shared developmental pathway. For example, each individual mammalian hair is considered homologous with all other mammalian hairs. Yet no *particular* hair of a cat can be confidently said to be the special homologue of a *particular* hair of a civet. However, the presence of hair in cats and

civets is a character that describes both a structural and developmental condition thought to have originated as a unique evolutionary novelty in the common ancestor of cats, civets, and all other mammals. Thus although particular hairs cannot be termed "special homologues" the presence of hair itself is a special homologue shared by all mammals. Botanists consider leaves (for example) to be such homonomies and usually apply the term "serial homologue" to such structures. Zoologists restrict the term "serial homologues" to those generally homologous structures repeated in animals because of metamerism (segmentation).

5. **Homoplasies.** Homoplasies are characters that display structural (and thus ontogenetic) similarities but are thought to have originated independently of each other, either from two different preexisting characters or from a single preexisting character at two different times or in two different species. Put simply, homoplasy exists if the two taxa showing the character have a common ancestor that does not have the character. Three character phylogenies shown in Fig. 1.2 illustrate simple patterns of homoplasous evolution. In Fig. 1.2a the character "triangle" in taxa A and B is a homoplasous similarity because the common ancestor of A and B had the character "square." Note that both "triangles" are homologous with "square" but that the "triangles" are not homologous with each other. Figure 1.2a illustrates homoplasy as **parallel development**, that is, the independent development of similar characters from the same plesiomorphic character. Figure 1.2b illustrates a similar condition using the homoplasous similarity "circle." Note that "circle" is homologous for A and B and that it is homologous for F and G, but it is homoplasous for A + B and F + G. Figure 1.2c illustrates homoplasy as **convergence**, that is, as the development of similar characters from different preexisting characters.

6. **Analogues.** As discussed in Chapter 5, analogues are functionally similar but structurally and developmentally different characters.

7. **Holomorphology.** The holomorphology of an organism is the total spectrum of characters exhibited by that organism during its lifetime. The holomorphology of a species is the total spectrum of the holomorphology of the individuals comprising that species.

8. **Epiphenotype.** The epiphenotype of an organism is its morphology at any particular time it is inspected during its life. The term is largely synonymous with the term **phenotype** although the connotation of the prefix implies that the epiphenotype is the result of an array of genetic and ontogenetic phenomena.

Terms for Classification

1. **Classification.** A classification is a series of words used to present a particular arrangement of organisms according to some principle of relationship thought to exist among the organisms.

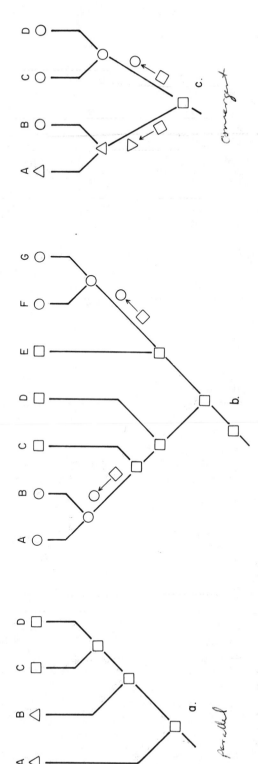

Figure 1.2 Concepts of homoplasy. Symbols are characters. Taxa are letters. Character evolution is indicated by arrows between characters.

13

2. Phylogenetic classification. A phylogenetic classification is a classification that presents the genealogical (historical, phylogenetic) relationship hypothesized to exist among a given array of organisms.

3. Category. A category is a class prefix name that denotes the relative subordination of a taxon in a Linnaean classification relative to other taxa. Assigning a categorical prefix name to a taxon has the effect of **ranking** that taxon relative to other taxa. In phylogenetic classifications that use the Linnaean convention, categorical ranks are a heuristic device for showing the phylogenetic relationships of natural taxa. And, within a monophyletic group, rank denotes relative time of origin. The following is an abbreviated list of categories used in this book:

    Kingdom
      Series
        Phylum
          Class
            Division
              Cohort
                Order
                  Family
                    Tribe
                      Genus
                        Species

This hierarchy is commonly expanded by using prefixes, such as "super-" and "sub-." This has the effect of expanding the number of categories. For example,

    Superclass
      Class
        Subclass
          Infraclass

As we shall see in Chapter 6, the Linnaean hierarchy is one of several different systems for translating a phylogenetic hypothesis into a phylogenetic classification.

Throughout this book the reader should remember that categories are classes and thus different from taxa. When I refer to "a family" I am referring to a taxon of familial rank *not* to "family" as a category.

Finally, botanists will immediately notice that "Division" is a quite different category than the botanical "Divisio," which is equivalent to the zoological "Phylum." I use Division at the rank shown earlier because it is extensively used at this level of the hierarchy by zoological phylogeneticists.

Systematics and Other Evolutionary Disciplines

Comparative biologists are interested in (1) discovering biological patterns and (2) elucidating the causal processes that produced these patterns. Different comparative biologists are, of course, interested in different patterns. An ecologist, for example, may be interested in patterns of predator-prey relationships or the ecological divergence that occurs during speciation. An evolutionary biologist (when working in the comparative mode) may be interested in genotypic changes during speciation. Whatever the interest, the comparative biologist can work most efficiently if he or she has a reference system to consult which will help to pick critical comparisons, that is, those species or populations that will provide the most critical data for the study.

Let us say that a worker is interested in studying genotypic changes which have occurred as a result of (or perhaps precipitated) speciation. At a minimum, the investigator should (1) pick a pair of species that are closest relatives (i.e., sister species), (2) know a minimum of one additional species that can reasonably be considered the closest relative to the two species compared, and (3) be reasonably certain that there has been no extinction of species which might be related to the three studied. The first and third pieces of information are necessary to insure that the investigator is studying a speciation event involving the two species actually compared rather than one of these species and an extinct relative. The second piece of information is necessary to establish an estimate of the ancestral genotype so that divergence can be measured. In other words, a simple accounting of gross genotypic differences will not yield the critical information needed to fully understand the phenomenon. Rather, the investigator should be interested in how much each of the species departed from the genotype of the ancestor if he or she is to approach some causal explanation or even a simple enumeration.

Systematics has the potential for furnishing the kind of information that other comparative biologists need for picking organisms for such comparisons. To produce such a reference system, systematists must (1) have a method for reconstructing the phylogenetic relationships among organisms and (2) produce classifications that reflect these relationships. Phylogenetic systematics is the systematic approach that accomplishes these tasks and thus it is in a unique position among the various disciplines comprising comparative biology. In other words, the phylogenetic system can furnish the baseline data from which others can begin their investigations.

PHILOSOPHY AND SYSTEMATICS

The philosophies of science are diverse and much is made by philosophers of rather minor differences between them. This is especially true of what is termed "logical empiricism." I do not propose to sort out these differences or to present one particular philosopher's ideas. I will outline briefly the major areas of philosophy and present two basic approaches to scientific reasoning. Neither section should be considered authoritative and those interested in further discussions should consult the writings of the philosophers themselves. A helpful survey of the whole field is Harré (1972).

Four Areas of Philosophy

Philosophy comprises four major fields. All enter the realm of science or affect scientists at one level or another.

1. Logic. Logic is the study of the principles of reasoning. It seeks to separate valid reasoning from invalid reasoning which might lead to faulty conclusions. As such, logic is an integral part of science. Harré (1972: 3) states:

> The written expression of scientific knowledge ideally takes the form of a reasoned and systematic exposition. Conclusions will be backed up by reasons. Hypotheses will be considered with respect to the balance of favorable and contrary evidence. Certain logical relationships hold between conclusions and the reasons for the conclusions. Other logical relationships hold between hypotheses and the reasons which call for their rejection, or modification. Relationships such as these are the stuff of logic. They [hypotheses and conclusions] must conform to the canons or principles of correct reasoning. Some considerations do support a conclusion; others do not.

2. Epistemology. Epistemology is the study of knowledge and knowledge acquisition. Genuine or valid knowledge must meet certain standards. Particular scientific methods may be applied to gather such knowledge. But all scientific methods have their limits beyond which their knowledge claims are not valid. In addition, what appeared to be genuine knowledge may be replaced by new knowledge, if that new knowledge is genuine and validly gained. There are several approaches to epistemology. **Phenomenalism** suggests that only observed phenomena can be considered genuine knowledge. Following this idea in its strictest sense, science would be reduced to the identification and classification of phenomena. **Fictionism** suggests that scientific hypotheses should be looked upon as fictions that give order and coherence to observations. In other words, theoretical statements do not really refer to or describe real things or processes but rather serve only to order observations (Harré, 1972). **Realism** suggests that certain theories and concepts actually exist in the real world. Harré (1972:99) lists three realist tenets: (1) some theoretical terms apply to hypothetical entities, (2) some hypothesized entities are candidates for existence, and (3) some candidates exist.

Each of these approaches to knowledge has been used in systematics in one form or another. Bridgman's concept of operationalism espoused by some pheneticists is a phenomenological approach. Nominalists who view species as convenient groupings made for ease of communication are fictionists. Systematists who view species as "real" (be they biological, evolutionary, or other types of species) are realists. I prefer the realism approach for systematics because it assumes that there are real patterns to be discovered in nature which can be used to study real processes.

3. **Metaphysics.** Metaphysics is the study of concepts and their relationships. To science in particular, metaphysics is concerned with the relationships between various concepts and scientific systems of knowledge acquisition. Some logical empiricists (for example, Popper, 1968a, b) suggest that metaphysical concepts lie outside the realm of scientific inquiry. Yet there is an interaction between the way we view the world (metaphysical), our collection of empirical data, and changes in our world view based on our data. In that certain metaphysical ideas limit the kinds of hypotheses we propose, there is at least a potential to overturn our metaphysical ideas by showing that these hypotheses are either too restrictive, too broad, or do not fit higher theories.

4. **Ethics.** Ethics is the study of values and moral evaluation. Logic, epistemology, and metaphysics directly participate in the larger aspect of science but ethics occupies a subtler position. Scientists are expected to display what is usually termed "professional behavior," and this behavior helps keep scientists directed toward scientific issues and not overburdening controversies with personalities and emotionalism. Being human, however, we frequently fail.

APPROACHES TO SCIENTIFIC REASONING

Every philosopher of science probably feels that his or her philosophy is sufficiently different to be termed a separate approach to scientific reasoning. In one sense each would be quite correct because each philosopher brings to his or her work a slightly different world view. Thus it is with some reservations that I discuss what I consider the two broad approaches, inductivism and logical empiricism, as if all inductivists share the same views or all logical empiricists agree on basic principles. The reader is encouraged to read widely. The selection of any two authors will dispel any thoughts of unanimity. An excellent start can be made by consulting Harré (1972).

Inductivism

Inductivism, improperly ascribed to Francis Bacon, is an extension of what has come to be known as "Mill's canons" (John Stuart Mill, 1879):

1. Canon of agreement. If two or more observations have a single circumstance in common, the circumstance is either the cause or the effect of the observation. This canon might also be termed the *canon of congruence*.

2. Canon of difference. If an instance in which a phenomenon under investigation occurs and another instance in which it does not occur have every circumstance save one in common, then that one circumstance is an integral part which is necessary for the phenomenon to occur.

Both canons are used in inductivism and logical empiricism (thus Mill was not an inductivist). In inductivism, however, these canons have been expanded to produce a complete philosophy of science. Harré (1972) lists its principles as follows.

1. Scientific knowledge grows by the accumulation of independent facts.

2. We may unequivocally infer general laws from particular facts.

3. The truth content of a general law may be judged from the number of favorable observations which conform to the law.

Inductivists recognize that there are an infinite number of hypotheses which might be derived from a particular set of perceived facts. To pick which of these is preferred, they rely on the principle of simplicity (parsimony) to pick the simplest hypothesis. (Logical empiricists also use simplicity.)

The problems with inductivism are many and I shall not dwell on them extensively. There are three common complaints: (1) Particular "facts" are actually perceived with all the bias of our previous experiences about the world. Thus, facts are dependent on our previous experience. (2) Science does not seem to grow with the accumulation of "facts" but with the forwarding of theories that cause "old facts" to be reinterpreted and that stimulate the collecting of observations that were not even thought necessary according to the replaced theories. (3) There does not seem to be a logically defensible link between observations of nature and the unequivocal construction of a theory. The very fact that some scientists are better at producing interesting theories suggests that there is a link between observation and theory making. But the fact that some scientists are terrible theory-makers seems to refute the idea that inductivism provides an adequate vehicle for proposing theories.

Logical Empiricism

Logical empiricists use what is known as the hypotheticodeductive method for scientific reasoning. My favorite philosophers utilizing this approach are Popper (1968a, b) and Hempel (1965, 1966). Each differs in emphasis of certain concepts, words utilized, and the inclusiveness of science. However, their similarities to my mind far outweigh their differences. The following are the basic tenets of logical empiricism:

1. Inductive principles fail to give adequate logical justification for proposing a scientific hypothesis. Further, it is too easy to dismiss contrary evidence if inductivism is employed.
2. Although such principles as Mill's canons may cause a hypothesis to be proposed, other and perhaps unanalyzable psychological processes may also result in interesting and informative hypotheses. Indeed, experience and cleverness may suffice. Thus, hypotheses that can be counted as scientific may come from induction, intuition, luck, or the experiences of the investigator.[1]
3. A scientific hypothesis must be structured in such a way that it has test implications; that is, it must predict something that may be actually or potentially observed. In many cases, the test implications follow from "covering laws" and ancillary principles under which the hypothesis is advanced.

HYPOTHESIS. *Fundulus nottii* is more closely related to *F. escambiae* than to *F. lineolatus*.

COVERING LAW. Evolution occurs such that species have certain phylogenetic relationships to each other.

ANCILLARY PRINCIPLE. Phylogenetic relationships may be reconstructed using characters of organisms.

Note that the hypothesis has a logical structure under the covering law and ancillary principle. In other words, if evolution has occurred and characters show this, then characters should help test the hypothesis. We would predict that *F. nottii* and *F. escambiae* would share one or more characters which demonstrates that they are more closely related to each other than either is to *F. lineolatus*.

4. The test to which a hypothesis is subjected may come from two basic sources. First, it may be examined internally to see if it is logically formulated and externally to see if it is in accordance with the covering

---

[1] The question, however, is if this hypothesis-making is a logical process. The answer is probably yes, but there does not seem to be a formal proof of the logic at this time. Nontrivial and informative hypotheses are probably produced by a combination of experience and logic.

law(s) and ancillary principle(s) under which it was proposed. Second, it may be examined in terms of its test implications to see if these implications are fulfilled in terms of data from observation or experiment.

5. Two or more hypotheses frequently compete against each other in explaining the same data. In such a case, the **principle of simplicity** (**parsimony**) is used to pick the hypothesis that explains the data in the most economical manner. Simplicity, economy, and parsimony all refer to the same general concept. However, the nature of this concept is not well understood and has been debated by philosophers without resolution for many years. Popper (1968a, b) views the simplest hypothesis as the most falsifiable hypothesis and ties simplicity with information content. Sober (1975) suggests that simplicity and information content are in some ways antagonistic—that we strive for a balance between parsimony and information content. Both agree that no really informative hypothesis can be limited by the data then available. (For a review of Sober's ideas and a contrast between Sober and Popper see Beatty and Fink, 1980.) For our purposes in phylogenetics the most parsimonious or simplest hypothesis is that with the fewest ad hoc statements that explains the full array of available data. And we shall prefer such a hypothesis over others that compete for the same data. This point is illustrated in Chapters 4 and 5.

I shall adopt the hypotheticodeductive approach throughout this book. I do not expect anyone to gain a thorough knowledge of it through the sketchy outline provided above. Rather, I recommend the following works: Hempel (1965, 1966); Popper (1968a, b); Harré (1972: an excellent introduction to all basic approaches). No matter what particular philosophical approach taken, I recommend the following points to be kept in mind:

1. *The philosophical bases on which a scientist operates should be clearly specified when those bases affect the conclusions drawn.* For example, a particular concept of "species" (Chapter 2) might affect certain basic conclusions concerning phylogenetic trees (Chapter 4).

2. *A scientist should adopt a philosophy that helps separate statements of observations, deduction, and speculation.* We must first train ourselves to understand the difference between hypothesis and speculation. We must then learn to write clearly enough so that the reader can tell what is deduced from observation and what is speculation.

3. *A scientist must be willing to submit ideas to criticism from other scientists.* Hypotheses must be articulated in such a form that they are testable or potentially testable. Statements and hypotheses not in such a form are not scientific. This does not mean that they are meaningless or illogical, only that they belong to certain areas of metaphysics.

4. *A scientist must take a critical view of the evidence that supports or rejects a hypothesis.* This applies to one's own work as well as the work of others.

Chapter 2

Species and Speciation

A basic task of the systematist is to estimate the diversity of species in the group or fauna investigated. The answer will never be known with certainty. How good the estimate is will depend on many factors. The most important is how closely the species concept applied by the investigator approaches species as they exist in nature. Other important factors include the experience of the investigator, the amount of previous work done on the group or fauna, the quality and quantity of the available material, and the biology of the organisms. Like many other problems in science, previous experience and the world view of the investigator will largely determine the estimate.

There are three major discussions in this chapter. First, I will discuss various concepts of species and expose my world view (or prejudice) on the species question. Second, I will discuss various modes of speciation. Particular attention will be paid to the assumptions of each mode and to the biogeographic implications of each mode. Third, I will present some recommendations for practical decisions concerning whether a particular population (or populations) should or should not be considered a distinct species.

SPECIES CONCEPTS

The word "species" carries a dual connotation in biological science. First, the species is a naturally occurring group of individual organisms that comprises a basic unit of evolution. Second, the species is a category within the Linnaean hierarchy governed by various rules of nomenclature. Certain evolutionary theorists and philosophers (see Ghiselin, 1966, 1969, 1974; Hull, 1976; Mayr, 1976a; Wiley, 1978) now tend to consider the taxon species as an individual, analogous to, but governed by different processes than, individual organisms (process = ontogeny/genetics) and demes (process = population genetics + ecology). But formal taxonomy, embodied in the various rules, recognizes the category species as a class among other classes in the hierarchy of rank. This basic difference has resulted in some problems in phylogenetic classification which I will discuss in Chapter 6. For now I will say that the concept of species as evolutionary entities takes precedence over species as taxonomic entities and that the formalism of taxonomy must be subservient to the demands of evolution.

Neither the formal rules of nomenclature nor the force of set theory nor other logical constructs outside the realm of biology demand that we have a particular concept of species. A systematist may produce descriptions, revisions, and faunal monographs by supposing that species are convenience classes, God-made and immutable, or the products of evolution. All the investigator must suppose is that there is an order in the world of organisms. Because we do observe order we suppose that this order was produced by a phenomenon or process which has worked on organisms to produce this particular kind of order. It is not necessary to understand the phenomena or process to recover the ordering. Linnaeus and Agassiz perceived the order of nature as confirmation of God's wisdom. Evolutionists perceive the same order as a historical by-product of genealogical descent with modification. If so, then why must we attempt to develop a particular species concept? Because systematics is part-and-parcel of an array of disciplines concerned with the study of evolution. Systematics is in the unique position of furnishing the raw data of the pattern of descent with modification. If that pattern is interpreted as a pattern of evolution, then it can be used to test and assess the significance of postulated evolutionary mechanisms. Thus, throughout this book the patterns of relationship among organisms are interpreted as phylogenies rather than dendrograms, character-state trees, or synapomorphy schemes.

The jump from observation to interpretation is a basic and indispensable part of all scientific endeavors. To jump from observation to interpretation we need what Hempel (1966) has termed "bridge principles." Two sets of bridge principles are needed in phylogenetic systematics, one set for entities and the other set for the characters exhibited by these entities. The bridge principle concerned with the entities is the species concept and will be dealt with here. (Bridge principles for characters exhibited by higher taxa will be dealt with in Chapter 3.) A species concept must provide a link that will result in a common basic language (or world view) for those interested in systematics on the one hand and those interested in such phenomena as speciation and species interaction on the other hand.

The Nature of Species

What is a species? Is it a unit of convenience, an imposition on nature, as Darwin (1859), Gilmour (1940) and Ehrlich and Holm (1963) maintain? If so, then speciation is also a man-made artifact (White,1978a). Or, is a species a real and discrete unit of nature as Huxley (1940), Meglitsch (1954), Simpson (1961), Mayr (1963 and elsewhere), Hennig (1966), Ghiselin (1966, 1969, 1974), Dobzhansky (1970), Grant (1971), Bonde (1975), Hull (1976), and White (1978a) maintain? I shall consider species to be natural units in the biosphere that can be objectively studied and that have certain characteristics. Thus, the *taxon concept* species is a universal (Wiley, 1978) and a class concept (Wiley, 1980) whereas a particular species is an individual (Ghiselin, 1966). I can agree with Mayr (1963: 29) that

> Whoever, like Darwin, denies that species are nonarbitrarily defined units of nature not only evades the issue, but fails to find and solve some of the most interesting problems of biology.

The development of a phylogenetically based species concept has a long and interesting history, too interesting to attempt a short synthesis in this book. Among those who have traced this history are Mayr (1957, 1963), Simpson (1961), Grant (1971). Much of Mayr's thought is contained in an anthology of his works (Mayr, 1976b). Slobodchikoff (1976) has collected other works. Grant's (1971) lamentation that species biology has received little attention is now being redressed by both philosophers and biologists. Of particular interest are the more recent works of White (1978a), Endler (1977), and Bush (1975a).

The Biological Species Concept

The biological species concept is perhaps the most popular of modern species concepts among vertebrate systematists. Its most persuasive proponent is Mayr (1957, 1963, 1969, 1976b) who defines **biological species** as "groups of interbreeding natural populations that are reproductively isolated from other such groups" (Mayr, 1969: 26). In other words, biological species are the largest basic reproductive units (the individual and the deme being smaller units). This concept is an ancient one derived from Cesalpino's observation that "like produces like" (Mayr, 1957; Grant, 1971).

Criticisms of this concept come from several sources. Operationalists criticize the concept for not providing a series of operations by which a biological species can be identified (Sokal and Crovello, 1970; Sokal, 1973; after the principles of Bridgman, 1945 and other works). Others have criticized the concept because it infers biological characteristics for species known only from phenetic evidence (Blackwelder, 1967; Sokal, 1962; Sneath and Sokal, 1973). In fact, this is a strength, not a weakness. The inference that a phenetic cluster is a biological species can be viewed as a prediction that stands or falls with increased data

(Hull, 1968, 1971). In spite of these criticisms, the biological species concept has enjoyed great popularity. No doubt, this is due to the fact that biologists perceive entities fitting the concept in nature. It would be surprising if the concept enjoyed this popularity without this perception having some truth.

Given the usefulness of the biological species concept, I have some reservations about its exclusive use, even for sexually reproducing species.

It is not clear that in all sexually reproducing species, reproductive ties (gene flow) between demes provide the major cohesive force. Although very little migration seems needed to prevent differentiation (Lewontin, 1974), at least some investigators suggest that migration between demes in some species may be limited to short distances (Erlich and Raven, 1969). Other authors, such as Eldredge and Gould (1972: 114), question the role of gene flow with regard to specific speciation models:

> The coherence of a species, therefore, is not maintained by interactions among its members (gene flow). It emerges, rather, as an historical consequence of a species' origin as a peripherally isolated population that acquired its own powerful homeostatic system.

Eldredge and Gould (1972) reject species as "ecological units of *interacting* individuals in nature" and accept that species are lineages bound by epigenetic homeostatic forces evolved from an ancestral population. I find this attitude to be a little extreme. I would prefer to think that various species show an array of interactions ranging from pure epigenetic homeostasis (asexual species) to an interaction between epigenetic homeostasis and reproductive ties between demes (the more usual Mayrian concept). However, this stance does help refocus certain attitudes toward the study of species. Mayr's (1970: 278) attitude that "the basic problem of speciation consists of explaining the origins of gaps between sympatric species" can be replaced by the attitude that to attempt to reach a better understanding of speciation we must first attempt to document historical patterns of genealogical descent and then attempt to examine those extrinsic and intrinsic mechanisms responsible for the speciation patterns observed.

The Evolutionary Species Concept

The evolutionary species is a lineage concept that avoids many of the problems of the biological species concept without denying that interbreeding among sexually reproducing individuals is an important component in species cohesion. It is compatible with a broader range of reproductive modes (Meglitsch, 1954; Simpson, 1961; Grant, 1971) and with all speciation models. Thus, it is the concept which provides the best bridge principle between evolutionary process and pattern of descent. The concept is defined thus:

*An **evolutionary species** is a single lineage of ancestor-descendant populations which maintains its identity from other such lineages and which has its own evolutionary tendencies and historical fate.* (Wiley, 1978; modified from Simpson, 1961)

By lineage I mean one or a series of demes that share a common history of descent not shared by other demes. Identity is a harder concept to define. Identity is a quality that an entity possesses which is a by-product of its origin and its ability to remain distinct from other entities. Concerning species, one aspect of estimating whether two forms have different identities is to examine various mechanisms that promote identity. The most important aspects for many species include recognition systems permitting individuals to discriminate between other members of their species and members of other species. Such recognition systems may operate on several levels. In sexually reproducing species such systems might include recognition because of phenotypic or behavioral character differences, differences in biochemical recognition systems, or any number of other things. In asexual species identity may be manifested only in phenotypic and genotypic similarity. In either sexual or asexual species identity may be manifested in distinctive ecological roles. Such qualities as distinctive phenotypic and genotypic differences are used by systematists to distinguish many species, but systematists should (and usually do) recognize that those qualities used by the species themselves may not be the qualities used by systematists. Systematists are aware of this fact and use such characters as those manifested in breeding systems and behavior to distinguish phenotypically similar but distinctive species. Such species are frequently given special names (cryptic species, agamospecies, semispecies, etc.) although their evolutionary status is the same as more distinctive species.

It is possible for species to lose their identities and thus become extinct. Reduction speciation (discussed later in this chapter) would consist of the loss of two species and the production of a third species. Species may be incorporated into other species. Ancestral species may become extinct during speciation events if they are subdivided in such a way that neither daughter species has the same fate and tendencies as the ancestral species.

A species may comprise a single deme, a series of demes bound by reproductive ties, or a locked-in epiphenotype of limited variation (i.e., by homeostasis or evolutionary stasis). Variation between demes may take the form of environmental fine tuning (manifested in geographic variation), genetic or epigenetic drift (manifested in intrademic variation which exceeds interdemic variation), or epigenetic factors correlated with environmental factors.

I have previously suggested that four corollaries can be derived from the definition (Wiley, 1978). Logical corollaries are statements that can be derived directly from a definition, hypothesis, or theorem.

Corollaries

Corollary 1. All organisms, past and present, belong to some evolutionary species. This corollary is derived from the observation that every organisms belongs to a lineage including at least its parents or parent (Vendler, 1975), and thus if species are lineages, all organisms must belong to one or another species. This makes the species a unique taxon. Whereas supraspecific taxa are collections of lineages, species are the lineages themselves (Wiley, 1978). These lineages are placed in higher groupings on the basis of hypothesized past linkages. Thus, natural higher taxa are purely historical constructs whose sole existence depends on how accurately they document the historical unfolding of lineage splitting (speciation). From this corollary we may deduce that (1) a phylogenetic tree is actually composed of evolutionary species, and (2) all terminal taxa are species, and linkages between terminal taxa are species. Therefore, all supraspecific taxa originate as species (Hennig, 1966; Bonde, 1975; Wiley, 1977a, 1978). There are two consequences of this line of reasoning. First, we must reject Løvtrup's (1974, 1979) antithesis to this corollary. He asserts that terminal taxa in the Linnaean hierarchy may be supraspecific taxa without also being species. The formalism of taxonomy must give way to the perceived realities of evolution.

Second, this concept of species coupled with the concept of higher taxa as collections of species imposes a certain view on the origin of higher taxa and thus directly affects taxonomic classification. Every organism belongs to some evolutionary species, but only the various rules of nomenclature (imposed by taxonomists on nature) demand that each species belong to the other "mandatory" categories. In other words, just because species A belongs to phylum X, it does not follow that it must belong to any other taxa within phylum X. For example, if species A was the ancestor of all other species in phylum X, then it can not belong to any subtaxon within X (Hennig, 1966). Only the various nomenclatural rules (if followed) would make it necessary to place species A in a genus, family, order, or other lower taxon. Such rules have no particular reality in nature and these redundant higher taxon names would be biologically meaningless. Taxonomy, that is, natural taxonomy, should proceed to group as nature has produced groupings. As many have asserted, evolution has proceeded from the apex to the base by lineage splitting (some proponents of this taxonomy are Rosa, 1918; Goldschmidt, 1952; Schindewolf, 1950; and Løvtrup, 1974.) This view led Hennig (1966) to assert that higher taxa arose with their stem species and not, as Simpson (1961) claimed, because of some subsequent evolution *within* these higher taxa. Hennig's assertion makes logical sense. No genus could have originated before the family in which it belongs has originated. That is, the stem species of the genus could not have originated before its own stem species. This is a logical deduction of the axiom of the continuity of genealogical descent and simply an affirmation that spontaneous generation does not occur. Thus, a family at the time of its origin was composed of a single evolutionary species (Hennig,1966;Wiley,1977a, 1978). Platnick's 1976b assertion that higher taxa only originate after the stem species split can not be supported because if we took this view of the origin of higher taxa the classification of ancestral species would be impossible, or it would multiply the number of necessary taxa by a factor of two.

Corollary 2. Species must be reproductively isolated from each other to the extent that this is required for maintaining their separate identities, tendencies and fates. The concept of interbreeding is a rather complex idea. The ability to interbreed depends on three factors: (1) the mode of reproduction, (2) developmental plasticity, and (3) geographic position. How one views the evidence of intergradation and hybridization depends heavily on how one views species and those three factors. For example, to Mayr (1942, 1970) speciation is the process of gaining reproductive isolation—no other criterion is applicable (of course, Mayr, 1970, notes that there are many ways to achieve this). Thus, many cases of secondary contact and hybridization would be considered *prima facie* evidence for incomplete speciation (Mayr, 1963: 502; 1970: 288).[1]

Botanists take a different view. Grant (1971: 41–42) suggests that the reproductive integrity depends heavily on reproductive mode:

> The unity and integrity of an evolutionary species depend on the likeness and relationship of its component individuals. . . . Existing plant species form a spectrum of breeding systems ranging from wide outcrossing at one extreme through various intermediate conditions to strict uniparental reproduction at the other extreme. It follows that the degree of integration reached in these species should vary also and, in a correlated manner, from a maximum in widely outcrossing species to a relatively loose condition in predominantly uniparental groups.

The more widely outbreeding a species, the greater the chance for hybridization between species. Botanists discuss hybridization in terms of reproductive modes rather than in terms of speciation, that is, hybridization as *prima facie* evidence for incomplete speciation. Grant

---

[1] This, of course, is not meant as a hard-and-fast rule, for Mayr, 1970, does discuss "old hybrid zones" between recognized species.

(1971: 39) following Simpson (1961) makes clear the attitude toward hybridization under the evolutionary species concept:

> Viewed from the standpoint of the evolutionary species concept, however, the important question is not whether two species hybridize, but whether two species do or do not lose their distinct ecological and evolutionary roles. If, despite some hybridization, they do not merge, then they remain separate species in the evolutionary perspective.

Davis and Heywood (1965: 446) have a similar view point:

> Despite the undoubted importance of intersterility in the evolution of species, crossability criteria must be assessed, for purposes of a general classification, like any other taxonomic character and not be unduly weighted.

One major problem with the exclusive use of the interbreeding community criterion is the tendency on the part of some workers to view speciation as incomplete until sympatry, and then if hybridization occurs to view this as incomplete speciation. This results in two problems with biological species. It makes one feel that he or she is committing heresy if he or she describes an allopatric or parapatric differentiated form as a species. After all, speciation is not "complete." The investigator then sagely refers these perfectly good evolutionary species to a single "polytypic" species composed of two or more subspecies. This invites confusion in two respects. First, it makes for underestimates of the number of independent evolutionary lineages present in a biota. Second, the "subspecies" as an evolutionary lineage will be confounded with the subspecies as a category of convenience—a variant population of an evolutionary species. This makes for confusion, especially for evolutionary theorists who are not sophisticated systematists. In addition, for those studying the genetics of species formation it is evident that most genetic differences develop in allopatry. *Drosophila* may not be the model on which to base all of species-level genetics, but it is the best we have. For the *Drosophila willistoni* group Ayala et al. (1974) present evidence that most genetic change develops in allopatry or parapatry and that such sister species differ as much in genotype as do sympatric sister species. Speciation, at least in this group, is completed in allopatry or parapatry and the eventual sympatry of two sister species may be irrelevant.

What, then, are we to think about zones of hybridization or intergradation (i.e., **contact zones** between differentiated populations or between species)? We must realize that the same populations may show both phenomena depending on the characters analyzed. Oligogenic and single locus characters (classic polymorphic characters) produce hybrid zones (i.e., those containing both parental types plus hybrid individuals).

Quantative characters produce zones of intergradation (i.e., zones where no parental forms are found, except perhaps at the extreme edges of the zones). We must also realize that such zones may be primary (developed *in situ*) or secondary (developed after a period of allopatry). We may separate primary and secondary contact only in the special circumstance that obtains when we have evidence that the two forms are more closely related to other species than to each other (i.e., they are not sister groups). In such a case, the zone is secondary and the fact that the two forms are interacting is of no relevence to a decision as to their species status. If, however, the two forms are sister groups, then we can not separate primary from secondary contact (as discussed by Endler, 1977). To assess the significance of the contact zone we would have to examine its characteristics. If the hybrid zone is narrow and there is evidence that it is old, then the forms are probably species because they are successfully maintaining their identities in spite of gene flow. The mode of speciation of such species may be allopatric or allo-parapatric (thus a secondary contact zone) or parapatric (thus a primary contact zone). In either case, the observation that such forms hybridize is no reason to reject their being evolutionary species. If the zone is wide, then the two forms are probably geographic variants of the same evolutionary species.

Although we might allow that two evolutionary species can hybridize without losing their identities, can we say the same thing for zones of intergradation? The answer is complex because intergradation may be actual or apparent. Actual intergradation results from gene flow between differentiated populations. Apparent zones of intergradation do not involve gene flow but are the result of residual geographic variation left from an ancestral species and either maintained by local selection or chance, or the zone is a transient phase which will disappear in the future. So far as I know, no one has explicitly recognized the possibility of apparent intergradation, although Grant (1975) alluded to it indirectly. The phenomenon should obtain between species that are reproductively isolated and yet show apparent intergradation. An example of the phenomenon may be seen in the water snakes *Nerodia sipedon* and *N. fasciata* (North American water snakes were formerly placed in *Natrix*.) These snakes inhabit rivers in the southern half of eastern North America. Studies on scutation, eye size, and color pattern in the eastern (Schwaner and Mount, 1976) and western (Blaney and Blaney, 1979) parts of their common range along the Gulf Coast resulted in data that were interpreted as evidence for intergradation. Electrophoretic studies point to a different conclusion. Schwaner et al. (1980) found that each species had unique alleles for certain loci (three in *N. fasciata*, one in *N. sipedon*). Specimens examined from the same study area as those examined by Blaney and Blaney (1979) showed no evidence of polymorphism at these loci. That is, all specimens were either fixed for the unique alleles of one or the other species and there were no allelic hybrids. This indicates that there is no interbreeding between the two forms and that the intermediacy shown for morphological characters in the two previous studies was apparent intergradation, probably representing residual geographic variation (perhaps reinforced by local selection) of the common ancestor of the two species. In the case of actual intergradation, we must recognize that although a zone of hybridization may be narrow, a zone of intergradation of the same populations may be very wide if the particular characters examined are not characters that reinforce species identity. Because identity has to do with many different aspects of species phenotype and genotype, two species can tolerate gene exchange for functional genes which do not effect that basic identity. In other words, if LDH1 is originally monotypic for species A and LDH2 is originally monotypic for species B, it makes little difference to either's evolutionary identity whether they introgress for these alleles so long as LDH is not one of the components that reinforce species identity. This may sound like pure heresy to some, but it explains much about identity. What happens if the two "species" swamp each other completely? Is this a case for (1) the observation that we originally had only one species or (2) the interpretation that we had a speciation event via fusion (i.e., a third species has evolved)? I do not know—one would have to look at the result and compare that result with the hypothesis of character distribution in the ancestor of the two introgressing species. If the introgressing forms were sister groups, then I would opt for (1) although I would not argue that my choice is critically testable. But, if the introgressing forms were not sister groups, then (2) would be the logical alternative.

This brings up the interesting point that the significance of both hybridization and introgression depends on phylogenetic position. The ability to breed successfully is, potentially, a plesiomorphic character. In swordtails (Teleostei, Poeciliidae) of the genus *Xiphophorus* this potentiality has become a reality. Rosen (1979) has studied both the phylogenetic relationships and interbreeding potential of four species of *Xiphophorus*. These species and their interrelationships are shown in Fig. 2.1 ("*X. sp.*" is an undescribed species). *Xiphophorus signum* is the most derived species of the group and is reproductively isolated from *X. helleri*, as shown by sperm competition experiments (a criterion well established for reproductive isolation among poeciliids). *Xiphophorus signum* is also reproductively isolated from *X. alvarezi*. But, *X. helleri* and *X. alvarezi* are completely interfertile and hybridize and introgress where their ranges contact in nature. We may conclude that *X. signum* has different and apomorphic features incorporated into its reproduction and development whereas *X. helleri* and *X. alvarezi* are similar and plesiomorphic for these features (at least to the point that development is not impaired). Their ability to interbreed is due to the retention of plesiomorphic features, not to the fact that they are closest relatives. One

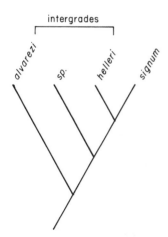

Figure 2.1 Phylogenetic relationships among four species of swordtails (*Xiphophorus*). Intergrades are found between *X. alvarezi* and *X. helleri* at specific geographic localities but *X. helleri* and *X. signum* are reproductively isolated. *Xiphophorus* sp. is an undescribed species. (Data from Rosen, 1979.)

should beware of the assumption that the ability to interbreed denotes a close phylogenetic relationship. In this case, that criterion would have led to a mistake in analysis.

Corollary 3. Evolutionary species may or may not exhibit recognizable phenetic differences, thus any investigator may overestimate or underestimate the actual number of existing independent lineages in a study. An over- or underestimate of the true number of species represents a source of error in any hypothesis about nature. These errors are similar to classic Type I and Type II errors (Wiley, 1978) and may stem from several sources including inadequate sampling or a lack of knowledge about the organisms. In the case of cryptic species, underestimates are inherent whenever morphology alone is used. It should be recognized that many such errors are largely beyond the control of the thoughtful and well trained investigator. We all work with what we have before us and mistakes regarding the number of species in a group do not necessarily denote bad systematic work. Indeed, the data available logically may lead an investigator to either type of error.

Inadequate data frequently lead to an overestimate of the number of species in a particular geographic area. For example, Coombs (1975) reinvestigated the species situation of chalicotheres (clawed perissodactyl mammals) from the Miocene Harrison Formation of western Nebraska. Previous investigators recognized two species of *Moropus*, one large species (*M. elatus*) and one small species (*M. petersoni*). Holland and Peterson (1914) discussed the possibility that these species were males and females of a single species but dismissed the idea because (among other things) immature specimens were known of both nominal species. However, with larger sample sizes Coombs (1975) demonstrated that samples from Agate Springs Quarry formed a continuous bimodal

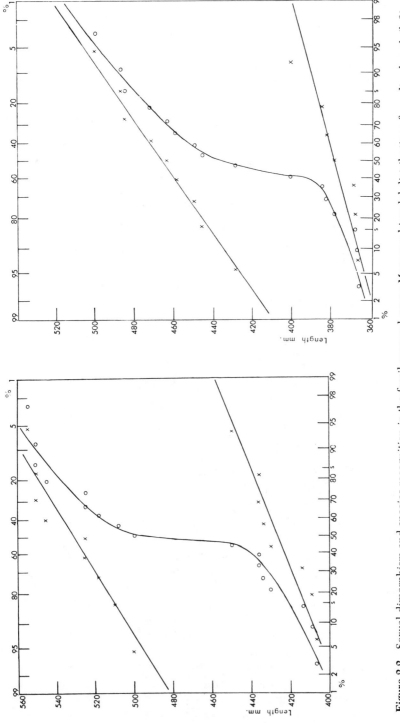

Figure 2.2 Sexual dimorphism and species recognition in the fossil mammal genus *Moropus*: bimodel distributions for radius length (left) and tibia length (right). Measurements are plotted on probability paper and bimodality is indicated by the sigmoid (rather than straight) curve fitted for the total population (shown by circles). Lines fitted to individuals identified by sex (indicated by the "x" plots) show male and female samples fitting a straight line, as shown above and below the sigmoid curve. (From Coombs, M. C., 1975. Used with the author's permission.)

distribution with about equal numbers of individuals comprising each mode (see Fig. 2.2). Thus, Coombs suggested that the simplest explanation was that there existed a single species of sexually dimorphic *Moropus*.

Another interesting example is the crested duckbill dinosaurs (hadrosaurs) of the Oldman Formation of Canada (Dodson, 1975). Previous workers recognized three genera and 12 species. Dodson collected data on the largest available sample ($n = 36$) of the 12 nominal species and analyzed a number of cranial characters. Using bivariate plots and comparing suspected allometry to a living analogue (crested cassowaries), Dodson suggested that only two genera were present. The first, *Corythosaurus*, was monotypic. The second, *Lambeosaurus*, contained only two species. The third previously recognized genus, *Procheneosaurus*, was found to contain juveniles of the other two recognized genera Variation in crest morphology, on which the 13 described species were largely based, was found to be related to growth and sexual dimorphism (see Fig. 2.3).

Underestimates are probably as common as overestimates. What were once thought of as relatively homogenous single species are frequently found to be composed of several distinct lineages when analyzed in

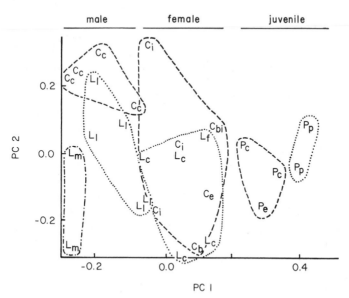

Figure 2.3 Sexual dimorphism and species recognition in lambeosaurine hadrosaurs: principle components analysis of the twelve nominal species interpreted as three sexually dimorphic species. Cb, Cc, Cbi, Ci, Pc, and Pe refer to *Corythosaurus casuarius* (circumscribed by dashes); Lf, Lc, Ll, and Pp refer to *Lambeosaurus lambei* (circumscribed by dots); Lm refers to *Lameosaurus maginicristatus* (circumscribed·by dots and dashes). (From Dodson, 1975. Used with the author's permission.)

detail. Such species are frequently termed "sibling species" (for reviews and examples see Mayr, 1963; Dobzhansky, 1970; White, 1978a;and Dobzhansky et al., 1977). For example, it was long thought that there was a single species of Western Grebe (Mayr, 1963; Mayr and Short, 1970) which was polymorphic for color phase (with dark and light morphs). Ratti (1979) and others have shown that the two "morphs" exhibit high assortative mating and are good species by either the biological or evolutionary species concept. Many polytypic species composed of discrete and easily diagnosable subspecies are now being broken into species.

Corollary 4.   No presumed separate, single, evolutionary lineage may be subdivided into a series of ancestral and descendant species. This corollary is self-evident and would hardly be worth discussing if it were not for the rather common practice among some paleontologists of subdividing a single lineage into a number of species (for example, Gingerich, 1979). Such subdivision is usually done to give recognition to a presumed series of ancestor-descendant populations undergoing anagenesis. These species are variously termed "paleospecies" or "successive species" (Simpson, 1961) or "chronospecies" (Mayr, 1942). They are overtly arbitrary (Simpson, 1961: 165) and delimited by rules of thumb. As such, they are unsuitable for critical evolutionary studies since one can never know from classifications containing successional species and true species which part of character evolution is correlated with cladogenesis and which is correlated with anagenesis. But, Simpson (1961) saw this arbitrary subdivision as necessary to escape an infinite regression in classification. Wiley (1978) pointed out that no such regression occurs because of the nature of the hierarchy and that arbitrary subdivision was not necessary. This renders the concept of successional species unnecessary.

Evolutionary Species as Ancestors

The evolutionary species concept provides the theoretical basis for interpreting a branching diagram as a phylogenetic tree. All terminal lineages are evolutionary species or descendant higher taxa represented by their ancestral evolutionary species. These terminal lineages are connected by common ancestral evolutionary species (lineages connecting the terminal lineages). Each branch, therefore, is the result of a speciation event. The concept does not preclude a particular ancestral species from surviving a speciation event. In cases where the mode of speciation is ecological—sympatric, via hybridization—or involves a small isolate, nonextinction of the larger population might be expected since the removal of these localized populations or the "wasting" of gametes might have little or no effect on the larger population. We can see that neither allopatric, parapatric, nor sympatric speciation nor speciation via hybridization necessarily renders the ancestral species extinct (Wiley, 1978). In the case of an allopatric model involving a small isolate we might question how large the isolate must become before the original ancestral species loses a sufficient chunk of its overall variation to be affected itself. I do not know, nor do I think an answer will be forthcoming until adequate criteria for recognizing ancestral species are formulated and applied to enough groups so that some statistical estimate can be made. Bell (1979) suggests that the survival of ancestral species might be quite common in some cases. I see no reason to doubt his assertion and I think his model (the *Gasterosteus aculeatus* complex) and additional examples (Gould, 1969) may very well be justifiable. Given this, the critical question is whether the evolutionary species concept, with its possibility of ancestral species survival, is compatible with Hennig's phylogenetics.

Hennig (1966) considered the extinction of ancestral species when they speciate to be a reasonable methodological rule given his perceptions of the inadequacy of paleontological data and the problems of classifying entities that he could see no possibility of adequately identifying. There are certain advantages to this view. For example, if taken as an axiom, a fully ranked hierarchy could be employed in which each clade was founded on a single *exclusive* ancestral species. Thus family A and genus X within family A would not have the same ancestral species. This methodological principle has been confused for a biological principle (see Hull, 1980, for discussion). Some opponents criticize and reject the phylogenetic system because they see this "biological principle" as incorrect and yet essential for the system. Adherents answer with more critical reasons why ancestral species can not be identified (Englemann and Wiley, 1977; Platnick, 1977b). Actually, Hennig's methodological principle is not essential to phylogenetic systematics regardless of the practicalities of ancestor identification (Wiley, 1979c; Hull, 1980). Ancestral species can be classified satisfactorily whether they become extinct or survive speciation events (as we shall see in Chapter 6). Thus, the evolutionary species concept is completely compatible with phylogenetic systematics and Hennig's methodological principle must be set aside.

Asexual and Unisexual Evolutionary Species

An **asexual species** is a species whose dominant or exclusive reproductive mode does not include mating between individuals (i.e., the exchange of parts of the genome). In other words, asexual species do not reproduce parasexually or sexually to the extent that these reproductive modes effect the spread of evolutionary novelties to the entire lineage. Asexual species are not facultative asexual propagators which go through cycles of sexual and asexual reproduction in response to a natural breeding cycle (such as many coelenterates, insects, and rotifers) or in response to environmental conditions. On the other hand, whether a particular protist species is an asexual or parasexual species is a question only an expert can answer. The important point for systematics is that asexual species do exist, are amenable to being characterized as evolutionary species and may be dealt with as real biological entities in reconstructing phylogenies and erecting classifications.

Meglitsch (1954) developed the concept of evolutionary species explicitly to include asexual lineages within the concept. [Simpson's (1951) first definition applied only to bisexually reproducing species—see Simpson, 1961, p. 162, footnote 11.] In contrast, the biological species concept, a special case of the evolutionary species concept, has always applied exclusively to bisexual species (Mayr, 1963). Those who hold to this concept as the *only* species concept view asexual species as pseudospecies (Dobzhansky, 1970; also see comments by Grant, 1971: 40). Yet, while some potential processes affecting asexual and sexual lineages are different, both types of reproductive modes result in lineages of organisms and thus both are amenable to the same higher species concept (Simpson, 1961).

Grant (1971) provides a good theoretical discussion of the similarities and differences between sexual and asexual evolutionary species. The cohesion of an asexual species depends upon likeness and the relationships among the component individual organisms. Sexual species have both these characteristics but also have the factor of interbreeding. Sexual species display two cohesive bonds, the vertical bond between parents and offspring and horizontal bonds between mates. Because of this, sexual species forming a series of local demes have the ability to maintain their species integrity in the face of the origin of evolutionary novelties given that mating bonds are not disrupted by these novelties and that demes show a certain amount of between-deme mating (what population geneticists term "gene flow"). Asexual species lack the mating bond and therefore likeness is dependent upon evolutionary stasis (or, perhaps more rarely, on parallel selective responses to common extrinsic factors).

I have called attention to the analogy between asexual species and completely allopatric demes of sexually reproducing species (Wiley, 1978). A sexually reproducing series of totally separate demes is like a series of clone lineages—neither the demes nor the clones have mating bonds. The similarity exhibited between two completely separate demes thus rests upon the same factors that give cohesion to an asexual species, likeness produced by a stabilized epiphenotype and parental links (albeit populational rather than individual parental links). Put simply, we group allopatric demes together in a single species because intrademic variation equals or exceeds interdemic variation and thus the evolution-

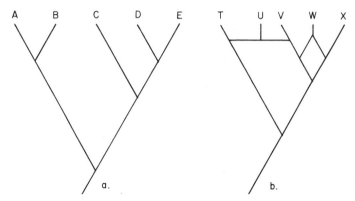

Figure 2.4 Divergent (*a*) and divergent plus reticulate (*b*) patterns of phylogenetic descent.

ary tendencies of the demes are hypothesized to be identical. The same can be said (on theoretical grounds) for the clones of an asexual species.

Phylogeny and Varying Modes of Reproduction

Phylogenetic descent is a relatively straightforward proposition within clades whose species reproduce by a single strategy. Complex patterns of descent occur when reproduction modes within a clade run the gamut from sexual to asexual modes overlain by simple to complex patterns of species origin by hybridization. In such situations, the resulting pattern of descent is reticulate (Fig. 2.4*b*). Such reticulate patterns are relatively rare in animals (1 in 1000 by White's, 1978, estimate) and more common in plants (Grant, 1971). (Both authors give excellent examples and discussions of the phenomena involved in producing these patterns.) Such complicated patterns cause no concern in the activity of classifying if the annotated Linnaean system is used (see Chapter 6). However, the nature of the evidence used to support hypotheses of such patterns is rather different from that encountered in the usual analysis of clades forming simple (nonreticulate) phylogenetic trees.

SPECIATION—MODES AND PATTERNS

Speciation is a general term for a number of different processes which involve the production of new evolutionary lineages (species). Except for the theoretically possible mode of reduction speciation (discussed later) and speciation by hybridization, speciation events couple lineage splitting with differentiation and this coupling results in two or more species where only one species previously existed. While all valid special species concepts may be gathered under the larger concept of evolutionary species, there are several ways that such species can gain their evolutionary independence. Thus, it is not possible to relegate all special modes of speciation to a single higher concept.

In this section I shall discuss the various modes of speciation and the patterns in space and time that they form. It is important to distinguish between these general *patterns* of speciation and the various genetic and epigenetic *mechanisms* peculiar to any one ancestral species' ability to differentiate. These mechanisms are the unique factors in the equation of speciation which differ from species to species and from clade to clade within a single geographic area. I term these the **intrinsic factors**. The other set of factors are the general modes of speciation which are mediated by **extrinsic factors**. Given the intrinsic ability to speciate (speciation potential), relatively unrelated species occupying the same geographical area will respond to the same extrinsic event (regardless of their intrinsic factors). That this is so is evidenced by the fact that many groups show similar biogeographic patterns regardless of the fact that they may use very different intrinsic genetic or epigenetic mechanisms to initiate and complete speciation.

The job of the systematist who wishes to contribute to speciation theory is to sort out those groups that demand particular explanations for their speciation patterns from those groups that demand a general explanation of their speciation patterns. Only when this is done can a scenario of particular mechanisms be invoked to explain speciation. Given a similar speciation pattern for groups with very different genetic systems, no such particular explanation is required because the causality must lie with the factors held in common between the groups. These factors are largely geographic—extrinsic events over time and space.

With this in mind, I shall discuss several modes of speciation and their affects on phylogeny and biogeography. Before doing this, I will examine an invalid speciation mode, "phyletic speciation."

Speciation vs. "Phyletic Speciation"

Many evolutionists view speciation as a dual phenomenon, the "dual dimensions" of speciation (Ross, 1974: 58). The first of these dimensions is lineage splitting. The second is said to be sequential production of species within a single evolutionary lineage. This has been variously termed "transformation of species in time" (Romanes, 1897), "phyletic evolution" (Simpson, 1961), and "phyletic speciation" (Mayr, 1963). Mayr (1963: 424–425) provides a hypothetical example:

> An isolated population on an island, for instance, might change in the course of time from species *a* through *b* and *c* into species *d* without ever splitting. (see Fig. 2.5)

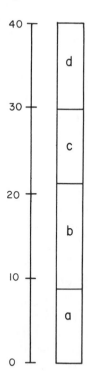

Figure 2.5   Mayr's example of allochronic speciation. A time axis in thousands of years is shown at the left. Letters are species. (Modified from Mayr, 1949.)

The new species produced in this fashion are variously termed "**successional**" species or "**paleospecies**" (Simpson, 1961) or "**allochronic**" species (Mayr, 1942). The extinctions caused by this speciation are usually termed "taxonomic extinctions."

The evolutionary species concept is, of course, diametrically opposite to the notion that phyletic evolution (in the sense of Lewontin, 1974) can produce new species. In other words, although I do not deny that phyletic evolution proceeds in lineages, I submit that such a process does not produce the wholesale transformation of the demes of one species into another species. My attitude is contrary to the general practices of paleontologists of subdividing hypothesized continuous lineages. Rather than simply be dogmatic and base my rejection on my species concept, I shall outline the reasons for my rejection in terms of generally accepted evolutionary theory and common sense. My reasons for the rejection of phyletic speciation (although not rejecting phyletic character transformation which would seem a common phenomenon) are taken largely from Wiley (1978, 1979b) and are listed:

1.   *The recognition of phyletic species is an arbitrary practice.*

That recognition of phyletic species is an arbitrary practice has been admitted by Simpson (1961: 165):

Certainly the lineage must be chopped into segments for purposes of classification, and this must be done arbitrarily ... because there is no non-
arbitrary way to subdivide a continuous lineage.

Mayr (1942: 153) substantiates this:

The delimitation of species which do not belong to the same time level
(allochronic species) is difficult. In fact, it would be completely impossible
if the fossil record were complete.

Mayr (1942: 153–154) goes on to state:

In the few cases, in which an almost complete record of a continuous line
is readily available, the paleontologist follows the reasoning of the taxonomist who is confronted with an unbroken intergrading series of geographic populations. He breaks them up for convenience.

This statement, of course, was made many years before Mayr's (1963)
book on species and I do not use it to attempt to show Mayr's present
attitude but to show what prior justifications were made to subdivide
continuous lineages and that the practice is admittedly arbitrary. From
Mayr's (1942) statement, which I fully agree with, we can see that the
ability to recognize allochronic species is a function of the amount of
data the investigator does not have. In other words, the less the data the
easier the decision. There is something terribly wrong with the proposition that the more data we have the more arbitrary we must become
in our decisions.

   2. *Arbitrary species result in arbitrary speciation mechanisms.* I
take the view that speciation is an array of natural mechanisms and not
a convenience existing in the minds of systematists. I quite agree with
White (1978a: 2) when he states:

Discussions of the nature of speciation assume that species have an objective existence. If species cannot be objectively defined, and are merely
artificial constructs or subjective figments of the imagination of taxonomists, then speciation can hardly be said to be a real process.

   In a phyletically evolving lineage where characters are being replaced
by their more apomorphic homologues, exactly where to draw the line
between two species is entirely a matter of opinion. How many of these
species to recognize is also a matter of opinion. Thus the number of
speciation events and the timing of these events is not a subject of
scientific inquiry but a matter of personal taste. From the standpoint of
the logic of science, such a practice is not acceptable. But more important
is the distortion such a practice produces in the study of the evolutionary

process itself. The estimations of the number of lineages and species important for gauging the mode and tempo of evolution can be vastly increased by the recognition of allochronic species. If another investigator does not have a clear idea of which species end as taxonomic extinctions then *real* extinction rates are obscured. If the same investigator cannot sort out allochronic species from evolutionary species, then the rate of lineage splitting cannot be determined and, furthermore, speciation events cannot be correlated with geographic events. The second point is vitally important. How can we determine historical biogeography and deduce general principles of earth history when rugose corals are arbitrarily divided into species in a way different than, for example, trilobites? In accepting phyletic speciation we must give up any hope of reaching a general synthesis of the evolution of biotas because speciation in one group can not be compared to speciation in another group.

3. *Phyletic speciation has never been satisfactorily demonstrated.* This point has been discussed extensively by Eldredge and Gould (1972). Put simply, the best cases for phyletic transformation of one species into another are refuted by a closer look at the data (For examples, see ibid. and Gould and Eldredge, 1977).

In summary, the phyletic transformation of one species into another species is contrary to the corollaries inherent in both the evolutionary and biological species concepts. This slow and steady transformation is what Darwin conceived as the major speciation mechanism. But Darwin was a nominalist who viewed species as arbitrary entities existing only in the minds of men. The modern view is that species are not arbitrary units, yet many still attempt to make them so. The problem is this: what Simpson (1961) saw as necessary for classification has somehow been transformed into an evolutionary process—that is, from the originally perceived necessary taxonomic convention we have a so-called evolutionary process of phyletic speciation.

Reductive Speciation

Given that phyletic speciation is invalid (while not denying phyletic evolution), we may now examine what may be termed "true" modes of speciation, or "speciation in its restricted modern sense" (Mayr, 1970: 250). Mayr (1963) suggests two major modes of speciation, reduction in the number of species (reductive speciation) and multiplication of the number of species (additive speciation). **Reductive speciation**, a theoretical possibility involves the complete fusion of two previously independent evolutionary species and is a hybridization/intergradation phenomenon. The complete fusion of two species over their entire range may be a rare phenomenon—I can think of no examples where such swamping resulted in a new species. (There may, of course, be examples of swamping of one species by another, but this usually results in one of the species incorporating the genes of the other, not in the origin of a third species.) The supposed examples of reductive speciation cited by Mayr (1963) involve only the reduction of the number of species at a particular locality or restricted geographic region (for example, the *Drosophila americana-texana-novamexicana* complex; Patterson and Stone, 1952). That hybridization does produce new species in plants is a well established observation (see Grant, 1971, for summary), but whether two evolutionary species have ever actually fused to produce a third species and themselves become extinct will probably remain problematical.

Additive Speciation

Additive speciation, as the name implies, is an umbrella term incorporating any mode of speciation which adds to the diversity of living organisms. With the theoretical exception of reductive speciation, all speciation is additive. A phylogenetic tree is a graphic representation of additive speciation and the pattern of branching and reticulation coupled with biogeographic distributions may be used in certain circumstances to discriminate between various modes of additive speciation. The following is a list of these modes (The organization of these modes largely follows Bush, 1975a and Endler, 1977; both authors provide excellent summaries of these phenomena).

1. **Allopatric speciation.** Lineage independence is achieved while two or more lineages are geographically disjunct (separated, vicariant).
2. **Allo-parapatric speciation.** Lineage independence is achieved while two or more lineages are geographically disjunct but their differentiation is enhanced after a period of parapatry.
3. **Parapatric speciation.** Lineage independence is achieved between geographically distinct lineages which maintain limited interlineage mating across a contact zone.
4. **Stasipatric speciation.** Lineage independence is achieved by major chromosomal rearrangements which give rise to postmating isolating mechanisms (White, 1978ab) while the population is in relative disjunction.
5. **Sympatric speciation.** Lineage independence is achieved without geographic separation by shifts in ecology, host, or timing of reproduction, or by hybridization or apomixia.

These five modes have been much discussed in the literature and I shall not attempt to duplicate others' efforts to characterize them fully (a sample of the literature is found in each discussion). In particular, I am not concerned with the mechanisms by which independence of evolutionary lineages is achieved (i.e., mechanisms that inhibit interbreeding between two evolutionary species or that permit discrimination between asexual lineages). The study of such mechanisms belongs to the larger field of evolutionary theory—the subject of interest here is recovering the *pattern* of speciation and not necessarily the causal intrinsic factors promoting evolutionary independence. Therefore, I will concentrate on factors that affect the pattern. And because patterns are replicated in many groups whose differentiation mechanisms are very dissimilar, I will assume that such factors exist beyond a simple recounting of examples of individual mechanisms.

Allopatric Speciation Models

Many discussions of allopatric speciation treat the phenomenon as a relatively monothetic subject. Actually, allopatric speciation can be divided into three major models. Two of these models (large scale subdivision/vicariance and peripheral isolation) are based on similar population structures but have quite different biogeographic characteristics in their more extreme cases. The remaining model, termed Model III, is based on a different population structure and has been frequently confused with the peripheral isolate model. The phenomenon of allopatric speciation has also been termed "geographic speciation" by many authors (e.g., Mayr, 1942, 1963, 1970; Cain, 1954; Grant, 1971). However, geography also plays an important part in the allo-parapatric and parapatric modes.

Historically, the notion of a correlation between geographic factors and speciation grew out of the observation that closest relatives tend to occupy separate but contiguous geographic regions (Wallace, 1855). Species in sympatry resembled each other less than species in allopatry. This was Wallace's major pre-1859 contribution to evolutionary theory, a fact little cited by modern evolutionists. Among other early workers who commented upon this phenomenon and contributed to the theoretical development of allopatric speciation are Wagner (1889), Romanes (1897), Karl Jordan (1896, 1905), and David Starr Jordan (1905). The phenomenon has been discussed extensively by (among others) Dobzhansky (1951, 1970), Dobzhansky et al. (1977), Mayr (1942, 1963, 1970), Stebbins (1950), Grant (1963, 1971), Croizat (1964), Hennig (1966), Lewontin (1974), Bush (1975a), Endler (1977), and White (1978a).

Allopatric Speciation, Model I—Vicariance or Geographic Speciation

Vicariance speciation is the physical separation of two relatively large populations of a single ancestral species and the attainment of lineage independence by these large populations. It is based on the assumption that the same processes of local adaptation and geographic differentiation (stochastically or in response to environmental factors) which affect local demes can also affect larger populations given a geographic disjunction. Grant (1971: 111) characterizes what I term Model I as:

> The theory in its modern form holds that geographic races are the pre-
> cursors of species in a continuous process of evolutionary divergence. . . .
> The spatial isolation of populations at the racial stage of divergence enables
> the separate populations to develop and maintain the gene combinations
> determining their distinctive morphological and physiological characters.
> The beginnings of reproductive isolation set in at some stage. Species
> formation is then an extension of these processes.

Model I allopatric speciation is based on three assumptions:
Assumption 1.   Demes tend to differentiate (within the limits imposed by developmental homeostasis) in response to stochastic and local extrinsic factors (i.e., selection).
Assumption 2.   Deme differentiation is countered by gene flow within the species range.
Assumption 3.   Between-deme differences are inversely proportional to interdemic gene flow and population size and directly proportional to selective differences between demes, the rate of origin of unique evolutionary innovations, and the initial geographic variation of the ancestral species.

These assumptions allow us to make certain deductions which may be stated in the form of predictions.
Prediction 1.   Given an initial amount of geographic variation, disjunction is likely to result in differentiation because gene flow ceases across the disjunction.

This prediction simply states that subdividing a variable species will result in differentiation of the two (or more) daughter species as a simple result of the cessation of the flow of genes across the geographic barrier. Before the disjunction complete differentiation would be inhibited by *species cohesion* resulting from gene flow and large population size. A species may be cohesive because of gene flow, as in this model, or it may be cohesive because of a very canalized developmental homeostasis which prevents even geographic variation. If a species is cohesive because of canalized homeostasis, we would expect it to be relatively monotypic—that is, to show little or no geographic variation and to show as much intrademic as interdemic variation. In the case where the species is held together by gene flow cohesion, we may see that any disruption of gene flow will result in differentiation that is directly related to the geographic differences between the two populations before the disjunction, the sizes of the two (or more) geographic units, and the acquisition of evolutionary novelties in one of them. Barring new evolutionary novelties arising in either daughter population, the time required for differentiation is a function of four factors:

1. The magnitude of geographic variation.
2. The magnitude of gene exchange between demes and the number of demes in each population.
3. The magnitude of selective differences among groups of demes.
4. The effective population sizes in the demes affecting the rate of genetic drift.

This exercise points to one interesting conclusion: one of the factors that provided cohesion for the ancestral species (gene flow) will help promote differentiation if a disjunction occurs.

Not all species show geographic variation, nor is evolution composed purely of the subdivision of previously existing geographic variation. If it were, evolution would rapidly run down. New characters may be considered evolutionary novelties—characteristics not present in the ancestral species. When these are considered, a new factor is introduced. Given a monotypic species and a disjunction which splits this species into two populations, the time required for differentiation will be a factor of the acquisition of a novelty in one population and its fixation (by drift or selection) throughout that population. Fixation depends on gene flow and selective differential. The acquisition of an evolutionary novelty in one of the subdivided populations will accentuate species differences and will be fixed independently of existing geographic variation.

Prediction 2. The relative apomorphy or plesiomorphy of a particular daughter species for a character that is polymorphic in the ancestral population and varies geographically will be determined by geography. Time to fixation will be determined by the number and size of the demes comprising the daughter species and the selection differentials between them.

This prediction is best explained by an example. Figure 2.6 portrays character evolution and speciation in the species C and its descendants C_1 and C_2. Species C was originally fixed for character 1. One deme in species C picks up the novelty 1' (Fig. 2.6a) which subsequently spreads from that deme and is found in increasing frequencies in other demes to the south as time progresses (Figs. 2.6b,c). Species C is thus polymorphic, with increasing frequencies of 1' in northern populations and none in southern populations. A geographic separation of C occurs south of the most southern population having 1' in any frequency. The new species C_1 becomes characterized by the autapomorphy 1' while its sister species, C_2, remains primitive for the character 1 (Fig. 2.6d).

This example involves an "either/or" character, a character sufficient to diagnose a species relative to all other species. However, this does not have to be the case. Many meristic and morphometric characters

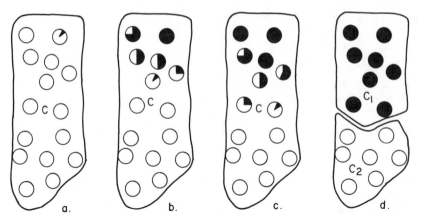

Figure 2.6   The spread of an evolutionary novelty and speciation of the ancestral species C. The circles are demes with the frequency of the evolutionary novelty indicated by shading. Descendent species C_1 is eventually fixed for 1^1 while C_2 retains 1.

show clear modal or mean tendencies but have ranges that overlap. We may use these means and modes in the same way that we use either/or characters and this model example provides the clear theoretical basis for their use.

Prediction 3.   The relative apomorphy or plesiomorphy of a particular daughter species for an evolutionary novelty after disjunction cannot be predicted and therefore can be assumed to be random (essentially, Wright's rule; Wright, 1955).

Prediction 4.   Time to differentiation may be relatively long and will depend on interdemic migration, deme size, the number of demes in the smaller of the two daughter species, and the selective advantage of different characters.

Phylogenetic Prediction 5.   A phylogenetic hypothesis of the relationships of species (as evidenced by synapomorphies) should reflect accurately the temporal geographic separation (vicariance) and speciation of these species to the extent that the character evolution of the group has been worked out correctly. Further, we should expect that this history of speciation will be largely (though not always) dichotomous because the chance that a multiple splitting and character divergence will occur simultaneously is relatively rare. This prediction follows from the deduction that the species involved will share direct genealogical ties with series of common ancestors, not a single common ancestor. Thus character evolution should be largely a linear process of picking up synapomorphies from sequentially evolved ancestral species. We shall see in the sections on peripheral isolates and island populations that the descendants of ancestral species exhibiting these modes of speciation are not expected to exhibit this type of character evolution, but are expected to have only autapomorphies and thus show polytomous relationships on the level of character analysis. Further, we should note that exceptions to this prediction will be encountered even with large populations when geographic areas are rapidly separated due to climatic or geologic changes and character evolution does not keep pace with these vicariance events.

Biogeographic Predictions. Certain biogeographic predictions may be deduced from Model I allopatric speciation.

Prediction 6. The range of the ancestral species may be estimated by adding the ranges of the daughter species. This prediction should hold for relatively young speciation events and for older events if the daughter species are poor dispersers.

Prediction 7. The geographic point of disjunction corresponds to the boundary between disjunct or contiguous daughter species. If dispersal occurs in some species pairs this prediction must be checked statistically. Many clades are required and some commonality arrived at by comparing many species borders and likely geographic events. Figure 2.7 dem-

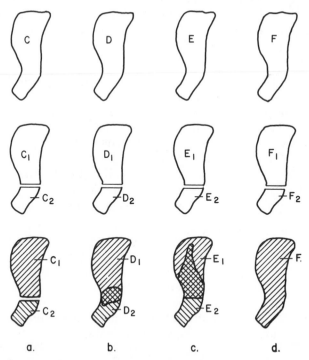

a.   b.   c.   d.

Figure 2.7 The outcomes of a single geographic subdivision on four potential ancestral species (C–F) inhabiting the same region: (*a*) the descendant species C_1 and C_2 show unaltered vicariance; (*b*) the descendant species D_1 and D_2 show limited sympatry; (*c*) the descendant species E_1 and E_2 show widespread sympatry; (*d*) the species F was unaffected by the geographic subdivision.

onstrates this principle with a simple four group example. We shall see in Chapter 8 that this forms the species-level basis for panbiogeographic track analysis.

Prediction 8. We should expect to see many clades that inhabit the same geographic area to show the same (congruent) phylogenetic and biogeographic patterns. (Methods for determining congruence of phylogenetic and biogeographic patterns are detailed in Chapter 8.)

Model I, Summary. Model I allopatric speciation is a model of large-scale geographic disjunction with (1) divergence of the resulting daughter species, in the absence of novel characters, dependent on the degree of geographic variation present in the ancestral species; and (2) time to differentiation dependent on the amount of migration between the demes within the smaller of the two daughter species. The model assumes that species cohesion is a function of developmental homeostasis and interdemic migration (gene flow). Evolutionary novelties may arise in a restricted part of the range of the ancestral species, thereby determining which of the daughter species will be apomorphic for the trait, or they may arise after disjunction. For monotypic species or species with geographic variation insufficient to produce differentiation, time to differentiation depends on the origin of at least one such novelty in one of the two daughter species.

Allopatric Speciation: Model II, Peripheral Isolates

The **peripheral isolates** model supposes that a new species arises in marginal habitats, usually on the boundary of a larger central population. In other words, it is a model that concerns a single deme or several closely situated demes on the outer range boundary of an ancestral species that is composed of many other demes. It is the mode termed "punctuated equilibrium" by Eldredge and Gould (1972) and is a favored model by many authors (see Mayr, 1963). Its assumptions may be stated briefly:

Assumption 1. Demes tend to differentiate (within the limits of developmental homeostasis) stochastically and in response to local factors of environment (selection) but interdemic migration is prevalent enough within the more central parts of the species' range to prevent differentiation of any one or any combination of demes.

Assumption 2. Interdemic gene flow is not strong enough at the periphery of a species boundary to prevent the establishment of new phenotypes at relatively high frequencies. Further, if interdemic migration is stopped (i.e., a disjunction occurs), one or more of these novel phenotypes may become fixed as new species.

The model predicts the following:

Prediction 1. New species will appear initially in marginal habitats, often, but by no means always, at the margins of the ancestral species' range. The place of appearance is causally related to a geographic disjunction which prevents interdemic migration, and the original range of the new species will be small relative to the ancestor.

Prediction 2. The peripheral isolate will be more apomorphic than the central population regardless of whether the central population becomes a species different from the ancestral species (i.e., Hennig's, 1966, deviation rule applies; one species will be more different from the ancestor). Further, the distribution of apomorphic characters cannot necessarily be predicted from the geographic variation of the ancestor.

This prediction requires some explanation. In regard to relative apomorphy, the model holds that the central population is necessarily conservative. Local demic differentiation is swamped by migrant phenotypes from neighboring demes that are not differentiated in the same way. Only in the peripheral isolate(s), so the model goes, are the constraints of interdemic migration slack enough to permit the establishment of uniquely different populational structures. In addition, because of the small size of the isolate, it would take much less time to establish an apomorphy throughout the range of the new species than it would throughout the range of the central population. (Of course, it is also less likely that such an apomorphy would arise—but given enough peripheral isolates, chances of *one* gaining an evolutionary novelty are greatly increased.)

Prediction 3. Wright's rule will apply. The divergence of the peripheral isolate will be random with respect to the ancestral array of epiphenotypes within the constraints of developmental homeostasis (see Wright, 1955; Eldredge and Gould, 1972; Gould and Eldredge, 1977).

Phylogenetic Predictions. There are two possible phylogenetic patterns, based on synapomorphies, that one might expect from the process of peripheral isolation.

Prediction 4. If the history of speciation of a group involved peripheral isolation, speciation, subsequent migration of the peripheral isolate, another peripheral isolation, and so on, such that a progression in time and space is the result, then we might expect the pattern of descent as evidenced by synapomorphies to be largely dichotomous. This progression in time and space forms Hennig's progression rule, a concept discussed more fully in Chapter 8.

Prediction 5. If the history of a group involved peripheral isolation and speciation of a number of peripheral isolates around the range of a single ancestral species, then we should expect the pattern of descent as evidenced by synapomorphies to be polytomous (a multiple furcation) because the descendant species would share with each other only the common characters also shared by the ancestral species and all other peripheral isolates.

Biogeographic Predictions. There are several biogeographic predictions which can be deduced from the peripheral isolates model.

Prediction 6. If the assumptions inherent in prediction 4 are to apply, dispersal through time and space must produce a pattern of biogeography concordant in every respect with the phylogenetic relationships of the species group. That is, one daughter species (or the ancestor itself) must be primitive in morphology and occupy the ancestral range whereas the other daughter species (or series of species) must be more derived (Hennig's deviation rule) and occupy newly gained geographic space or its original (small peripheral) range (Hennig's progression rule).

Prediction 7. If the assumptions inherent in prediction 5 are to apply, then dispersal through time and space should not follow either the deviation rule or the progression rule.

Prediction 8. Due to differing dispersal capabilities, the biogeographic patterns of different species groups (i.e., different clades) inhabiting the same geographic range will not be expected to show similar biogeographic patterns under the assumptions inherent in either prediction 5 or 6. Congruence of tracks (see Chapter 8) would be expected only in the original areas of peripheral isolation (Fig. 2.8) when that area in-

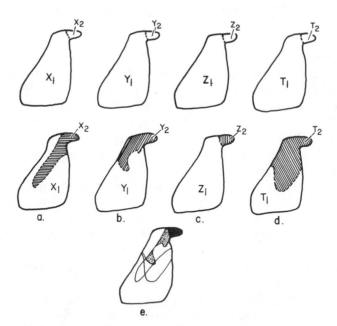

Figure 2.8 Peripheral geographic isolation of four ancestral species and subsequent speciation. (*a–d*) isolation (above) and subsequent speciation and range alteration (below) of the descendant species X_2, Y_2, Z_2, and T_2; (*e*) patterns of distribution of all four descendant species plotted on the same map. Note in (*e*) that the only common ranges shared by all four decendant species is their original area of peripheral isolation (the shaded area), whereas three of the four species are found in a somewhat larger (stippled) area.

volves isolates of two or more species groups, or, it should involve only those species which have similar habitat requirements.

Model II, Summary.   Model II, the peripheral isolate model, presupposes that only demes at the outer limits of an ancestral species' range or demes in marginal habitats will escape the homeostatic restraints of the ancestral species' phenotype and maintain enough difference so that disjunction, when it occurs, will rapidly precipitate speciation. It should not be thought that this precondition for speciation exists at only one geographic locality; most workers assume that many peripheral demes are preconditioned for differentiation and will differentiate if disjunction occurs. Speciation is more rapid than in Model I because the peripheral isolate is smaller. Therefore, preexisting differences will be consolidated more rapidly and evolutionary novelties, if they occur, will be fixed in a shorter time period.

Two additional features of Model II should be mentioned. First, there is a contrast between predictions 4 and 5. This leads to different biogeographic predictions (6 and 7). Prediction 4 states that the phylogenetic relationships of taxa involved in geographic progression should show dichotomous phylogenetic relationships. This is the same prediction as in Model I. However, we may distinguish between Model I and Model II prediction 4, because we would expect that under Model I speciation there would be a number of clades having the same biogeographic patterns, even clades with different ecological requirements. However, under the assumptions of Model II and prediction 4 we might expect to see only those clades with the same ecological requirements and similar dispersal capabilities showing the same geographic patterns, given that the speciation event is relatively young. Prediction 5 of Model II carries different connotations than Model I in that we would expect multifurcations.

Finally, it should be noted that clades exhibiting both Model I and Model II speciation may show particular patterns of phylogenetic and biogeographic relationships. For example, if a clade alternated between models, such that the first speciation event involved vicariance, the second peripheral isolation, the third vicariance, and so on, the phylogeny would be dichotomous. But when we compared the pattern to other clades, we would expect to find congruence involving only the vicariance speciation events, not the peripheral isolation events. A practical example that might involve such a sequence is discussed in Chapter 8 and involves the speciation history of Middle American sword tail *Xiphophorus* and *Heterandria* fishes.

Allopatric Speciation: Model III (Unnamed)

Model III is similar to Models I and II in that Wright's rule prevails, and it is similar to Model II in that speciation events are postulated to involve single demes or a series of closely situated demes that exchange genes. It differs from Model II in one important aspect. Model III assumes that gene flow is so restricted that it can not effectively provide for species cohesion. In other words, both Models I and II assume that there are two major factors effecting species stability—homeostasis and gene flow. Model III differs in that it assumes the overriding force promoting maintenance of species identity is evolutionary stasis (homeostasis) (Eldredge and Gould, 1972; Gould and Eldredge, 1977). Thus Model III can also be applied to asexual species. Indeed, it is *the* model of speciation in asexual clades. The assumptions of this model may be stated briefly:

Assumption 1. Species composed of two or more demes are genetically and epigenetically similar because of homeostasis (Gould and Eldredge, 1977) and not because of gene flow. In other words, similarity of epiphenotype is a historical by-product of descent from a common ancestral deme with the same epiphenotype.

Assumption 2. Gene flow is not a significant factor promoting the cohesion of the ancestral species (Ehrlich and Raven, 1969; Eldredge and Gould, 1972).

Assumption 3. Any variation between demes is due to chance (drift) or parallel development of some (but not all) of the demes in response to selection. Geographic variation may be present if selection gradients are present over the geographic range of the species.

Like Model II, Model III is a model of rapid speciation—rapid, that is, on a geologic time frame. But the predictions of Model III are considerably different from the predictions of Model II.

Prediction 1. Any autapomorphies shared by the demes of a species were developed in the original species deme before dispersal and retained in all subsequent demes. The alternate is the wholesale production of a parallelism in response to some overriding selection. Barring such parallelisms, we should expect intrademic variation to equal interdemic variation.

Prediction 2. Wright's rule applies—that is, the direction of differentiation is random with respect to the ancestral genotype and epiphenotype.

Prediction 3. The original ancestral species is unaffected by the speciation event in terms of both its genotype and epiphenotype. Thus, an ancestral species would be expected to survive a speciation event.

Prediction 4. Speciation may occur at any point within the range of the ancestral species. That is, speciation is not primarily an event that concerns peripheral populations.

Phylogenetic Prediction 5. As under the assumptions inherent in prediction 6 of the peripheral isolates model, we should expect that three or more descendants of a single ancestral species would show a polytomous relationship to one another when analyzed by apomorphic characters. Dichotomous phylogenetic hypotheses would be expected only when there is a sequence of speciation events involving extinction of a previous ancestral species followed by the speciation of a new species from the species that caused the extinction of the original ancestor, and so on. However, such a case would not produce the same biogeographic pattern as would be expected with a peripheral isolates model, especially when several species groups inhabiting the same geographic range are analyzed.

Biogeographic Predictions.
Prediction 6. The geographic character variation of a species bears no particular relationship to the geographic character variation of its sister species. Biogeographic patterns between clades or species should be due entirely to chance; that is, there should be no necessary underlying

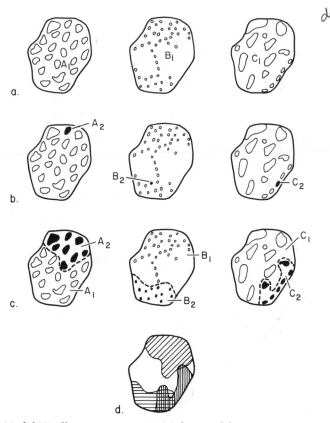

Figure 2.9 Model III allopatric speciation: (*a*) demes of three species, A, B, and C, that inhabit the same geographic area; (*b*) speciation in one deme of each ancestral species resulting in the origins of A_2, B_2 and C_2; (*c*) dispersal of descendant species and their displacement of ancestral species in part of their original ranges; (*d*) patterns of distribution of descendant species plotted on the same map. Note in (*d*) that there is no congruence of ranges for all three descendant species.

reason for distributional correlations (see Fig. 2.9) between monophyletic groups.

Prediction 7. The area of disjunction of the ancestral species cannot be determined by inspecting the ranges of its descendants nor by inspecting several groups to determine a likely point of common disjunction.

Allo-Parapatric Speciation

Allo-parapatric speciation occurs when two populations of an ancestral species are separated, differentiate to a degree that is not sufficient for lineage independence, and then develop lineage independence during a period of parapatry (limited sympatry). This model has been discussed by Mayr (1942), Dobzhansky (1951), Key (1968; termed "stasipatric speciation"), and Endler (1977). Endler (1977) points out that the major differences between allo-parapatric and allopatric speciation are (1) that speciation is completed after a period of sympatry and (2) that the process of attaining lineage independence is potentially reversible because it is possible that the two partly differentiated populations could form a single evolutionary lineage showing clinal variation after they meet rather than the period of sympatry reinforcing the differences between them. This is an important distinction because the allopatric models assume that the attainment of lineage independence occurs in allopatry. Thus, cases of "failed" allopatric speciation are in reality cases of incomplete allo-parapatric speciation.

Allo-parapatric speciation shares the same assumptions of population structure as Models I and II allopatric speciation and the predictions that follow the assumptions depend on whether speciation involves large or small populations.

Parapatric Speciation

Parapatric speciation occurs when two populations of an ancestral species differentiate despite the fact that no complete disjunction has occurred.[2] In other words, the daughter species may share a small fraction of their respective ranges and interbreed within this narrow contact zone and yet still differentiate. Although this mode of speciation has been specifically repudiated by some (e.g., Mayr, 1970), it has been shown to be reasonable by others (Endler, 1977). We shall see from the predictions that it cannot be discriminated from allopatric speciation and secondary contact except in certain circumstances.

Assumption 1. Demes tend to differentiate in response to stochastic

---

[2] Parapatric speciation should not be confused with parapatric distribution, which is the condition where two species' ranges abut but do not overlap, and may be caused by several modes of speciation.

processes and local selection within the limits of developmental homeostasis.

Assumption 2. Deme differentiation tends to be countered by interdemic migration within the range of the ancestral species.

Assumption 3. The individual members of the ancestral species have relatively low vagility (Endler, 1977) and thus local differentiation is pronounced.

Assumption 4. Given a decrease in fitness of a heterozygous epiphenotype along a geographic variation gradient (a cline), assortative mating may occur such that speciation may go to completion.

Prediction 1. If there is competition among the new sister species, then a narrow zone of sympatry will be established. Owing to the assumptions dealing with interdemic migration, we would expect this situation given that one daughter species does not outcompete and thus eliminate the other.

Prediction 2. The phylogenetic pattern of parapatric speciation as evidenced by apomorphic characters is similar to that expected from vicariance allopatric speciation. That is, we would expect a largely dichotomous pattern given correct analysis of characters.

Prediction 3. The relative apomorphy or plesiomorphy of any one of the daughter species will be determined, given no evolutionary innovations, by the geographic distribution of characters in the ancestral species. The origin of any particular evolutionary novelty after establishment of the contact zone (effectively after onset of speciation) cannot be predicted.

Stasipatric Speciation

Stasipatric and parapatric speciation are frequently considered synonyms (Endler, 1977; White, 1978a and references therein). However, stasipatric speciation as envisioned by White (1978a,b) concerns chromosome modifications and its phylogenetic and biogeographic implications are quite different from the usual parapatric scenario as discussed by Bush (1975a) and Endler (1977). It is, in fact, a special case of Model III allopatric speciation involving a particular mechanism (chromosome mutation). Thus, most of the assumptions and predictions follow Model III. White (1978b) gives the best summary of this speciation mode. A spontaneous chromosome rearrangement arises in a deme of an ancestral species of limited vagility. This chromosome mutation may be a translocation, a fusion, or a pericentric inversion. The resulting chromosome arrangement must be fully viable in the homozygous condition but of reduced viability in the heterozygous condition. Such an arrangement is established in the deme by inbreeding, drift, or meiotic drive. A deme fixed for the new chromosome arrangement (i.e., a deme homozygous for this new arrangement) is established. This results in a species of limited range within the range of the ancestral species which, with the dispersal of the new species out of the original area, will result in a parapatric contact zone with heterozygotes (for the chromosome rearrangement) being at a selective disadvantage. The narrowness of the contact zone depends on the vagility of the organisms and on selection of the two chromosome types. The assumptions of this mode may be summarized thus:

Assumption 1.   The members of the ancestral species have low vagility; that is, interdemic migrations are of relatively low frequency, thus gene flow is low.

Assumption 2.   Chromosomal rearrangements are sufficient to cause lineage splits, with or without phenotypic differences.

Assumption 3.   The new chromosomal rearrangement results in a more competitive phenotype in the homozygous condition, at least within the geographic region of its origin. The heterozygous condition is either inviable or of lower fecundity.

Prediction 1.   We may expect the origin of a new species via stasipatric speciation to be a random phenomenon with respect to variation within the ancestral species.

Prediction 2.   Because the phenomenon involves small demes, the ancestral species may not be expected to become extinct at the speciation event.

Prediction 3.   In cases where the dominant mode of speciation is stasipatric, we should expect that the daughter species of an ancestral species will show a polytomous relationship when analyzed by apomorphic characters. Dichotomous phylogenetic hypotheses should only be expected when one species which itself has originated via stasipatric speciation gives rise to another species, and so on. Put simply, the phylogenetic patterns of stasipatric speciation should be similar to that expected for Model III allopatric speciation.

Prediction 4.   Because stasipatric speciation is a random phenomenon there should be no biogeographic correlation between clades that exhibit stasipatric speciation in the same geographic region.

Sympatric Speciation

Sympatric speciation is an array of mechanisms that produces one or more new species from one or more ancestral species with no geographic segregation of populations. Although most authors (e.g., Smith, 1966; Bush, 1975a,b; White, 1978a) focus on ecological segregation as the major phenomenon of sympatric speciation, there are at least three other phenomena which result in sympatric speciation. These will be listed first, followed by a short discussion of ecological sympatric speciation.

   1.   *Speciation by hybridization.*   In most sexually reproducing or-

ganisms the origin of a species by hybridization of two parental species is usually sympatric (Grant, 1971).

2. *Speciation by apomixis.* Asexual species derived from a biparental ancestral species may arise in place and there is no *a priori* reason to believe that such speciation must involve geographic barriers because "reproductive isolation" is instantaneous.

3. *Speciation of a single local deme in a population which is composed of demes with rare or no interdemic migration.* This is Model III allopatric speciation with the isolate deme moved inside the range of the species. Such a situation would look for all intents and purposes like sympatric speciation whereas it may be considered as allopatric or microallopatric.

4. *Ecological sympatric speciation.* Grant (1971) terms this "sympatric speciation" in the restricted sense. Most discussions about sympatric speciation center around this mechanism (Thoday and Boam, 1959; Smith, 1966; Bush, 1969, 1975a,b; Richardson, 1974; White, 1978a). The mechanism is based on segregation of habitat. Bush (1975b) presents a detailed picture of sympatric speciation in phytophagous parasitic insects, and Richardson (1974) a convincing case for habitat segregation and assortative mating in two species of Hawaiian *Drosophila*. Smith (1966) discusses the whole problem in detail from a theoretical perspective. In spite of Mayr's (1963) objections that the majority of cases for ecological sympatric speciation can be interpreted allopatrically, I believe that convincing evidence (theoretical and empirical) has been presented for such speciation.

Sympatric speciation via ecological segregation requires a very specific phylogenetic hypothesis. The postulated species must be each others' closest relatives. If one of the species is actually more closely related to a third species, then the sympatric model is falsified. In the case of Richardson's fruit flies, if *Drosophila mimica* and *D. kambysellisi* are sister species, then sympatric speciation is a viable alternate to allopatric or stasipatric speciation. However, if one is more closely related to some other *Drosophilia* species, then sympatric speciation involving the pair is falsified. This is shown in Fig. 2.10.

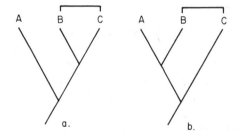

Figure 2.10 Two phylogenetic hypotheses involving species of supposed origin via sympatric speciation: (*a*) a tree compatable with the hypothesis that B orginated via sympatric speciation from C; (*b*) a tree incompatable with the hypothesis of sympatric speciation.

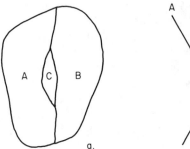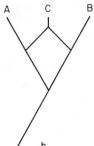

Figure 2.11 The ranges of three species (*a*) and their phylogenetic relationships (*b*). A pattern such as this might be expected to be repeated in groups that show hybridization and subsequent speciation along suture zones.

Biogeographic implications of sympatric speciation are diverse. The following is a list of some implications:

1. *Hybrid species.* A hybrid species forms a phylogenetic triad (reticulation) with its parental species. A pattern of such triads will be uncorrelated with other biogeographic distributions involving species of nonhybrid origin. However, along a suture zone of sister species produced by normal allopatric species, we might expect a common biogeographic pattern of parental-hybrid triads (see Figure 2.11).

2. *Apomictic species.* Apomictic species are the result of unique intrinsic events (i.e., by events having no potential influence on other species in the same area). I would suspect that apomictic species would not show congruent biogeographic patterns with other clades inhabiting the same region.

3. *Ecological sympatric species.* Congruent biogeographic patterns would be restricted to those few clades which respond to the same ecological situation. I would expect little or no congruence.

DECISIONS ON THE SPECIES LEVEL

Theoretical discussions are valuable in our considerations of the relevance of species to the evolutionary process and their relationship to the phylogenetic trees we attempt to produce. However, such discussion can only partly guide an investigator in solving an immediate problem. This section addresses the practical aspects of determining when an investigator is justified in naming a population as a distinct species. This implies that the discussion involves very closely related organisms (i.e., probable sister groups) and the terminology is presented with this in mind. It should be viewed as a series of suggested strategies and is not meant to exhaust all possibilities. Experts with different groups of organisms frequently employ different criteria in determining the species status of populations. They do this because the biological attributes of various groups may be quite different. What may seem like local demic variation to an ichthyologist may indicate distinct species to an entomologist (or vice versa). Thus, *there is no substitute for knowing as much about the biology of the organisms as possible.* In spite of this, and in the spirit of realizing that everyone will find something wrong with this section, the beginning systematist might benefit from the general suggestions outlined in the following guidelines.

Distributions

Strategies for making decisions on the species level involving very closely related populations revolve around the distributions that the organisms exhibit.

1. **Sympatric distribution.** Two populations occupy largely the same geographic area (see Fig. 2.12*a*).

2. **Partly sympatric distribution.** Two populations are found together in one part of the total range and apart in the remaining portion of the total range (see Fig. 2.12*b*).

3. **Allopatric distributions.** Two populations are found in different geographic ranges (see Fig. 2.12*c,d*). When the ranges abut they are frequently termed **contiguous** or **parapatric distributions** (Fig. 2.12*c*, but not necessarily a result of parapatric speciation). I prefer the term "contiguous" because it does not imply a particular speciation mechanism. When the ranges are separated they are termed **disjunct distributions** (Fig. 2.12*d*).

In addition, there are two special terms for sympatric distributions (Rivas, 1964).

1. **Syntopic distributions.** Two phenotypes are found in the same

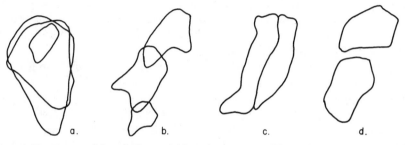

Figure 2.12 Types of distributions: (*a*) largely sympatric; (*b*) partly sympatric; (*c*) contiguous allopatric or parapatric; (*d*) disjunct allopatric.

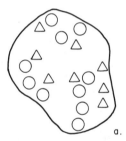 

Figure 2.13 Two types of sympatric distributions: (*a*) allotopic and (*b*) syntopic. Circles and triangles are samples of each of two species.

habitat; that is, they are microsympatric. In fishes, syntopic phenotypes are caught in the same seine haul (see Fig. 2.13*b*).

   2.   Allotopic distributions.   Two phenotypes are found in the same geographic range but in different habitats (i.e., they are microallopatric, see Fig. 2.13*a*).

Mode of Reproduction

In many groups decisions on the species level are tied to reproductive mode as well as morphology. Angiosperms frequently have apomictic species complexes composed of one bisexual and several apomictic species. The apomicts may be cytologically distinguishable but morphologically indistinguishable. Decision on the species level involving various modes of reproduction have been excellently discussed by Ross (1974) from whom much of this discussion is taken.

   1.   *Asexual species.*   Functional asexual species show no sexual differences between members of the population. Many are parasexual rather than strictly asexual. They may reproduce by a variety of means depending on the group. Decisions on the species level usually involve a decision by the investigator about the distinctness of the clonal populations. Where two clones can be adequately diagnosed, the naming of species is reasonable. Where two clones cannot be diagnosed but are differentiated only on a statistical level, the naming of species may be arbitrary and I would not advocate it.

   2.   *Agamospermic and parthenogenic species.*   Rather than asexual, these species are unisexual. Such species are derived from bisexual species. There may be evident chromosomal differences between such populations. In plants, an asexual species is frequently the sister group of an agamospermic species (Grant, 1971). Both may be found sympatrically with a closely related (and even ancestral) bisexual species. Both agamospermic and parthenogenic populations represent independent evolutionary lineages and thus are evolutionary species.

3. *Gynogenetic species.* These species are largely unisexual and if males are present they tend to be infertile. The female uses the sperm of another species to activate the egg to develop, but no genetic introgression occurs.

4. *Bisexual species.* Bisexual species reproduce by fusion of egg and sperm. Many bisexual species have complex life histories involving sexual and asexual phases, as in the alteration of generations in plants. Many coelenterates and lower plants furnish other examples of complex life histories. Before the intricate nature of these life histories were known, different life history stages of the same species were frequently placed in different taxa. This is still true in some parasites. Among elopiform teleosts, many larvae are classified as *Leptocephalus* simply because their adult species are not known.

In cases where life histories are sufficiently well known so as not to be the source of confusion, decisions on the species level may be divided into two major areas—decisions concerning populations that are (1) sympatric and (2) allopatric. The decisions involve two or more epiphenotypes. I use the term *"epiphenotype"* to connote morphological, ontogenetic, and genetic components of organisms (see Chapter 1). The terms "morphotype" or "phenotype" could be used, but their implications are more limited. I assume that we must have two or more epiphenotypes because if we only have one and no evidence that cryptic species are involved, there will not be a need for a decision.

Bisexual Species—Sympatric Occurrences

Complete or More-or-Less Complete Sympatry (see Fig. 2.12a)

1. *Specimens from various localities are not distinguishable as distinct epiphenotypes although they may fall into integrated size classes.* This is a case in which intrademic variation equals or exceeds interdemic variation. I would conclude that one species is present unless there was additional evidence pointing to cryptic species.

2. *Certain epiphenotypes predominate in part of the range while other epiphenotypes predominate in other parts of the range.* In this situation the investigator should fill in the unsampled area between sampled demes that show the distinct epiphenotypes. If additional samples show that the distinct epiphenotypes are connected by a continuum of variation that cannot be ascribed to hybridization and if the distinctive epiphenotypes are closest relatives, then I would conclude that one species, exhibiting geographic variation, was present. If additional sampling shows that the epiphenotypes are actually allopatric, then consult the section on allopatric distributions which follows. If the epiphenotypes are allotopic or syntopic and do not form a continuum of variation, then I would conclude that they are separate species.

3. *Epiphenotypes occur syntopically but each comprises a separate sex in mature individuals.* A single sexually dimorphic species is present. Frequently, the immature specimens will be similar to one or the other sex.

4. *Epiphenotypes are syntopic, one composed of sexually mature males and females, the other(s) composed of sexually immature specimens.* A single species is present. Growth patterns are frequently distinct because of discrete life history stages.

5. *Mature specimens fall into two epiphenotypes, each represented by both sexes, immature specimens fall into one or more epiphenotypes.* Two sexually monomorphic species are present. In cases where the immature specimens fall into one epiphenotype, they may resemble one of the two epiphenotypes present in adults. If so, that mature epiphenotype is likely plesiomorphic.

6. *Epiphenotypes are syntopic, one composed of diploid and one of haploid individuals.* A single species showing alternation of generations is present. One must, of course, match the haploid with the diploid where more than one species showing this life history is present in the same locality.

7. *Epiphenotypes are syntopic or allotopic, reproductive modes differ.* This example covers apomictic and asexual complexes. See the earlier discussion on asexual species.

8. *Epiphenotypes are syntopic or allotopic, basic chromosome number differs.* If the epiphenotypes form a natural group (a clade), the epiphenotypes are probably separate species. However, some species show chromosome polymorphisms and some polyploids occur as natural polymorphisms in a single species (reviewed by Davis and Haywood, 1965), so care should be exercised.

Partial Sympatry (see Fig. 2.12b)

1. *Each geographic population is a distinct epiphenotype, no intermediate epiphenotypes are found in the area of sympatry.* The two epiphenotypes are distinct species. Sympatry with no evidence of interbreeding is *prima facie* evidence for separate species. The situation is somewhat complicated by sexual dimorphism where the females are very similar. In such cases, the female epiphenotype may be considered a plesiomorphic similarity and should not influence the decision.

2. *Each geographic population is a distinct epiphenotype in allopatry, sporadic hybridization or local (limited) introgression in part(s) of the area of sympatry.* I would conclude that the populations are separate species. Localized hybridization or introgression occurs regularly in some groups. Species that do not hybridize or introgress under normal circumstances may do so in disturbed habitats (see reviews by Stebbins, 1950; Hubbs, 1955; Mayr, 1963; Grant, 1971; White, 1978a;

important primary literature includes Hubbs et al., 1943; Hubbs, 1961; Anderson, 1949, 1953; Brooks, 1957; Mecham, 1960, 1975). Either may also occur under special ecological circumstances (Hubbs, 1955).

3. *Each population is a distinct epiphenotype in allopatry, but there is complete introgression throughout the zone of sympatry.* Introgression is an interesting phenomena. Whether two introgressing populations warrant species status depends on (1) the phylogenetic relationships of the populations and (2) the width of the zone or intergradation. Three cases will be discussed:

a. *Introgression occurs over a wide geographic area and the populations are nearest relatives (see Fig. 2.14a).* I would tend to consider such populations part of a single species showing local adaptation and, consequently, introgression between the locally adapted epiphenotypes. At the same time, it is quite possible that parapatric speciation is occurring or that the zone of introgression is becoming narrower. However, given the inability to diagnose two forms without resorting to geographic

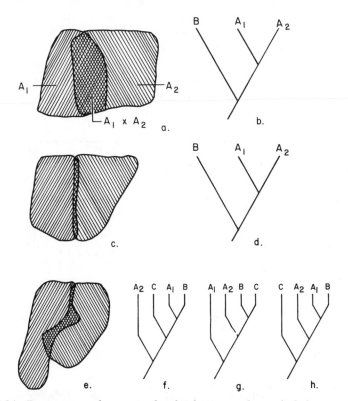

Figure 2.14 Decisions on the species level I, biogeographic and phylogenetic patterns involving contact zones: (*a, b*) species A_1 and A_2 hybridize widely and are closest relatives; (*c, d*) species A_1 and A_2 hybridize in a narrow zone and are closest relatives; (*e–h*) species A_1 and A_2 hybridize but are not closest relatives.

data and the inability to diagnose a significant geographic population (the introgressed individuals), I do not see any great advantage to naming the two parental populations as distinct species.

b.   *Introgression occurs between nearest relatives only in a narrow sympatric zone, parental epiphenotypes predominate outside the zone* (see Fig. 2.14c,d).   The investigator is justified in naming the parental epiphenotypes as species. This situation is evidence for species-distinct characters being selected outside the zone of sympatry. It is evidence that both epiphenotypes show independent species cohesion. Whether characters that do not contribute to species cohesion are introgressed is not relevant. I know of no example from the literature, but it is quite possible for characters that do not contribute to species cohesion to introgress widely without effecting the taxonomic conclusion that A_1 and A_2 (in our example) are distinct species. In cases where certain features of parental epiphenotypes determine assortative mating, it is quite possible for introgressed individuals possessing these features to mate back only with one parental type (A_1) carrying fully one-half of the non-essential genes of the second species (A_2) into the A_1 species. Since these genes do not affect the species cohesion of A_1, they are free to introgress widely without disturbing species identity or cohesion.

c.   *Introgression occurs over either a wide or narrow geographic area, the populations are not closest relatives* (see Fig. 2.14e–h).   Such cases demand more study than is usually accorded them. Introgression is fre-

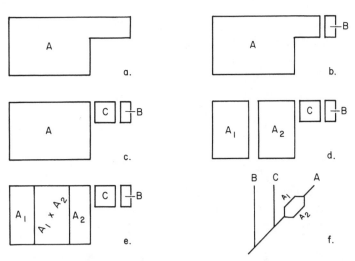

Figure 2.15 Decisions on the species level II, biogeographic and phylogenetic patterns involving peripheral isolates, large scale geographic subdivision and secondary contact zones: (*a*) original range of ancestral species A; (*b*) speciation of B; (*c*) speciation of C; (*d*) vicariance of A_1 and A_2; (*e*) widespread interbreeding of A_1 and A_2 preventing speciation; (*f*) a phylogenetic tree of the relationships of A, B, and C.

quently interpreted as evidence that two populations are nearest relatives. The tacit assumption is that interbreeding populations must be
the same species, or at the most, closest relatives. However, Rosen (1979)
has made the point that the ability to interbreed may be a plesiomorphic
character. Thus, the ability to interbreed cannot be considered *prima
facie* evidence for a sister group relationship. In each phylogenetic hypothesis shown in Fig. 2.14*f–h*, the investigator is faced with one of two
choices. First, the investigator can consider A_1, A_2, B, and C as members
of a single species. Second, the investigator can consider each a separate
species. Given that there is evidence for the separate species status of
B and C, then A_1 and A_2 are species regardless of their introgression
unless it can be demonstrated that A_1 and A_2 are two populations of the
ancestral species that gave rise to B and C. Such situations are possible
given a complex biogeographic history involving peripheral isolation
(see Fig. 2.15). If such is the case, similar tree topologies would be
expected in at least some other groups inhabiting the same biota.

Bisexual Species—Allopatric Occurrences

Because sister species are frequently allopatric, critical analysis of species status must be conducted without the aid of the interbreeding
criterion. Laboratory studies are frequently illuminating (see later discussion) if the organisms are suitable for such analysis. But the majority
of decisions must be made without the benefit of lab studies. This may
be due to the interest of the systematist, but usually it is the nature of
the systematic material one has to work with. The number of different
situations involving allopatric populations are many. I have attempted
to organize them into three general cases.

1. *A single epiphenotype (or two if sexually dimorphic) found in two
or more disjunct areas.* A single species is probably present. The inability of an investigator to diagnose the disjunct populations is evidence
that they are the same species. Variation within each disjunct population
would be expected to equal or exceed variation between the populations.
Consistent differences should be on a statistical level only. Multivariate
analyses (Chapter 10) would *not* be expected to cluster individuals into
consistent clusters corresponding to the *a priori* designated disjunct
populations.

2. *Two disjunct populations are obviously differentiated but no one
character or combination of characters can be used to diagnose them
as separate entities.* Such a situation may involve one or two species
and further study is warranted. Further study may be carried out at
several levels. One might look for less obvious but diagnostic characters
(in many fishes subtle but constant pigment patterns have frequently
been used). One might also look for karyotypic differences. Another

strategy might involve the use of canonical variates analysis. If individuals are consistently classified correctly in their *a priori* designated group, the hypothesis that two species are involved is given some support. Another way of approaching the problem might involve a canonical variates analysis with local deme samples and not the two suspected species being designated as *a priori* groups. For example, in Figure 2.16, there are two suspected species, A_1 and A_2. Rather than lump population samples 1–7 into A_1 and 8–13 into A_2 and perform a discriminant analysis, the investigator performs a canonical variates analysis using 1–13 as *a priori* groups. If 1–7 separate into one cluster (the original A_1) and 8–13 into another (the original A_2), then the hypothesis that A_1 and A_2 are separate species is given support (see Fig. 2.16*b*). If the samples are

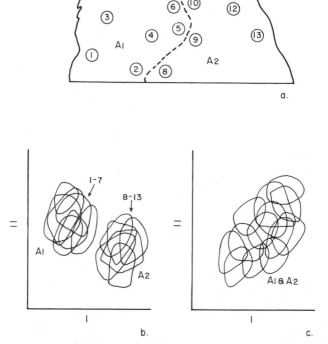

Figure 2.16 Decisions on the species level III, the use of canonical variates analysis (multiple discrimination analysis): (*a*) the distributions of A_1 and A_2 showing numbered local population samples (1–13) of each form; (*b, c*) canonical analyses with each local population identified as a group but neither A_1 nor A_2 identified as a group. In (*c*) there is no clustering of populations belonging to A_1 or A_2 into discrete groups. In (*b*) each of the presumptive forms is seen by the ability of the algorithm to discriminate the two forms but not to discriminate any of the populations withing each form. In (*c*) there is no reason to suppose, based on the available data, that two species are present, whereas in (*b*) there is evidence that two species are present. See Chapter 10 for a real example of this use of canonical variates analysis.

mixed (see Fig. 2.16c), then the hypothesis of two species is not supported (see Chapter 10 for further discussion). Finally, if the organisms are suitable for lab study, breeding and behavioral characteristics can be studied.

3. *Both populations are diagnosable without recourse to geographic data.* This situation indicates two species unless the investigator can ascribe the observed differences to ecophenotypic variation. Given that ecophenotypes are not involved, consistent differences between two allopatric populations is indicative of independent evolution of one or both from the original ancestral epiphenotype. This is reinforced if the populations are contiguous but show no signs of intergradation.

Biogeographic Data and Allopatric Decisions

Elements (clades) of entire biotas frequently show correlated speciation patterns, with sister groups in one clade having species boundaries that correlate with those of quite unrelated clades (Rosen, 1978). The most plausible explanation for such patterns is that a common speciation event effected these diverse clades (Croizat, 1964, and others). Thus, an ichthyologist can benefit from knowledge about the speciation patterns of insects and the entomologist can benefit from knowledge of the speciation patterns of fishes. Of course, population structure, species cohesion, and vagility vary from clade to clade. Vagile organisms or species with tight developmental homeostasis may not be affected by the same vicariance event that results in speciation in other clades. But, where biogeographic patterns are evident, they may be regarded as clues in solving species questions in other clades. For example, given the result in Fig. 2.16b that two populations (A_1 and A_2) can be distinguished, we might look at their phylogenetic relationships to B and their biogeographic relationships to other clades. If several groups recognized as "good" species show a similar biogeographic pattern this is some support that A_1 and A_2 are also species. In Fig. 2.17, for example, X and Y as well as N and O are "good" species and it is likely that the same vicariance event which separated A_1 and A_2 also separated X–Y and N–O. Given the morphological and biogeographical information, I would call A_1 and A_2 two species.

Experimental Data and Species Decisions

It is frequently possible to reinforce a decision based on morphology and biogeography by performing experiments in the lab or garden. If members of two presumptive species are successfully crossed and the progeny are inviable, sterile, or largely sterile then the hypothesis of species status is reinforced. However, some animals are hard to breed in the lab and a failure to mate may be due to lab conditions or to species

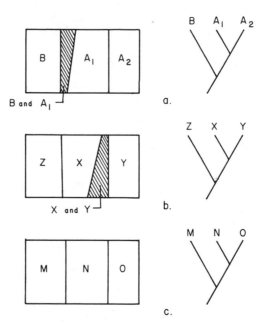

Figure 2.17 Decisions at the species level IV: biogeographic and phylogenetic relationships of three clades inhabiting the same areas. See text for discussion.

distinctions. Behavioral information may resolve the situation if it can be shown that mating behavior differs in the two presumed species so that premating isolating mechanisms are working (frequently the case in *Drosophila*). Often, however, these isolating mechanisms break down when a female of one species is offered only the male of another species. And, many perfectly good species may freely interbreed in the lab or garden but not in nature because they exhibit ecological differences (perhaps they are allotopic) or, in the case of plants, they are pollinated by different animals. Thus, successful crossing is subject to many interpretations, only one of which is the conclusion that the two populations belong to the same species.

Cryptic or Sibling Species

Cryptic species (or sibling species, Mayr, 1942) are species that cannot be diagnosed by morphology but that act as independent evolutionary lineages in nature. A classic zoological example is *Drosophila pseudoobscura* and *D. persimilis* which occur sympatrically in western North America (Dobzhansky and Epling, 1944; Mayr, 1963, reviews the occurrence of sibling species of animals). A classic example in plants is the *Gilia transmontana* species group which consists of three diploid species and two tetraploids derived from hybridization (Day, 1965; see summary by Grant, 1971). Day's careful morphological study showed that each species was somewhat morphologically distinct, but this conclusion followed after the lineage distinctness of the five forms were recognized (by Grant and Grant, 1960; Grant, 1964).

Cryptic species are good evolutionary species and deserve species names. The practical problems of keying such species complexes should not override their biological importance (Ross, 1974). Where sibling species occur, they should be keyed to species group or superspecies with an indication that several cryptic species may be represented in the sample. Preserved specimens may be labeled in the same manner.

In this chapter I have attempted to develop the concept of species-as-taxa as being evolutionary units and components of phylogenetic trees. Some modes of speciation result in different patterns of descent and some of these patterns of descent result in different biogeographic patterns. In the next chapter we shall be concerned with collections of species, supraspecific taxa.

Chapter 3

Supraspecific Taxa

In Chapter 2 I explored the nature of species. In this chapter I will discuss the nature of groups of two or more species—supraspecific taxa. Such groupings may be associated with a name and rank in a formal hierarchy (in the Linnaean hierarchy; order Diptera and genus *Drosophila* are examples). Others may remain unnamed but potentially nameable (as in a sequence convention, see Chapter 6). Still others may be named informally (e.g., the *Fundulus nottii* species group).

Literally any proposed and named assemblage of two or more species is a supraspecific taxon. Most of the mathematical permutations of the possible array of supraspecific taxa are not acceptable to any taxonomist (the number of possible taxa is vast: Felsenstein, 1977). We may infer from this that some supraspecific taxa are better than others. One of the purposes of this chapter is to discuss the idea of a natural phylogenetic system composed of only those taxa that have an objective basis in evolutionary history. Such taxa can be considered best of the astronomical number of possible taxa. *It is the objective of phylogenetic systematists to discover these taxa and either name them or make their presence immediately apparent.*

NATURALNESS AND SUPRASPECIFIC TAXA

"Natural" is an elusive term when it is applied to systematics. Almost everyone claims that their system is natural. Perhaps this is because "natural" is such a pleasing word. A minority, such as Mayr (1969), have abandoned the term because it is plagued with the weight of historically different uses and, perhaps, misuses. There is much to agree with in Mayr's attitude. Yet there does not seem to be a better term than "natural" for the concept I wish to develop in this chapter. So I shall use the word in a very specific context and contrast it with other usages.

Historical Concepts of Naturalness

The word "natural" has been applied to three basic systems of classification. **Aristotelean naturalness** is the oldest of the three. A taxon exhibits Aristotelean naturalness when the things placed in that taxon agree in characters that embody the essence of the group. In short, these characters are both necessary and sufficient to demonstrate group membership. The members of such a group can also be expected to resemble each other in additional characters (Crowson, 1970).

What we may term **phenetic naturalness** is the second oldest of the three. A taxon that exhibits phenetic naturalness is composed of members that resemble each other (are similar) more than they resemble any nongroup member. In other words, all of the members are more similar to each other than to anything outside the group (Davis and Heywood, 1965; Crowson, 1970; Sneath and Sokal, 1973).

The youngest concept is **phylogenetic naturalness**. The members of a phylogenetically natural group share a common ancestor not ancestral to any other group (Crowson, 1970); that is, they are thought to be phylogenetically related (Davis and Heywood, 1965).

When we examine the development of recent taxonomic thought it becomes apparent that the concept of Aristotelean naturalness is no longer taken seriously. Aristotelean naturalness was based on two principles: the principle of logical division of Theophrastus (c. 370–287 B.C.) founded on a set-theorylike rule of dichotomy (an object is either a member of X or a nonmember), and the assumption that the members of a group could be characterized by a few essential characters. Essential characters were those thought important to the functioning organism. Thus, Linnaeus classified plants on the basis of reproductive morphology in his *Species Plantarum* (1753). After all, what is more important to a plant than perpetuating itself?

Aristotelian naturalness is a much maligned concept and yet many of the groups named by essentialists are still used today and are considered phylogenetically natural. Aves is a taxon considered natural by all but the most ardent pheneticists (who would classify *Archaeopteryx* as a reptile by any criterion of overall similarity). Yet Aves is based on few characters (feathers, for example). We shall see later that its naturalness stems from the fact that the "essential" characters used by Linnaeus to define the group are the very characters (synapomorphies) that demonstrate its phylogenetic naturalness.

Many taxonomists had trouble with the Aristotelean concept of naturalness. Such workers as Adanson (1763), de Jussieu (1789), and de Candolle (1813) moved away from Linnaeus's sexual system with its "essential characters" and began to classify plants taking into account total morphology (in the United States the champion of such new systems was Asa Gray). The natural system began to be recognized as a system of general similarity based upon different life stages with a focus on constant characters within groups. Degree of similarity became the criterion for defining natural groups. No "essential" characters were necessary or sufficient—group inclusion is based upon the maximum sharing of all possible characters. Numerical taxonomists (as exemplified by Sokal and Sneath, 1963; the up-dated Sneath and Sokal, 1973; also Davis and Heywood, 1965) claim their roots with Adanson but have a slightly different concept termed "Gilmour naturalness." This concept will be discussed in Chapter 6 because it relates to the controversy surrounding numerical taxonomy and phylogenetics.

With the rise of evolutionary thinking, concepts of naturalness began to be linked with phylogenetic naturalness. As Mayr (1942) states:

> the puzzle of the high degree of perfection of the natural system (was solved) in a manner that was as simple as it was satisfactory: The organisms of a "natural" systematic category agree with one another in so many characteristics because they are descendants of one common ancestor! The natural system became a "phylogenetic" system. The natural system is based on similarity, the phylogenetic system on the degree of relationship.

The Natural Taxon

One could argue that a natural taxon is one that competent taxonomists agree is natural. Or one could argue that a natural taxon is one that can be defined by a set of operations. But neither of these approaches is really based on what phylogeneticists perceive as the connotation of the word "natural." When applied to science, naturalness carries the connotation of "existing in nature, neither artificial nor man-made." In the spirit of this connotation we may define a natural taxon in the following manner:

> A **natural taxon** is a taxon that exists in nature independent of man's ability to perceive it.

This definition of a natural taxon carries several specific connotations.

1.   Natural taxa exist whether or not there are any systematists around to perceive or name them.

2.   Because they exist in nature, natural taxa must be discovered, they cannot be invented.

3.   Natural taxa originate according to natural processes and thus must be consistent with these natural processes.[1]

---

[1] It should be noted that it can work the other way. Natural taxa may precipitate the fall of a currently accepted natural process in favor of a new natural process. Nevertheless, it must be consistent with a perceived natural process, new or old.

4. When we propose natural taxa, that is, when we hypothesize that a particular grouping is natural, we invoke all of the connotations implied in 1–3.

The idea that natural taxa are groupings of organisms existing in nature is neither new nor novel. We could, no doubt, trace the origin of this concept back to the classical Greeks. Among recent authors, several have stated one or the other connotations. Simpson (1961: 55) states:

The taxa of natural classifications must have some relationship ... with *groups* of whole organisms really existing in nature.[2]

Crowson (1970: 275) states:

a perfectly natural classification of plants and animals might even be considered as objectively existing, and thus requiring to be discovered rather than invented.

Hennig (1966: 77–83) clearly held an identical concept of natural taxa when he described the groups of the phylogenetic system as groups characterized by individuality and reality. Bock (1977: 868) agrees with connotation 3 in stating that "Natural has always had this meaning of being in agreement with the theory (underlying it)."[3]

We may now turn our attention to the question of what exists in nature. We might say that things we suspect to exist in nature as discrete entities are things we can explain or describe in reference to perceived natural laws or processes. For example, physicists believe that the various elements are natural entities because they and their interactions are understood (albeit incompletely) by certain physical laws and the processes these laws describe. At the biological level of complexity there are many kinds of natural entities, or I should say entities we believe are natural because we can explain their characters and interactions in terms of scientific theories. Among those entities not dealt with by systematics in a strict sense are cells and their subcellular parts (cell theory), individual organisms (reproductive theory and, in the case of multicellular organisms, ontogeny), and ecosystems (various ecological theories). We shall see in the next section that supraspecific taxa are natural entities, but of a different sort from those just mentioned. And we shall see in

[2] Simpson also states that there need not be a one-to-one ("point for point") correspondence between the taxa and the groups existing in nature. This robs his particular concept of any critical meaning, for if there is no one-to-one correspondence, then what correspondence should there be (Hull, 1964)?

[3] Bock, an evolutionary taxonomist, provides the very basis for rejecting his system with his own concept of "natural" because, as we shall see in Chapter 7, evolutionary taxonomic classifications *are not in agreement with current evolutionary theory.*

subsequent sections that natural supraspecific taxa are monophyletic groups sensu Hennig (1966).

The Natural Higher Taxon

To understand the nature of natural higher taxa we must consider the differences between classes, individual entities, and supraspecific taxa. Although Hennig (1966) seemed to have carefully distinguished between these concepts, Michael Ghiselin (1966, 1969, 1974, 1980) independently resolved the ontological status of individuals and classes as they relate to biological taxa. Ghiselin's ideas have profound implications for systematics and have greatly influenced (indeed, guided) my thinking about species and natural taxa. The following distinctions between individuals and classes are taken from the excellent discussion by Hull (1976) and references therein.

A **class** is a construct that has members. Membership in the class is determined by the class definition or class concept. Classes are spatiotemporally unbounded things constructed such that any entity that fits the definition of a particular class belongs to that class regardless of its historical origin or its location in time and space. For example, "helium" is a class defined as containing entities that have an atomic number of two. An atom or group of atoms all with the atomic number of two are helium. This is true throughout the universe, even when the helium atoms have quite different historical origins. The evolutionary species concept is also a class construct. Any unit of evolution which fits the concept is an evolutionary species. Thus we might expect to find evolutionary species anywhere in the universe where organic evolution has occurred.

In contrast to classes, **individuals** are restricted to particular spatiotemporal frameworks and have both cohesion and continuity. Because individuals are spatiotemporally restricted, they have particular and unique places in history. Cohesion and continuity imply that individuals are actual participants in natural processes (classes do not so participate). A particular helium molecule is an individual and participates in various processes. Particular individual species (*Homo sapiens* for example) are individuals which participate in evolutionary processes (and many other processes). While a group of sapient organisms on Earth and another group on a distant planet may both comprise evolutionary species, they are not members of the same species (unless they are historically connected). This is true even if they look exactly the same, because they have quite different origins and thus different insertions in history.

We may now consider supraspecific taxa. A natural taxon comprising two or more species is a spatiotemporally bounded entity. In being so, it is distinctly like an individual. However, supraspecific taxa that are natural differ from species and other individuals in several respects.

Cohesion in a species is maintained by reproductive ties (in the case of sexual species), evolutionary stasis (asexual and sexual species), and similar responses of the component organisms of the species to extrinsic factors of evolution. In contrast, there is no active cohesion within a natural supraspecific taxon because it is comprised of individual evolutionary units which have the potential to evolve independently of each other. Another point should be mentioned. Sexually reproducing species have continuity based on reproductive ties among the individual organisms that comprise the species. But natural supraspecific taxa have only a historical continuity of descent from a common ancestral species. In other words, species show both historical and ongoing continuity whereas supraspecific taxa have only historical continuity. These important distinctions result in a simple characterization of species and higher taxa; species are units of evolution, and higher taxa containing more than one species are not units of evolution (Wiley, 1978); rather, natural supraspecific taxa are units of history.

As units of history, natural supraspecific taxa display the individuallike quality of being spatiotemporally bounded entities but the classlike quality of being entities that do not participate in natural processes while at the same time containing individual entities that do participate in such processes. Wiley (1980) concluded that natural supraspecific taxa are neither individuals nor classes. Rather they are **historical groups** derived from individuals.[4] This concept has some important implications, five of which are listed below.

1. There is no ongoing process which gives a natural higher taxon cohesion nor is there a process by which such taxa arise which can be divorced from speciation.

2. Therefore, supraspecific taxa must be historical units that have resulted from speciation. To put it another way, there is no origin of supraspecific taxa except through the origin of species because species are the largest units of evolution.

3. We may conclude that genealogical lineage splitting and other

[4] Patterson (1978) and Ghiselin (1980) suggest that natural higher taxa are individuals. Both authors are correct in rejecting such taxa as classes, and it is true that higher taxa are individuallike (Hennig, 1966). I do not think that my conclusions differ that much from Hennig's, Patterson's, or Ghiselin's. The distinctions are more likely to be of interest to philosophers than to systematists. The important point of agreement between these authors and myself is that natural higher taxa are real entities with histories and not classes. Two other points may be mentioned. First, Wiley (1980) used the term "historical entity" to refer to natural higher taxa. The term "historical group" is preferred because species and individual organisms are also "historical entities" in that there is a historical component to their identities. Second, Ghiselin (1980) has made the important point that paraphyletic and polyphyletic groupings are classes, not historical groups or individuals. As classes, such groupings cannot be considered natural because the individuals comprising these classes cannot be defined in terms of natural processes.

speciation processes are both necessary and sufficient conditions for the origin of natural supraspecific taxa. Those taxa that do not accurately document these necessary and sufficient conditions cannot be natural taxa.

4. A natural supraspecific taxon cannot overlap another at the same level of universality. By this I mean that when we have two supraspecific taxa of the same categorical rank one cannot contain species that are more closely related to species placed in the other supraspecific taxon than to the species placed in its own supraspecific taxon. However, one natural supraspecific taxon may include others within it. As an analogy, two cells cannot overlap but they may contain within them other individuals (mitochondria, for example). This is one of the more important individuallike qualities of higher taxa. It means, for example, that the definitions of two supraspecific taxa cannot overlap. If these definitions are cast in terms of characteristics, then we may conclude that any single character may only be used once to define a particular historical group and may never be used again to define any entity (either a lower supraspecific taxon or a species) included within this taxon.

5. Although species as individuals have no adequate definition except that of their insertion in history, historical groups must be justified by an investigator in terms of evidence. Thus, species may be "christened" (Ghiselin, 1974), but supraspecific taxa must be justified by characters that demonstrate their status as natural groups (i.e., by synapomorphies).

Monophyly

The arguments presented above outline why unique genealogy on the species level is the necessary and sufficient condition for recognizing a historical group, a natural higher taxon. We shall term such taxa **monophyletic taxa** following Hennig (1966). Hennig gave two explicit definitions of monophyly which will be discussed later in the chapter. For purposes of discussion, I shall now define a monophyletic group (taxon) following the definition of Farris (1974):

> A **monophyletic group** *is a group of species that includes an ancestral species* (known or hypothesized) *and all of its descendants.*

I shall have more to say on other supraspecific groupings in later sections. For now I would only like to stress that only monophyletic groups sensu Hennig (1966) are natural taxa and historical groups. We shall see in the next section that monophyletic taxa must be justified in terms of characters, but that this character justification is not enough in itself to provide a necessary and sufficient condition for a group's status. Rather, characters must be tied to genealogy for a complete and coherent system.

We shall see that this system results in natural taxa that are monophyletic in the sense just discussed.

Characters and Genealogy

Systematists attempt to discover natural groups through the use of character analysis. Characters are the only available evidence we can apply to our efforts, yet from the previous discussion we can see that characters are not the primary basis for natural groupings. Indeed, I previously suggested that characters are neither necessary nor sufficient *by themselves* to justify a group's natural status (Wiley, 1979c). This is because organisms exhibit two phenomena that preclude their use without some bridge principles that specifically tie the characters to the process of genealogical descent. These phenomena are ontogeny and character modification during genealogical descent (Wiley, 1979c).

Ontogeny

Multicellular organisms exhibit a range of characters during development. Frogs, for example, do not exhibit the tetrapod limb until rather late in development, yet during their entire development each individual is a member of Tetrapoda and one or more of the various frog taxa down to species level. A cranial capacity of around 1200 cc is said to be a synapomorphy of *Homo sapiens*. Yet no *Homo sapiens* is born with a 1200-cc brain. At conception both the individual frog and the individual human have only two of the characteristics of their respective array of hierarchial groups. First, each has a unique genealogical descent, particular to the individual within the species and particular to that species compared with other species. Second, most *but not all* individuals have the genetic and cytogenetic characteristics which will guide their development. The first characteristic is both necessary and sufficient to place any individual *Rana pipiens* or any individual *Homo sapiens*. The second is neither necessary nor sufficient because there are always inviable organisms that are members of their species, genus, family, and so on. This inviability is, of course, the direct result of *not* having those factors that will produce the characters used to classify the species. The extent to which the embryo or juvenile has developed will determine the level to which it can be classified on the basis of characters alone.

Character Modification During Evolutionary Descent

If natural taxa exhibited Aristotelean essences, then characters could be used as necessary and sufficient criteria to place all adult organisms, regardless of their insufficiency to place some or most embryos and young of a species or higher taxon. Because of character modification during phylogenetic evolution, however, such a criterion fails in many instances. For example, is the character "tetrapod limb present" both necessary and sufficient to place an organism as a tetrapod? I would certainly say that it is sufficient; I know of no organism that has a tetrapod limb that would not be classified in a phylogenetic system as a tetrapod. But, is it necessary to have a tetrapod limb to be a tetrapod? No, many adult snakes and legless lizards have no vestige of a tetrapod limb. They are tetrapods nevertheless. The absence of tetrapod limbs in snakes and legless lizards are separate subcharacters of the presence of a tetrapod limb. But we can draw this conclusion only because all the other knowledge we have indicates that both snakes and legless lizards descended, separately, from legged ancestral species.

Genealogy, Characters, and Bridge Principles

We may conclude that characters alone are insufficient to define a natural taxon. What I mean is that characters cannot be divorced from evolutionary principles that tie characters to genealogy. These evolutionary principles may be termed "bridge principles" (Hempel, 1966). I have previously suggested two such principles (Wiley, 1979c):

Bridge Principle 1. The hypothesized defining characters of a taxon may be used to justify the naturalness of that taxon if it is also hypothesized that these characters indicate that the members of the taxon are genealogically more closely related to each other than to any other organism outside the taxon.

Bridge Principle 2. The hypothesized defining characters of a proposed natural taxon may be present only in certain stages of ontogeny or they may be modified during the subsequent evolution of subtaxa of the proposed natural taxon.

These bridge principles require only two assumptions about the evolutionary process: (1) genealogical descent occurs in nature and (2) heritable characters may be modified during genealogical descent. They allow us to escape a real dilemma—how to proceed from characters that are observable but neither necessary or sufficient to justify the naturalness of a taxon to genealogy that is necessary and sufficient to justify this naturalness but is not directly observable.

I draw three basic conclusions from this discussion about natural taxa:

1. Natural supraspecific taxa are genealogical entities and they are historically unique.

2. Characters hypothesized to justify unique genealogical groups must be those which indicate that the descendant members of the group share a common ancestral species not shared with another taxon. The

only characters that justify such groups are synapomorphies employed at their correct level of universality (Wiley, 1975, 1976, 1979c).

3. Unique genealogical groups corroborated by synapomorphies are monophyletic groups sensu Hennig (1950, 1966). *Thus, monophyletic groups sensu Hennig are natural taxa.*

The Insufficiency of Overall Similarity

I have tried to establish two points. First, natural biological groups dealt with by systematists are genealogical groups and inferred or observed genealogy is both necessary and sufficient to include an entity in a group. Second, certain characters (apomorphies) are biologically connected to the concept of genealogy and thus can provide justification for group membership in the absence of directly observed genealogy. In this section I will take up the insufficiency of overall similarity in justifying natural taxa.

Modern concepts of overall similarity have been developed by pheneticists. Pheneticists, like phylogeneticists, claim that the best system of classification is one that contains natural taxa. Thus, both groups are striving for an estimate of the natural order of organisms. But, pheneticists and phylogeneticists have radically different opinions as to the relationship of characters to the natural system.

To understand the numericist's concept of natural taxa, it is necessary to understand the types of organismic groupings distinguished by Sneath and Sokal (1973: 20–21), monothetic and polythetic groups. The monothetic group is characterized by having a number of characters exhibited by members of the group which are both necessary and sufficient to place a particular taxon in that group. Examples are Aves and Mammalia. Sneath and Sokal (1973) consider such groups "rigid," primarily because they are defined on the basis of a few characters rather than many characters. In contrast, they consider "polythetic groups" as natural. To understand what a polythetic group is to Sokal and Sneath, I will introduce its definition and contrast it with monothetic and fully polythetic groups following the definitions of Beckner (1959: 22; the same definitions used by Sneath and Sokal, 1973: 21).[5]

G is a group (set) of entities.

g_1-g_n are the entities (subsets) belonging to G.

C_1-C_n are the group characters of G.

For example, if G is the group Mammalia, then g_1 might be monotremes,

---

[5] The concept of monothetic and polythetic espoused by Sneath and Sokal and by Beckner differs from the concepts held by others.

g_2 therians; C_1 might be presence of hair, C_2 presence of mammary glands, and so on. We may now make a formal distinction between monothetic and polythetic.

   1.   A set, G, is defined by characters C_1–C_n.
   2.   If all of the subsets of G (g_1–g_n) have all of the characters C_1–C_n then the set G is **monothetic**.
   3.   If each g_n has a large (but unspecified) number of C_1–C_n, but *not all* C_1–C_n, and *if* each C_n is possessed by a large (but unspecified) number of g_n, but *not all* g_1–g_n, then the group is **polythetic**.
   4.   If no one character of the arrary C_1–C_n is possessed by *all* of the subgroups g_1–g_n, then the group is **fully polythetic**.

Sneath and Sokal (1973: 21) then make a simple assertion—"in a polythetic group, organisms are placed together that have the greatest number of shared character states." From this assertion they then justify overall similarity as being the criterion for natural taxa. Because polythetic groups are based on overall similarity, and because polythetic groups are natural, natural taxa are based on overall similarity. This is a bold assertion and conclusion because *if* it can be shown that polythetic groups are not definable on the basis of overall similarity, but only by special similarity, then we may conclude that phenetic taxa are not necessarily natural. Later I will show that polythetic groups cannot be defined on the basis of overall similarity. Further, I will show that there is a distinction between the monothetic and polythetic groups of pure mathematical set theory and the mono- and polythetic groups of biological classification.

Are polythetic groups based on overall similarity? To put it another way, are the C_1–C_n group characters all of the potential characters observed in the organisms of G (or some average of this similarity)? My answer is no; the reasoning is rather simple.

If our set, G, needs to be defined by characters, it needs to be defined because there is another group in the universe containing G. In terms of set theory, there must be a G and a not-G (or, $\overline{G}$) which comprise the universe, U, we are considering. This is expressed in the basic set formula:

$$G + \overline{G} = U$$

In the universe of organisms on earth, we may examine several examples and their defining characters. Before we do, please note that the formula exists only at a single particular level of universality (Wiley, 1975) and that this single level of universality depends entirely upon the organisms included in the set G. At this particular level of universality there are four types of homologous characters (admitting evolution as an axiom of the system). The first type of homologous character is a homologue found in G and all of $\overline{G}$. The characters "presence of carbon, oxygen, and hydrogen" and "presence of hereditary material in the form of RNA and/or DNA" are examples of such characters. These are defining characters of neither G nor some portion of $\overline{G}$. They are defining characters of U. Since all life exhibits certain of these characters, we may make our first conclusion: there are no fully polythetic groups of organisms (Sneath and Sokal, 1973, alluded to this conclusion). The second set of homologues are those found in some subgroups of $\overline{G}$ and in some or all subgroups of G. These characters can define neither G nor U. Rather, they define a subgroup of U composed of some (but not all) subgroups of $\overline{G}$ plus G. In other words, they are synapomorphies of a monophyletic group existing at a higher level of universality than the problem at hand. For example, the presence of a tetrapod limb in mammals defines neither Mammalia nor "life," but Tetrapoda, a group intermediate in universality between the set and the null set (between G and $\overline{G}$). The third set of homologues is found only in G and not in any subgroup of $\overline{G}$. Only these characters can define G in reference to $\overline{G}$. Thus the true (hypothesized true) group characters of G are synapomorphies at the level of universality of G in relation to U. Examples of this type of character are hair in Mammalia and the tetrapod limb in Tetrapoda. The fourth set of homologues are those found in some subgroups of G, but not in all subgroups of G. These are synapomorphies which define a subset within G (at a lower level of universality than the problem at hand). For example, the lack of a tetrapod limb in snakes defines the subgroup Serpentes (Ophidia) within the Tetrapoda. But, some average value of the lack of the tetrapod limb and the presence of the tetrapod limb within the Tetrapoda does not provide a valid group character (a hypothesized true C_n) for Tetrapoda. I will summarize my conclusions in a series of statements:

1. All characters shared among the subgroups of $\overline{G}$ and G are group characters (C_s) of the universe to which G and $\overline{G}$ belong. Therefore, the group characters of G cannot be determined without reference to $\overline{G}$. This provides the basic rationale for the principle of the out-group comparison in phylogenetic systematics.

2. Characters shared among the subgroups of G and some, but not all subgroups of $\overline{G}$ are hypothesized group characters (C_s) of a group containing G and those subgroups of $\overline{G}$ which display the character. An array of such characters may define (purport to define) varying subgroups of U of intermediate levels of universality between U and G.

3. From the previous two conclusions I draw the deduction that attempts to define G without reference to $\overline{G}$ by the use of some measure of overall similarity cannot be justified. The concept of overall similarity cannot discriminate the hypothesized true group characters from other

characters. Thus, measures of overall similarity cannot discriminate between a monothetic and a polythetic set. Some of the supposed C_s purporting to define G may actually define either subgroups of G or a group composed of G and a subgroup of $\overline{G}$. It follows that an average of such an array of characters cannot define G in relation to $\overline{G}$.

4. I conclude that when a group based on overall similarity corresponds with a group based on synapomorphies, this correspondence is due to the fact that the noise of irrelevant characters is not loud enough to influence significantly the performance of the true group characters employed. I also conclude that groups based on overall similarity cannot be evaluated critically because it is not knowable from the results of such analyses which of the so-called "natural" groups are based on relevant characters and which are not. *If we worked with only the hypothesized relevant characters the phenetic system would collapse into a pure phylogenetic system in which sets were analogous to monophyletic taxa.*

5. This line of reasoning also has relevance to the claims of evolutionary taxonomy. If measures of overall phenetic similarity cannot be used to defined groups, then measures of overall genetic similarity cannot be used to define natural groups. Additionally, paraphyletic groups cannot be defined on the basis of overall similarity any more than overall similarity can be used as a general paradigm for classification.

Pheneticists are to be congratulated for attempting to search for a natural system of classification. I believe that Sneath and Sokal (1973) were largely justified in viewing with suspicion groups based on few (and often unanalyzed) characters. But one *true* synapomorphy is enough to define a unique genealogical relationship. The problem is that the synapomorphies we hypothesize may or may not be true synapomorphies. Thus, the more shared derived characters we have to corroborate a group's purported natural status, the better our hypothesis. I suggest that numerical taxonomy's major failing lies in the fact that the characters exhibited by organisms apply to grouping of organisms at varying levels of universality. Thus, estimates of overall similarity include numbers of characters which do not apply at the level of universality of the problem to be solved.

TERMS FOR SUPRASPECIFIC TAXA

As I stated in the introduction to this chapter, any combination of two or more species given a name constitutes a supraspecific taxon. I have equated only one type of supraspecific taxon, the monophyletic taxon, with the concept of the natural taxon. This follows directly from Hennig (1966). Hennig (1966) distinguished two additional classes of supraspecific taxa which he considered nonnatural, *paraphyletic* and *poly-phyletic* taxa. There has been a controversy over what these terms should mean and the purpose of this section is to examine this controversy and arrive at definitions for these terms. The controversy revolves around four sets of definitions, Hennig's (1966), Aslock's (1971), Nelson's (1971), and Farris's (1974). For clarity of discussion all four sets of definitions are presented in Table 3.1. One additional set of definitions has been presented by Simpson (1961). Simpson did not recognize paraphyletic groups, but his concept of monophyly results in paraphyletic groups like those of Ashlock's. I shall not take up the claim that paraphyletic groups are monophyletic in this section, that claim is discussed in Chapter 7.

Monophyletic Groups

Hennig (1966) equated monophyletic group with natural taxon as an individuallike entity. There is general agreement among phylogeneti-cicts that Hennig's concept is biologically relevant. The two definitions employed characterize a particular monophyletic group in terms of the relationships between ancestors and their descendants (dfn. 1) and between descendants in relation to taxa classified outside the group (dfn. 2). "Relationship" in this context means genealogical or blood relation-ship—the relationship between parents and their children or the rela-tionships between an ancestral species and its descendants. In reference to Fig. 3.1, Miidae is more closely related to Xidae than to Yidae because Miidae and Xidae share a common ancestral species that is not shared by Yidae. Hennig (1966: 71) made it clear that a monophyletic group must include the ancestor and all descendants (notwithstanding the opinions of Tuomikoski, 1967, and Ashlock, 1971).

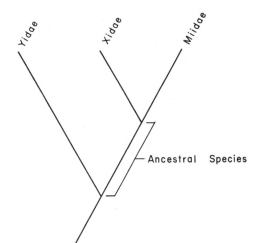

Ancestral   Species

Figure 3.1 Phylogenetic relation-ships of three hypothetical families and one ancestral species.

Table 3.1 Definitions of Monophyly, paraphyly, and polyphyly

Author	Monophyly	Paraphyly	Polyphyly
Hennig (1966)	1. A group of species descended from a single ("stem") species and which includes all species descended from this stem species 2. "A group of species in which every species is more closely related to every other species than to any species that is classified outside the group" 3. (Characterization)—a group based on synapomorphous similarity	1. "A group of species that has no ancestor in common only with them and thus no point of origin in time only to them in the true course of phylogeny" 2. (Characterization)—a group based on symplesiomorphous characteristics	1. (By inference)—a group in which the ancestor is not included in the group 2. (Characterization)—a group based on convergent similarity
Ashlock (1971)	1. A group whose most recent common ancestor is cladistically a member of the group	1. A group that does not contain all of the descendants of the most recent common ancestor	1. A group whose most recent common ancestor is not a member of the group
Nelson (1971)	1. A group into which have been placed all species or groups of species that are assumed to be descendants of a single hypothetical ancestral species, that is, a complete sister-group system	1. An incomplete sister-group system lacking one species or one monophyletic species group	1. An incomplete sister-group system lacking two species or monophyletic species groups that together do not form a single monophyletic group
Farris (1974)	1. *A group that includes a common ancestor and all of its descendants* 2. *(Algorithm dfn.)—a group with unique and unreversed group membership characters*	1. *A group that includes a common ancestor and some but not all of its descendants* 2. *(Algorithm dfn.)—a group with unique but reversed group membership characters*	1. *A group in which the most recent common ancestor is assigned to some other group and not to the group itself* 2. *(Algorithm dfn.)—a group whose membership characters are not uniquely derived*

Hennig (1966) viewed his definitions of monophyly simply as critical restatements (improvements) over less restrictive definitions that had come into vogue since Haeckel (1866) coined the term. Some critics have complained that Hennig's concept of monophyly is nontraditional, or even "contradicted by common sense" (Mayr, 1969: 75). I reject the "nontraditional" argument because tradition has no particular place in science. We may judge whether Hennig's concept is contradicted by common sense by examining other definitions and characterizations of monophyly.

Mayr (1942: 280) characterized a monophyletic group in this way:

> We employ the term monophyletic as meaning descendants of a single interbreeding group of populations, in other words, descendants of a single species.

Hennig (1966) criticized such definitions as being incomplete because they did not specifically state that all of the descendants must be included in the group. Earlier definitions seem more consistent with Hennig's when all of their logical connotations are examined. For example, Wernham (1912) considered a group monophyletic if the ancestor was included in the group. This fits well with the definition of Heslop-Harrison (1958). Both definitions logically produce the same results as Hennig's definition. For example, if Mammalia is to be considered monophyletic we would have to dismember the Reptilia because, as it now stands, the stem mammal is, by inference (Reptilia gave rise to Mammalia), a reptile. Further, each succeeding relative of mammals would have to be classified in such a way that ancestors could be placed with descendants. This would result in a system of classification that is phylogenetic sensu Hennig. One might argue that Reptilia is monophyletic because the stem reptile is placed within the group. But such a claim could only be supported if Aves and Mammalia were also placed within Reptilia. This taxon already exists in the phylogenetic system, it is termed Amniota.

It is not my intention in bringing up Wernham or Heslop-Harrison to defend Hennig's concept as traditional. Rather, I wish to point out that some older definitions have the same logical connotations as Hennig's and that Hennig had good historical precedent for formulating his definitions in the manner he did. To say that Hennig has "created enormous confusion by adding to the traditional definition" (Mayr, 1974: 104) is simply wrong. Traditional definitions result in Hennig's concept if their logical connotations are followed. Hennig simply took these logical connotations and formulated a definition that made them explicit. In doing so, he fostered clarity, not confusion.

The question of whether Hennig's concept of monophyly corresponds in some critical way to the basic principles of evolution is a more interesting matter, one that I have dealt with in previous sections. It would seem to fulfill the two criteria advocated by Mayr (1974: 104) concerning monophyly:

1. It results in groups whose "component species, owing to their characteristics, are believed to be each other's nearest relatives."[6]

2. It results in groups whose members "are inferred to have descended from the same common ancestor (ancestral species)."

Paraphyletic Groups

In Table 3.1 you will note that all four sets of definitions concerning monophyly are rather similar. This is not true for the various definitions of paraphyly. Hennig's principal criterion for distinguishing paraphyletic groups was based on characters. Consider the genealogical relationships shown in Fig. 3.2. If A and B are grouped together on the basis of a plesiomorphic similarity, then the taxon AB is paraphyletic. The following indicates the groupings and their character justifications:

Classification	Character Justification
ABCD	6'
AB	3, 4, 5
CD	3', 4', 5'

Hennig (1975) considered this definition to exist purely on a methodological level and he considered paraphyletic groups one of two types of nonnatural groups. But definition 1 in Table 3.1 can clearly be related to the definitions provided by Ashlock and Farris. Thus, Hennig's genealogical concept of paraphyly closely resembles all other definitions except Nelson's (1971). To clarify Nelson's definition we must first consider polyphyly.

Polyphyletic Groups

There is agreement among both evolutionary taxonomists and phylogeneticists that polyphyletic groups are nonnatural. Hennig's characterization of polyphyly (Table 3.1, dfn. 2) results in the concepts espoused by both Ashlock and Farris. Nelson's definition is rather different. Nelson considers a group paraphyletic if one sister-group system (monophy-

---

[6] Mayr (1974) equates "relationship" with "genetic relationship" and asserts that genetic relatives are not necessarily genealogical relatives. In Chapter 7 we shall see that this assertion is not supported by the facts presently available.

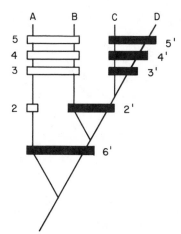

Figure 3.2 Phylogenetic relationships among four hypothetical taxa. Synapomorphies are black bars, plesiomorphies are open bars.

letic subgroup) is removed from a taxon and he considers a group poly-phyletic if two or more sister-group systems are removed from a group that are not themselves sister groups (i.e., the groups in question must not form a monophyletic subgroup within the larger taxon). For example, all of the authors cited in Table 3.1 would consider the dicots a para-phyletic group. All authors except Nelson would consider Reptilia a paraphyletic group. Nelson's definition renders Reptilia polyphyletic since both Aves and Mammalia are excluded from the group.

Discussion of Paraphyly and Polyphyly

I shall use Farris's (1974) definitions of paraphyly and polyphyly throughout this book. It is important to understand why this is desirable and why the other definitions will not do.

Hennig (1975) considered the distinction between paraphyletic and polyphyletic to be useful as reflective terms to characterize the groupings of other authors. If an investigator justified a taxon with plesiomorphies then that taxon is paraphyletic, if with homoplasies (convergences or parallelisms) then the taxon is polyphyletic. The weakness of this ap-proach is that the application of this distinction may result in identically circumscribed taxa being either paraphyletic or polyphyletic and yet have exactly the same genealogical tree topology. Examine Fig. 3.3. Taxon A–F is a monophyletic group. If taxon ACE is justified on the basis of a plesiomorphy, then ACE is paraphyletic. The alternate group-ing, BDF, may be rendered polyphyletic simply by making its justifi-cation a convergence. Both groupings have the same tree topology. And either group might be termed polyphyletic if the right characters were selected. While Hennig's (1966, 1975) distinction between paraphyletic

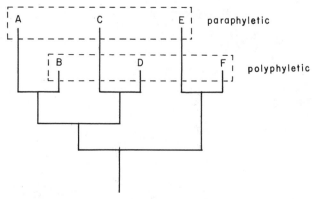

Figure 3.3 Hennig's concepts of paraphyly and polyphyly; the group ACE is paraphyletic because it is based on plesimorphic characters whereas the group BDF is polyphyletic because it is based on a convergent character. (Redrawn from Hennig, 1975. Used with permission of the editor, *Systematic Zoology*).

and polyphyletic groups certainly has utility when discussing characters, it simply will not do as a universally satisfactory distinction.

Ashlock's (1971) definitions are closely akin to Farris's and to the concept of Hennig (1965). The problem with Ashlock's concepts is that they depend on actual identification of ancestral species to fully distinguish paraphyly and polyphyly in every case. We may consider Fig. 3.4 to examine this point. This figure is identical to Ashlock's (1971) Fig. 1 except that the ancestral species are considered lineages rather than nodes. Figure 3.4a is Ashlock's concept of "holophyly" (monophyly sensu Hennig), 3.4b his concept of paraphyly, and Fig. 3.4c his concept of polyphyly. The following are the classifications resulting from these concepts. Note that the classifications containing a paraphyletic group and a polyphyletic group are exactly the same.

	Classifications	
Figure 3.4a	Figure 3.4b	Figure 3.4c
ABCD	ABCD	ABCD
AB	A	A
A	BC	BC
B	B	B
CD	C	C
C	D	D
D		

Suffice to say, Ashlock (1971), like Hennig, did not solve the problem of distinguishing paraphyletic from polyphyletic groups at the level of

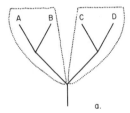
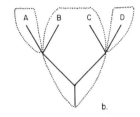
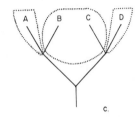

Figure 3.4 Ashlock's concepts of (*a*) holophyly, (*b*) paraphyly, and (*c*) polypyly. (Redrawn and modified from Ashlock, 1973.)

actual classification (also see Nelson, 1971; Colless, 1972, Farris, 1974; Hennig, 1975; Platnick, 1977a; and Wiley, 1979c).

Nelson's definitions are very different from the others because polyphyly and paraphyly are defined not by the inferred inclusion of an ancestral species but by the overt omission of certain numbers of species from a group which includes that ancestor. Farris (1974) chose to lay aside Nelson's definitions because:

> It is, however, useful to distinguish between the concepts of a paraphyletic Reptilia—a group united by a series of amniote characters—and a polyphyletic Reptilia—a group distinguished by characteristics which evolved from "amphibian" to "reptilian" stages of expression in two or more independent phyletic lineages.

I agree with Farris (1974) and believe that his definitions provide a more satisfactory basis for distinguishing these various groupings.

Examples of Monophyly, Paraphyly, and Polyphyly

In adopting the set of definitions of Farris (1974) presented in Table 3.1, it is important to distinguish between the set of definitions based on genealogy *per se* and the set of definitions that define genealogy in terms of group membership characters and the algorithm that accompanies these operational definitions. The purpose of this section is to distinguish these concepts and to utilize the algorithm on some simple examples.

The algorithm Farris (1974) suggested for checking the status of any proposed group is based on the assignment of **group membership characters** followed by a check as to whether the group membership characters are unique and unreversed or, if reversed, how many times they are reversed. A group membership character is nothing but a number assigned to each member of a group. For example, if we wanted to check the status of the Reptilia (Fig. 3.5), we would begin by assigning all members of the Reptilia the group membership character 1. Then all nonreptile groups at the level of the group to be evaluated would be assigned the alternate group membership character 0, which means "not-

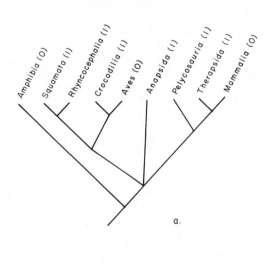

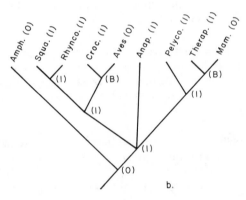

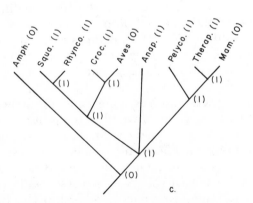

Figure 3.5 Steps in implementing the algorithm of Farris (1974) in determining the paraphyletic nature of Reptilia. See text for further explanation.

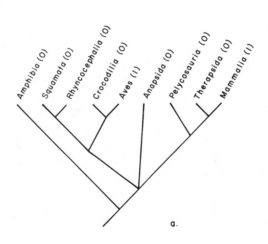

a.

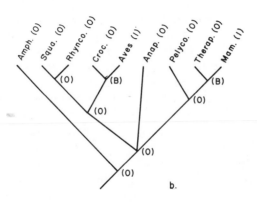

b.

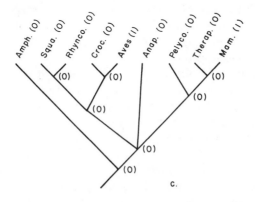

c.

Figure 3.6 Steps in implementing the algorithms of Farris (1974) in determining the polyphyletic nature of Homothermia. See text for further explanation.

reptile." These assigned numbers are then placed on a cladogram of relationships based on actual characters showing the sister group relationships of the taxon under consideration and its relatives. This has been done in Fig. 3.5a. Group membership characters are then assigned to the hypothetical ancestors. If both descendants have the same group membership character number, then their immediate ancestor is assigned the same number. For example, the common ancestor of rhyncocephalians (= 1) and snakes and lizards (= 1) is scored "1." If the two groups are scored differently, as are the crocodiles and birds, then the common ancestor is provisionally given a noncommittal designation "B" which means "either 1 or 0." Ancestors whose descendants are scored as either "B" or "1" are then scored "1." For example, the common ancestor of squamates (rhyncocephalians, lizards, and snakes), crocodiles and birds is scored "1" because one of its descendants (the common ancestor of squamates) is scored "1" while the other descendant (the common ancestor of crocodiles and birds) remains "B." The common ancestor of all groups in the analysis is then scored "0" because the base of the cladogram represents all other nonreptiles not included in the analysis. This stage of the evaluation is shown in Fig. 3.5b. The investigator then goes up the tree assigning all B-scored ancestors the value of its own immediate common ancestor. For example, the B value of the common ancestor of crocodiles and birds would be reassigned the value of 1 and likewise for the immediate ancestor of therapsids and mammals. At the end of this part of the analysis there would be no more B values assigned to any of the ancestors on the cladogram, all hypothetical ancestors would either be assigned a 1 or a 0. This is shown in Fig. 3.5c. We may now begin the actual evaluation of the group in question. Reading up the tree, if we go from $0 \rightarrow 1$ only once and there are no reversals from $1 \rightarrow 0$, then the group in question is monophyletic. If, however, there is one or more reversals from $1 \rightarrow 0$, then the group is paraphyletic. We can see from the example that the Reptilia is paraphyletic because, although we went from $0 \rightarrow 1$ only once we also had two reversals from $1 \rightarrow 0$; one at the level of Mammalia and one at the level of Aves.

We now take the example of the "Homotherma" to show its polyphyletic nature by Farris's definition. In polyphyletic groups, the group membership character is not uniquely derived. In Farris's algorithm this is reflected by finding the transformation $0 \rightarrow 1$ *more than once*. If we go through the same procedure with the "Homotherma" as we did with the reptiles, we will arrive at a series of cladograms with group membership characters like those shown in Fig. 3.6a, b, and c. In this particular case the $0 \rightarrow 1$ transformation has occurred twice. Therefore, by Farris's criteria, the "Homotherma" is a polyphyletic group.

In conclusion, Farris's definitions lead to unambiguous results concerning the nature of particular supraspecific groupings. They would also seem to conserve the original intent of Hennig. More importantly, they provide a method for distinguishing natural and nonnatural groups.

Chapter 4

Phylogenetic Trees

Phylogenetic trees are diagrams depicting the evolutionary descent of whole organisms or groups of whole organisms. The linkage between these organisms is genealogical descent. This descent appeals to one of three processes. Genealogical descent within populations appeals to the process of reproduction for its justification. Linkage between populations within species may appeal to both geographic subdivision and reproductive ties. Linkage between species appeals to speciation for its justification. Since I shall be talking primarily about species and supraspecific groupings I shall restrict my discussions about phylogenetic trees to species and speciation.

Haeckel (1866) is usually regarded as the first evolutionary biologist to publish an explicit phylogenetic tree of real organisms although the honor should probably go to Lamarck. By phylogenetic tree I mean a diagram explicitly denoted as the *evolutionary* relationships among diverse forms. However, diagrams which look for all intents and purposes like phylogenetic trees have been around at least since Agassiz (1844; see Patterson, 1977) and Patterson (1977) points out that they look strikingly like those of later authors such as Romer (1966).

Phylogenetic trees have been depicted in various forms for various purposes. Haeckel's (1866) diagram was actually a tree. Romer's (1966) diagram of vertebrates incorporates an absolute time frame and an estimate of the numbers of species of each group at any one time (see Fig. 4.1). Milne and Milne's (1939) tree of caddis flies is three dimensional and incorporates such factors as habitat and casing construction. Phylogenetic trees may have characters placed on them (see Fig. 4.2a) or they may be bare (see Fig. 4.2b). In short, phylogenetic trees come in many shapes and forms, each with a slightly different emphasis on the subject.

The most basic kind of phylogenetic tree portraying the relationships between species and supraspecific taxa, and the one with the most inherent information is a tree of species. In such a tree, the lines portray evolutionary species and the branches are graphic representations of

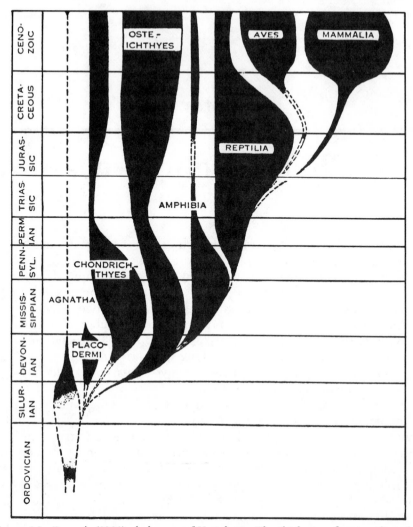

Figure 4.1 Romer's (1966) phylogeny of Vertebrata. The thickness of various branches indicates relative abundance for a clade at any particular time. (From *Vertebrate Paleontology* by A. S. Romer. Copyright 1966 by the University of Chicago Press. Used with permission.)

speciation events. All specialized trees are reducible to such a diagram. For example, the type of tree invented by Agassiz and used by Romer that estimates numbers of species over time may be translated directly from a branching tree of species (see Fig. 4.3*a,b*). I shall term a branching tree of species a **basic phylogenetic tree** because it is the ultimate source from which all trees of supraspecific taxa are derived. It is the basic phylogenetic tree which I wish to discuss in this chapter.

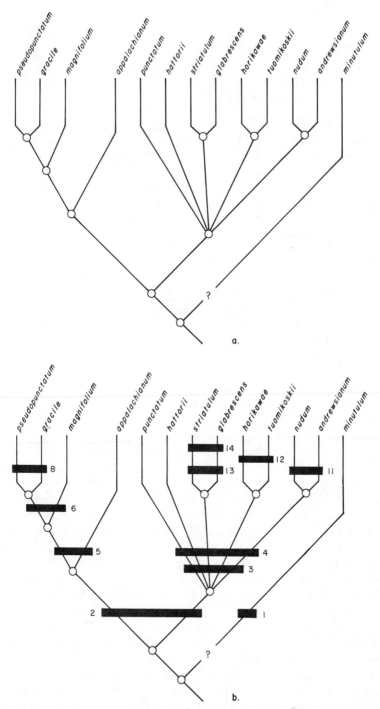

Figure 4.2 Interrelationships of *Rhizomnium* after Koponen (1973): (*a*) the phylogenetic relationships of species; (*b*) the same phylogenetic relationships with characters added which justify the relationships. (*b* redrawn and modified from Koponen.)

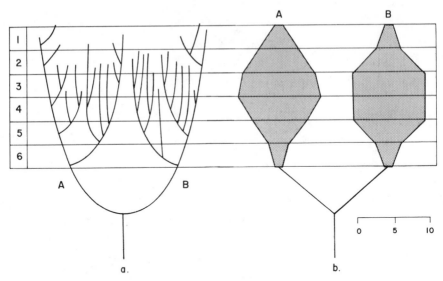

Figure 4.3 Different phylogenetic trees: (*a*) a basic phylogenetic tree of species in taxa A and B through six geological periods; (*b*) a tree of the same groups showing relative abundance but not species interrelationships. The scale to the lower right is in number of species.

PHYLOGENETIC TREES

All basic phylogenetic trees are composed of two things, species and speciation events. The species are evolutionary species, genealogical continua of populations. The speciation events may be of any sort that will produce species. Supraspecific taxa are often included on phylogenetic trees. When included, supraspecific taxa are considered to be represented by their ancestral species. This is, perhaps, intuitive, but it makes a real difference theoretically. As we saw in Chapter 3, a natural supraspecific group is produced when an ancestral species splits. As we shall see, the characters that the daughter species develops before it splits are the synapomorphies of its own group of descendants. Thus, the critical evolutionary divergence between sister groups is that divergence displayed by their separate ancestors from their common ancestor. In Fig. 4.4 this divergence is displayed graphically. Common ancestral species *a* speciates, giving rise to species *a'* and species *b*. Divergence of *a'* and *b* proceeds unequally because of unequal rates of character evolution. The critical difference in evolutionary divergence between clades X and Y concerns only the differences displayed between *a'* and *b*. The fact that a later ancestral species in clade X (ancestral species *b'*) evolved further is of no relevance to the origin and diversification of *b* *relative to a'*. Of course, it is of relevance to the total evolutionary change which has taken place *within* clade X, but not to the basic divergence *between* clades X and Y. Thus, X can be represented by its ancestral species, *b*, in a diagram including Y without loss of information pertaining to the divergence of X and Y. The same is true of Y. And, no relevant evolutionary information will be lost from the diagram concerning the origin of either X or Y.

From this we may say that *all phylogenetic trees are composed of evolutionary species or higher taxa represented by their* (hypothesized or actual) *ancestral evolutionary species.*

Dendrograms and Phylogenetic Trees

Any branching diagram may be termed a dendrogram (Mayr et al., 1953). The nature of the entities and the justification for linking these entities determines the nature of the dendrogram. A phylogenetic tree is one kind of dendrogram. Similar dendrograms have been termed "cladograms" (Mayr, 1969) or "phylograms" (Nelson, 1979) depending on the particular point of view of the author employing the various terms available. There are, of course, other types of dendrograms. The definitions provided below will serve to define various branching diagrams as I use the terms in this book.

1. **Dendrogram.** A branching diagram containing entities linked by some criterion.

2. **Cladogram.** A branching diagram of entities where the branching is based on the inferred historical connections between the entities as evidenced by synapomorphies. That is, a cladogram is a phylogenetic or historical dendrogram.

3. **Phylogenetic tree.** A branching diagram portraying the hypothesized genealogical ties and sequence of historical events linking individual organisms, populations, or taxa.

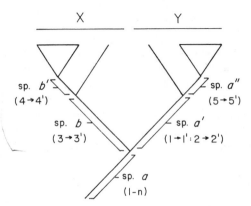

Figure 4.4 The relationships and character evolution among various ancestral species within the taxon XY.

4.   Phenogram.   A branching diagram linking organisms by estimates of overall similarity as evidenced from a sample of characters.

Cladograms and Phylogenetic Trees

The nature of cladograms and their correspondence to phylogenetic trees has been discussed extensively in the recent literature. Until about 1976, most phylogeneticists considered cladograms as phylogenetic trees justified by synapomorphic characters. Since ancestral species could not be identified, such trees were taken as incomplete; that is, they were considered as portraying only sister group relationships. Gareth Nelson, in a manuscript that was widely circulated but never published, challenged all of this by suggesting that cladograms were not phylogenetic trees but dendrograms portraying patterns of unique characters. This distinction was discussed by such workers as Tattersall and Eldredge (1977) and Platnick (1977b). Cladograms to some workers became branching diagrams of organisms without any *a priori* connotation of the special biological relationships of the groups classified. Such cladograms were said to have an array of possible trees (as many as six per cladogram; Platnick, 1977b). Phylogenetic trees were said to be branching diagrams that specify the biological relationships between the groups considered (sister group or ancestor-descendant relationships) and interpret the unique characters as evolutionary innovations. I see nothing wrong in making this distinction. However, some workers, including Cracraft (1974a), Harper (1976), Nelson (ms.), and Platnick (1977b), have come to the conclusion that such a large array of phylogenetic trees exists for each possible cladogram that tree analysis may be prohibitive.[1] I have attempted to show that there is a one-to-one relationship among all dichotomous cladograms and trees with identical topologies (Wiley, 1979a). Of course, this requires that the entities comprising both have specific evolutionary connotations. But evolutionary history is what we are in business to investigate.

Populations, Cladograms, and Phylogenetic Trees

Populations within species are evolutionarily interdependent groups of individual organisms evolving together or in a state of relative evolutionary stasis. Some individuals of a population at a particular time give rise to all the individuals comprising the population at a later time. If we take sectional samples at various time planes we get a succession of populations in ancestor-descendant relation to each other *without* branching. Such a populational genealogy is shown in Fig. 4.5a. A single

[1] The number of trees for each of the four cladograms of three taxa has been variously set at between 13 and 22 (*ibid*).

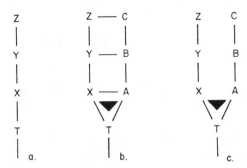

Figure 4.5 Population genealogies: (*a*) the relationship among four populations of the same infraspecific lineage; (*b*) the relationships among populations of two intraspecific lineages which have mating ties; (*c*) the relationships among populations of two intraspecific lineages which do not interbreed. Arrows indicate reproductive ties and triangles represent geographic separations from the ancestral population T.

unbranched line connects the populations because it is supposed that the genealogical relationship is based on a continuum of reproduction without differential divergence. Two or more populations may also be portrayed using a phylogenetic tree. If these populations share reproductive ties, then the relationship between populations may be shown as in Fig. 4.5*b*. If they do not share reproductive ties, then the relation is as in Fig. 4.5*c*. Given these purely genealogical relationships, we may now consider character change.

For a single population, we may have a genetic character evolution such that a character is changed over time. In reference to Fig. 4.5*a*, let us say that character 1 is being replaced by character 1' and that 1 is fixed for X and 1' is fixed for Z. Considering our three samples we might expect a cladogram of these populations like that shown in (Fig. 4.6*a*). Note that the cladogram is different from the tree. But this does not matter because no evolutionary connotations will be drawn from the cladogram. In other words, because it is a cladogram and not a tree, the branch points do not represent anything but linkage of similar phenotypes. We may suspect that because Y is polymorphic for 1 and 1' that it represents a population rather than a species. A rational approach would be to take samples of intermediate populations to see what percentage of individuals had which character. With enough intermediate samples the investigator could establish a statistically insignificant continuum from population to population, establishing the evolutionary unity of this series of ancestor-descendant populations.

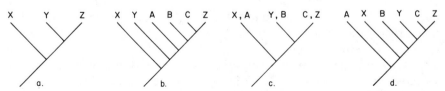

Figure 4.6 Population cladograms interpreting character modification inferred from samples taken from the populations shown in Fig. 4.7. See text for further explanation.

   The same reasoning is applied when examining geographic series of populations existing at the same time period (specifically, for example, the Holocene) to establish that variation observed in a sample organism can be ascribed to geographic variation rather than interspecific variation (for a review of geographic variation see Gould and Johnston, 1972).

   Two (or more) populations through time require both geographic (lateral) and vertical analyses of character variation. In reference to Fig. 4.5b, let us say that character 1' arises in population A, but not in population X. Because the two series of populations occasionally share members through migrations, we may expect the populational lineage XYZ to lag behind the populational lineage ABC in the relative frequency of 1'. A plausible cladogram, with relative apomorphy being expressed as the percent of individuals having 1', is shown in Fig. 4.6b. This cladogram requires explanation as an evolutionary species composed of two interdependent populations because if this were not so, the populations would have to play geographic and temporal leapfrog to produce the observed cladogram.

   The phylogenetic tree shown in Fig. 4.5c assumes one of two things. First, both populations could be in evolutionary stasis. If so, then no branching cladogram could be drawn on the basis of morphology. One could be drawn on the basis of time, as shown in Fig. 4.6c. But only one evolutionary species would be recognized. Second, we could have parallel changes in morphology such that 1' was being fixed at either the same or different rates in the two populational lineages. One possible cladogram of this type is shown in Fig. 4.6d. It would require consideration of a single evolutionary species unless two populations are, again, playing leapfrog with each other's ranges.

   There is, of course, a third possibility in regard to Fig. 4.5b,c. Either one or the other populational lineage could speciate. Such a situation would sort out on the cladogram level but I shall need another example to make this clear.

   Consider the two populational lineages whose population-level phylogeny is shown in Fig. 4.7a. I shall assume that (a) the evolutionary innovation 1' arises among individuals in population A, (b) there is migration between populations, and (c) that 1' is selected for increase in frequency in the A lineage or that it will drift to fixation given a sufficiently frequent 1' in the population (whether 1 or 1' will be fixed given $f(1') = .50$ is a 50–50 proposition). Now let us say that a barrier arises between descendant populations O and D (the stippled triangle), so that no increase in $f(1')$ occurs in the L lineage because it no longer receives migrants from the A lineage. Further, let us say that 1' drifts to extinction or is selected against in the sequence N → O → P while it drifts or is selected to fixation in the C → D → E lineage. Using increasing frequency of 1' as the apomorphy and disregarding stratigraphic position or geography, a cladogram of such samples would look like Fig. 4.7b.

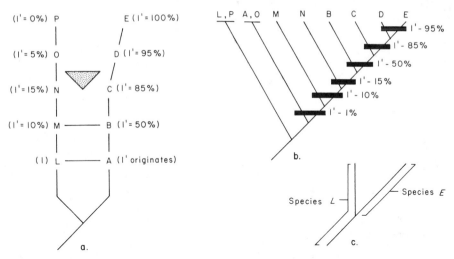

Figure 4.7 Character evolution, population cladograms, and species-level trees I: (*a*) a genealogy of populations with anagenetic evolution in the transformation series 1 ---→ 1′; (*b*) a population cladogram based on increasing frequency of the apomorphy 1′; (*c*) a phylogenetic tree of species based on the population cladogram, biogeography, and temporal sequence of the populations.

The only phylogeny of this cladogram which can be drawn with the additional knowledge of stratigraphy and biogeography is Fig. 4.7c.

There is another possible complication. Although it may be true that 1′ acts as it does, 1′ is only a single character. What happens if the N → P lineage picks up an evolutionary innovation of its own? There are two additional considerations. First, it is possible that this character (2′) arises before the speciation event but acts in the opposite manner to 1′ (i.e., it becomes fixed in N → O and lost in C → E). Second, it could arise after the barrier disrupted migration.

In the first case, we would have a tree as shown in Fig. 4.8*a*. Considering the character evolution of both 1′ and 2′ together, Fig. 4.8*b* shows the cladogram for character 1′ and Fig. 4.8*c* shows the cladogram for character 2′. The two cladograms are not easily reconciled and, in fact, I have had difficulty in trying to put the two together into one harmonious cladogram. The sequences M, N, O, P and B, C, D, E could be characterized by increasing frequencies of 2′ and 1′ respectively. But trying to fit both character transformation series onto a single cladogram seems stilted, and would involve reversals which we know from the (given) model did not occur. Given the geographic and stratigraphic information inherent in the sample itself, coupled with the character analysis, the species-level tree in Fig. 4.8*d* is the logical choice. Note that the absolute barrier was not necessarily the initiation of speciation. In other words, speciation was initiated by whatever factor restricted gene flow (i.e., a

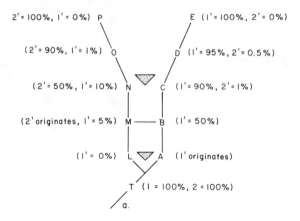

2' = 100%, 1' = 0%) P   E (1' = 100%, 2' = 0%)

(2' = 90%, 1' = 1%) O   D (1' = 95%, 2' = 0.5%)

(2' = 50%, 1' = 10%) N   C (1' = 90%, 2' = 1%)

(2' originates, 1' = 5%) M —— B (1' = 50%)

(1' = 0%) L   A (1' originates)

T (1 = 100%, 2 = 100%)

a.

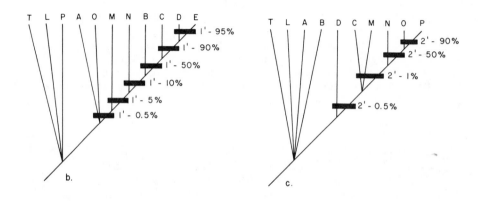

T L P A O M N B C D E

1' - 95%
1' - 90%
1' - 50%
1' - 10%
1' - 5%
1' - 0.5%

b.

T L A B D C M N O P

2' - 90%
2' - 50%
2' - 1%
2' - 0.5%

c.

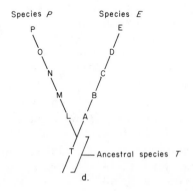

Species P   Species E

P   E

O   D

N   C

M   B

L   A

T —— Ancestral species T

d.

Figure 4.8 Character evolution, population cladograms, and species-level trees II: (*a*) a genealogy of populations with anagenetic character evolution in the transformation series 1 ----→ 1' and 2 ----→ 2'; (*b*) a population cladogram based on increasing frequency of 1'; (*c*) a population cladogram based on increasing frequency of 2'; (*d*) a species-level phylogenetic tree based on both transformation series, biogeography, and temporal sequences of the populations.

parapatric model). However, in this particular example the geographic barrier was necessary for the completion of speciation because if it was not present the sharing of 1' and 2' might have continued until both populations were fixed for both characters.

We may now consider the second case, where 2' is picked up after migration ceases (i.e., after the barrier). This case is shown in Fig. 4.9a.

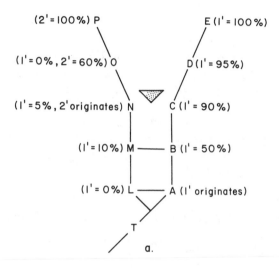

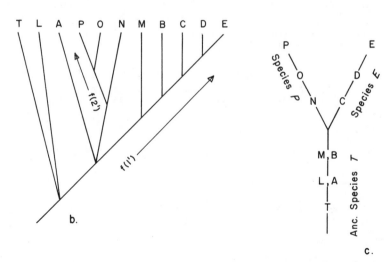

Figure 4.9 Character evolution, population cladograms, and species-level trees III: a genealogy of populations with anagenetic character evolution in the transformation series 1 ———→ 1' and 2 ———→ 2'; (b) a population cladogram based on increasing frequencies of 1' and 2'; (c) a species-level tree reflecting both character transformations, biogeography, and the temporal sequences of populations.

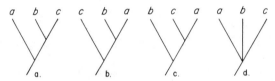

Figure 4.10 Species-level trees I: Four alternate and mutually exclusive hypotheses of the phylogenetic relationships among the species *a*, *b*, and *c*.

A cladogram of populations without regard to geography or stratigraphy but expressing increasing trends in 1′ and 2′ is shown in Fig. 4.9*b*. A tree that takes all these factors into consideration is shown in Fig. 4.9*c*. *Note that the tree in Fig. 4.9c is exactly the same species-level tree as that in Fig. 4.8c.* This leads to an important summary: *Whether an ancestral species speciates by allopatric or parapatric or sympatric speciation, a cladogram of populations will show an identical species-level phylogenetic tree when the additional factors of time of occurrence of samples and geographic positions of samples are incorporated into the analysis.*

Species, Cladograms, and Phylogenetic Trees

Species-level analysis implies a different array of phylogenetic trees than population analysis because the theoretical bases for the relationships between species are different (Chapters 2 and 3). When we consider three evolutionary species there are four possible character cladograms (see Fig. 4.10) and seven possible phylogenetic trees (see Fig. 4.11;

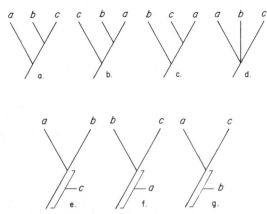

Figure 4.11 Species-level trees II: Seven alternate hypotheses of the phylogenetic relationships among species *a*, *b*, and *c*. Trees (*e*), (*f*), and (*g*) are possible alternative hypotheses of tree (*d*) given that an ancestral species is present in the analysis.

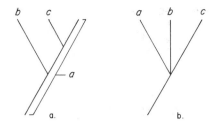

Figure 4.12 Species-level trees III: Interpretation of a dichotomous tree (*a*) as trichotomous tree (*b*) because character evolution did not occur in the ancestral species *a*.

Wiley, 1979a). Each of the dichotomous cladograms implies only a single trees. This is not to say that the phylogenetic tree is correct, but that it is the only alternate tree which can be derived from the data at hand without ad hoc assumptions. The trichotomous cladogram is another matter. Without rejecting some characters as synapomorphies, a dichotomous phylogenetic tree can never be trichotomized. But, a single new character may dichotomize a trichotomy. Thus, most phylogeneticists view trichotomous topologies with the suspicion that they are unresolved dichotomies. However there are situations where we would expect a character cladogram to be trichotomous.

1. *One of the three species is an ancestral species.* Descendant species which share homologous characters share them because they are descendants of an ancestral species which had the characters. It is not necessary for the ancestor to have the character during all stages of its existence, only before it splits. Given that the ancestor was sampled during a time period in which it had fixed the character, we would expect an unresolved trichotomy on theoretical grounds.

2. *The actual lineages are dichotomized but character evolution did not keep step with the speciation.* This situation might be expected whenever the ancestral species survives the initial speciation event. Consider, for example, Fig. 4.12*a*. Species *a* is ancestral to both *b* and *c* but has remained essentially unchanged. Because *b* and *c* share no characters not shared with *a*, the cladogram would continue to show a trichotomy (see Fig. 4.12*b*).

When working with extinct ancestral species there is another possibility, which I will now discuss.

Ancestor-Descendant Relationships as Dichotomies

When one is working with extinct taxa with associated stratigraphic ranges there are other problems. In Fig. 4.13 we have an ancestral species (*Dus aus*) that is evolving anagenetically, replacing the morphological character 1 with 1'. By the time it speciates into *Dus bus* and *Dus cus*, 1' will have been fixed so that we would interpret 1' as a synapo-

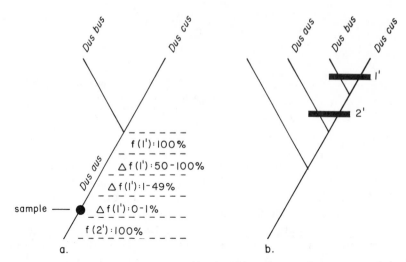

Figure 4.13 Species-level trees IV: Interpretation of a dichotomous tree (*a*) as dichotomous tree (*b*) because the only sample take from the ancestral species was taken before the synapomorphy 1^1 was frequent enough in the population series to be sampled.

morphy of *D. bus* and *D. cus*. If we took a single sample of *Dus aus* from the strata before the innovation occurred, or, if we sampled *Dus aus* when 1′ was so rare that no specimen in our sample happened (by chance) to have 1′, then we would conclude that *Dus aus* was the sister group of *D. bus* + *D. cus* and not the ancestor (see Fig. 4.13*b*). Put simply, our cladogram would be in error. Hopefully, we would sample *Dus aus* throughout its stratigraphic range over several localities and see the anagenesis. This may not be possible. however, and we might continue to assume that two speciation events had taken place rather than one.

Whether ancestral species can ever be identified is a question that has been addressed by several phylogeneticists, and the answer is usually "no" for one reason or another (Hennig, 1966; Engelmann and Wiley, 1977; Platnick, 1977b). However, when morphology, stratigraphy, and biogeography are all applied to groups with a good fossil record, it would seem entirely plausible to postulate ancestral species. Bretsky (1975) suggested this strategy. Wiley (1979b) commented on the possibility that biogeography might help in producing testable hypotheses of ancestry. More recently, Prothero and Lazarus (1980) used all three lines of evidence to discuss an ancestor-descendant relationship between two species of the radiolarian genus *Eucyrtidium* hypothesized by Hays (1970) to be ancestor and descendant. The argument, summarized by Prothero and Lazarus (1980), that *E. calvertense* is ancestral to *E. matuyamai* and that *E. calvertense* survived the speciation event that produced *E. matuyamai* is convincing, for it combines morphological analysis, stratigraphic occurrence, and biogeography in such a way as to make it reasonble to postulate that *E. matuyamai* is the product of a peripheral isolation event that involved a restricted portion of the range of *E. calvertense*. Prothero and Lazarus (1980) concede that the quest for testable hypotheses of ancestor-descendant relationships in many groups will probably fail, but they stress that in at least some groups testable hypotheses can be produced. I agree. And this supports Hull's (1980) assertion that the difference between testing ancestor-descendant hypotheses and sister-group hypotheses is a matter of the quality of the data needed and not a matter of any inherent difference between the two types of relationships. Thus, Bretsky (1975) seems to have been right and I must recant some of my earlier assertions concerning the seemingly impossible task of testing ancestor descendant hypotheses (i.e., Engelmann and Wiley, 1977; Wiley, 1979a). Nevertheless, ancestor-descendant hypotheses require additional data and thus are harder to assess than sister group relationships and this must be realized when one is attempting to postulate such relationships.

Higher Taxa, Cladograms, and Phylogenetic Trees

As species differ from populations in the array of trees which may be inferred from character analysis, so higher taxa differ from species. Species may stand as ancestral to other species or to higher taxa. But, higher taxa cannot be ancestral to other higher taxa because higher taxa are not

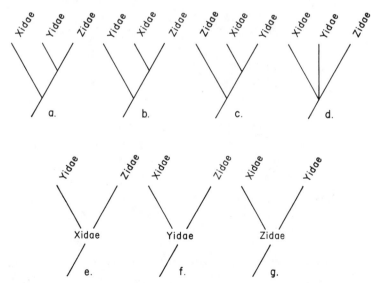

Figure 4.14 Four possible (*a–d*) and three prohibited (*e–g*) phylogenetic trees of the relationships of three higher taxa.

units of evolution but historical units composed of separately evolving species (Wiley, 1978, 1979a,b). Given three higher taxa there are the same possible character cladograms as for three species (see Fig. 4.14a–d). But, there are not seven possible phylogenetic trees. There are only four possible phylogenetic trees. Trees such as those shown in Fig. 4.14e to g are prohibited by the phylogenetic concept of monophyly (see Chapter 3) because supraspecific taxa cannot be ancestors.

Species and Higher Taxa

When we have species mixed with higher taxa it is possible that one species is ancestral to another of the taxa. For example, with two higher taxa and a single species, there are five possible phylogenetic trees (Wiley, 1979a), as shown in Fig. 4.15. A trichotomous character cladogram, identical in topology to tree 4.15d may be (1) an unresolved dichotomous tree, or (2) an unresolvable character cladogram representing a true trichotomous speciation event, or (3) a tree in which *Yus aus* is the ancestral species of families *Xidae* and *Zidae* (Fig. 4.15e).[2]

For two species and one higher taxon there is one additional possible tree. In Fig. 4.16 we can see that one or the other species may be an ancestor (both cannot be because of the prohibition of phyletic speciation).

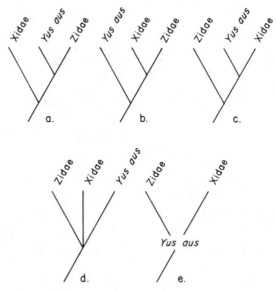

Figure 4.15 Five alternate phylogenetic trees of the relationships among two higher taxa and a species.

---

[2] I will discuss in Chapter 6 the conventions for classifying such trees.

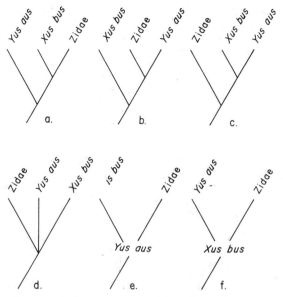

Figure 4.16 Six alternate phylogenetic trees of the relationships among two species and one higher taxon.

TESTING PHYLOGENETIC TREES

We shall see in Chapter 5 that the results of character analysis are the major bases for accepting one phylogenetic tree over another. Further, biogeographic analysis may be of some help in resolving questions relating to trichotomous situations (see Chapter 8). For now, I would like to discuss what might be termed the "mutual incompatibility of trees."

That there is only one phylogeny of living organisms is usually taken as self-evident by most systematists (it is incorporated as an axiom of the phylogenetic system by Wiley, 1975). This means that there is only one sequence of speciation in nature. We may deduce that there cannot be two trees of the same taxa that are both correct. I do not mean that one must be correct, only that both cannot be correct. It is this incompatibility of trees which stands at the base of phylogenetic testing. Hypotheses of relationship compete for attention and character analysis points to that hypothesis an investigator feels most justified as a best estimate of truth. Investigators, of course, make mistakes, and I shall discuss these mistakes in a later section. For now I shall discuss the basic system of trees dealt with.

1. *Two-taxon trees.* Two taxon trees are solved. Given no other taxon the two included in the diagram are related. Thus, two taxon trees are inherently uninteresting.

2. *Three-taxon trees.* Three-taxon trees are basic phylogenetic statements. Four to seven hypotheses compete in the arena of character analysis. Given some internal criterion to determine relative apomorphy, for example, an ontogenetic criterion (Chapter 5), these competing hypotheses may be tested to see which is justified as a best estimate of truth. However, three taxon statements do not appeal to the concept of *overall parsimony.* Let me explain.

The whole phylogeny of life is a conjunction of three-taxon statements taking the form "A shares a common ancestor with B not shared by C" (Hennig, 1966). Engelmann and Wiley (1977) point out that any three-taxon statement accepted by an investigator must be externally parsimonious. It must rest comfortably with the entire history of life and not simply be the most parsimonious solution to character distribution *within* the three-taxon statement.

3. *Three-taxon trees and the out-group.* This is the basic system dealt with in testing trees. We shall see why in the next section.

Parsimony and the Need for the Out-group

The distribution of characters provides potential tests of the relationships of taxa showing the characters. Each character is a potential test so long as it is shared by two and only two taxa in a three-taxon system (or $n - 1$ taxa in an n-taxon system). For simplicity, I shall talk only about three-taxon systems. A character shared by all three taxa is consistent with any of the four to seven possible ways these taxa are related. A character unique to one taxon implies that its alternate is shared by the other two. A valid test must be capable of falsifying a hypothesis, or in failing to do so, justifying that hypothesis. Falsification of the phylogenetic hypotheses that B and C share a common ancestor not shared by A (see Fig. 4.17a) is provided by a character shared by taxon A and C (7', Fig. 4.17b) or a character shared by A and B (8', Fig. 4.17c) because either character distribution is inconsistent with the expected distribution if the hypothesis (see Fig. 4.17a) was correct (i.e., we would expect B and C to share the character). Corroboration of Fig. 4.17a would be provided

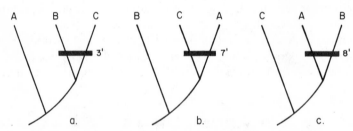

Figure 4.17 Three phylogenetic hypotheses with their character justifications. See text for discussion.

by finding a character shared in B and C (3′) but not found in A. Such a character would not only corroborate Fig. 4.17a, it would also refute Fig. 4.17b and c.

Note that no consideration has been given to whether any of these characters are apomorphic or plesiomorphic. This is because, as Engelmann and Wiley (1977) point out, our example is a *closed system*. Without referring to taxa outside the three-taxon problem, or without referring to some other criterion (such as an ontogenetic rule), the *a priori* designation of a particular character as apomorphous or plesiomorphous would be ad hoc, rendering the character itself invalid as data in an empirical system. Why, then, must we make such designations at all? Cannot we simply work with pure similarity to produce a system or relationships? We could, of course (and pheneticists have done this for a number of years). But, we would like to have a system whereby we can sort out characters that are directly applicable to solving the problem from characters that are not. When character distributions conflict within this closed system such that a number of phylogenetic hypotheses have "corroborating" data we must assume that some of these characters provide valid tests and other characters do not provide valid tests because there is only one true phylogeny of the three taxa we are dealing with and we can therefore logically assume that there should be only *one* justifiable phylogenetic hypothesis *if* we have correctly applied valid characters to the alternative hypotheses.

In a closed system (a phenetic system) we are forced to apply the parsimony criterion to select that hypothesis which minimizes character conflicts. The hypothesis that does this requires the fewest ad hoc hypotheses about invalid character tests. In other words, we have no external criterion to say that a particular conflicting character is actually an invalid test. Therefore, saying that it is an invalid test simply because it is unparsimonious is a statement that is, itself, an ad hoc statement. With no external criterion, we are forced to use parsimony to minimize the total number of ad hoc hypotheses (Popper, 1968a: 145). The result is that the most parsimonious of the various alternates is the most highly corroborated and therefore preferred over the less parsimonious alternates. For example, in Fig. 4.18a B and C share five characters (1′–5′), A and C two characters (6′, 7′), and B and A one character (8′). We accept the first hypothesis because it requires only three ad hoc hypotheses (i.e., that characters 6′, 7′, and 8′ are invalid tests). Acceptance of the second hypothesis (see Fig. 4.18b) would require six ad hoc hypotheses, and acceptance of the third (see Fig. 4.18c) would require seven ad hoc hypotheses.

The problem of phylogeny reconstruction of living and extinct organisms involves far more than three taxa. As Hennig (1966: 88) has stated, the larger problem of the phylogenetic history of life is a single large statement made up of conjunctions of three taxon statements. The test

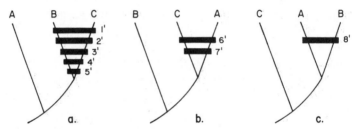

Figure 4.18 The resolution of character distributions using parsimony but not using out-group comparison. Only shared characters are plotted. See text for discussion.

of this larger hypothesis includes all of the characters known to biological science. No low level three-taxon statement can be set apart from the larger phylogeny of life. Because of this, the most parsimonious solution of a given three-taxon problem taken in isolation is not necessarily the most parsimonious solution when this three-taxon problem is placed within the context of the larger phylogeny. To achieve **overall parsimony** (parsimony within the context of the larger phylogeny), the investigator may accept an alternate phylogeny which is less parsimonious than the most parsimonious solution taken without reference to the larger phylogeny of life. In other words, a phylogenetic hypothesis which is an *internally* less parsimonious solution. As Engelmann and Wiley (1977) state, the investigator is faced with an apparent dilemma. An attempt to solve the total phylogeny of life simply to resolve on three-taxon problem the investigator is interested in would be completely impractical. Yet, to disregard some information for the purpose of expediency would be working within a closed system and would introduce ad hoc parsimony statements which are contrary to an empirical approach.

This apparent dilemma is avoidable. For the purpose of working on a limited problem, the investigator assumes the most highly corroborated higher level phylogeny to be true. Such an assumption is permissible because the higher level phylogeny is, itself, open to testing at a higher level of universality than the problem at hand. Of course, the better corroborated and more detailed the higher level phylogeny, the better the results within the more limited problem at hand. This acceptance of a higher level phylogeny does not change the basic criterion for a valid test of phylogeny. That is, the valid character must still be shared by two and only two of the three taxa in question. But, the acceptance of the higher level phylogeny does allow the investigator to resolve character conflicts and to choose the most parsimonious hypothesis within the context of the higher level phylogeny. It does this by identifying characters as relevant or irrelevant to the immediate problem at hand, the three-taxon problem. Reference to the higher level phylogeny is summarized for the problem at hand by hypothesizing characters as apomorphic or plesiomorphic. Synapomorphic characters are valid tests of the problem at hand because they are relevant tests of unique common ancestry and because their distributions as shared characters are parsimonious within the context of the higher level phylogeny. Symplesiomorphic characters are not valid tests of the problem at hand because if they were taken as evidence of relationship this would result in an unparsimonious overall solution. In terms of the parsimony criterion, plesiomorphies are ad hoc hypotheses of homology taken as true to achieve overall parsimony. Ad hoc hypotheses cannot test other hypotheses.

This discussion implies that we may think of apomorphies and plesiomorphies as purely formal constructs—a by-product of a purely logical system. We shall see, however, that this formal reasoning also appeals to concepts implicit in evolutionary descent with modification (Chapter 5). In other words, they appeal to natural process.

Errors in Tree Analysis

Making mistakes is inherent in scientific investigation. We attempt to minimize our mistakes by developing critical methodologies and grounding ourselves on firm theoretical grounds. Mistakes in any science can be ascribed to two basic causes: (1) getting the wrong "facts" or (2) interpreting the right facts in an illogical manner or in a logical manner from the wrong assumptions. Either may lead to the same types of errors which plague any testing scheme which is deductive. These errors are usually termed Type I and Type II errors. I shall apply both types of errors to tree analysis and character analysis because they are interlocking. In terms of trees, we may define these errors as below.

Type I Error. *Accepting a false phylogenetic hypothesis as being a true phylogenetic hypothesis.* Put simply, Type I errors occur when we accept the wrong alternate tree as a correct tree.

Type II Errors. *Rejecting a true phylogenetic hypothesis in favor of a false phylogenetic hypothesis.* Put simply, if one makes a Type I error, then one will also make a Type II error.

The major difference between phylogenetic testing and, for example, statistical testing concerns probability levels. Statisticians have worked out the probabilities of making their own Type I errors by empirically analyzing known populations of measurements and counts. Given known populations of measurements, one can determine empirically whether one should calculate the standard error by n or $n-1$ to estimate a parameter from a subsample of that population of measurements.[3] There are no known models of phylogenies with which the phylogenetic systematist can empirically calculate the chances for either types of errors.[4] This forces us to use nonstatistical inference at the "100 percent" level and parsimony must take its course. The best a phylogeneticist can do is to attempt to make the analysis as complete as possible by attempting to explain, in a way grounded in evolutionary theory, all of the character distributions. In other words, conflicting characters must be explained as plesiomorphies or nonhomologies. I have confidence that this is sufficient because if you do not catch your mistakes, someone else is bound to catch them for you!

[3] Statisticians have also been successful in determining probabilities for Type II errors. However, such errors are not frequently reported because to determine the level of probability one must know the alternate hypothesis as well as the null hypothesis. Whereas the null hypothesis is given, the alternate hypothesis is known only in hypothetical data sets or data subsets where the larger set of measurements are taken as true.

[4] Models have been taken as true to see which of the three general systems estimates the models best. In most cases, some form of phylogenetic analysis performs better (Mickevich) and Johnson, 1976; Farris, 1977; and Mickevich, 1978).

Chapter 5

Characters and Reconstruction of Phylogenies

The previous chapters attempted to convey the notion that there is a real phylogeny of organisms and that natural taxa make up this phylogeny. The job of the phylogeneticist is to attempt to discover parts of this phylogeny. All available relevant evidence should be used to accomplish this goal. Much of this evidence involves attributes which are inherent to the organisms themselves, that is, the intrinsic characters of the organisms, and by inference, of the species and higher natural taxa. These may be contrasted with extrinsic characters, such as geographic range or ecological niche.

In this chapter we shall explore the nature and general kinds of intrinsic characters, termed "characters" for the sake of simplicity. This is followed by a discussion of homology and the particular kinds of homologous and nonhomologous characters. I will then outline the criteria which have been used to apply characters to the reconstruction of phylogenies. Finally, some practical examples of phylogenetic analysis are presented to show complete analyses using many characters. The various categories of characters, such as characters of external morphology, physiology, and so on, are discussed in Chapter 10.

CHARACTERS

The term "character" is used by every taxonomist. Further, I have talked to no taxonomist who did not have an intuitive idea about what characters are and there has been little or no confusion in taxonomists communicating their ideas about characters. Yet the term is usually defined in an inadequate manner. I do not think this will be easily remedied because I suspect that "character" is a primitive term to systematists like "set" is a primitive term to mathematicians. This does not matter so much as long as we are speaking the same language.

Many taxonomists have attempted to characterize "character," some have tried to define it. Cain and Harrison (1958) considered characters as "anything that can be considered as a variable independent of any other thing considered at the same time." Davis and Heywood (1965) considered characters as attributes of organisms separated from these organisms for the purpose of rational discussion. On the practical level they considered characters as the attributes of organisms that could be adequately described or assessed. This is an important practical concept. If one worker can communicate to another worker about an attribute or feature of an organism or group of organisms, then the colleague is likely to be informed and will consider the character as "real." This, of course, does not mean the colleague will consider the character significant. The significance of a character will probably depend on the research interest of the other investigator. Taxonomists frequently refer to "taxonomic characters" as being of interest. Mayr (1969: 121) defines such a character as "any attribute of a member of a taxon by which it differs or may differ from a member of a different taxon." The problem with this definition is that similar characters are used for taxonomic groupings at high levels. However, at any single level (genus to genus within a family, for example) the definition does have merit. Mayr simply means that you need differences to detect taxonomic diversity. And, Mayr does acknowledge the fact that similarities are important to assess relationships. Thus, his discussion is worthwhile.

Much has been made of characters and the states of characters. For example, Davis and Heywood consider the character as an abstraction and the character expression or character state as that which is actually described or discussed. Mayr (1969) disagrees and terms the abstraction the feature and the expression of that feature the character. Bock (1977) agrees and denies that a distinction should be made between characters and character states. Hennig (1966) does not make such a distinction, and his idea of "character" (ibid., p. 7) is closely akin to Mayr's (1969) "taxonomic character."

My opinion is that much of the earlier discussions of characters and character states are largely semantic. Systematists have an excellent intuitive idea about the attributes of organisms and have not confused the practical taxonomic literature by confounding the concept of character. I define a character in the following manner:

A **character** *is a feature of an organism which is the product of an ontogenetic or cytogenetic sequence of previously existing features, or a feature of a previously existing parental organism(s). Such features arise in evolution by modification of a previously existing ontogenetic or cytogenetic or molecular sequence.*

A character in this sense is not an abstract entity (Davis and Heywood, 1965), it is one product of the evolutionary process. Such characters do not have to be independent of each other (Cain and Harrison, 1958). The presence of neural arches is not independent of the presence of a notochord. In fact the presence of neural arches is closely linked to the presence of a notochord because a notochord is a necessary ontogenetic precursor of neural arches. Both are characters of vertebrates. One (the notochord) is a synapomorphy of Chordata, which includes vertebrates. Such characters linked by ontogeny may be termed "character complexes." Independent characters, in the sense of Cain and Harrison (1958), also exist. They are parts of different ontogenetic pathways present in the same organism and thus are free to respond to evolution differently. "Independence" used in this manner does not imply absolute independence because all characters are linked by being a part of the whole organism. Some, however, may change while others remain the same or change in different ways. For example, the number of teeth which border the snout of gars and the length of the lower jaw which opposes the snout are linked in a character complex to the elongation of the ethmoid region of the skull. But, the number of caudal fin rays is not linked to ethmoid elongation and thus may be considered independent.

Types of Characters

No attribute of an individual organism is exactly similar to the same attribute of another individual organism. This is true even of twins or clones when quantitative characters are considered. When we speak of an attribute of one organism being the same character as that of another organism, we make this statement under certain specified or intuitive conditions. We may set these conditions quantitatively or qualitatively depending on the nature of the character we wish to express and the purpose of the study. Considering only attributes inherent to the organisms themselves, there are at least three major types of these characters:

1. **Structural characters.** Two attributes are similar in basic or detailed structure in such a way that parts are directly comparable. Structural characters may be phylogenetic (see below) or they may be nonhomologous. Because they display similar structure they are usually called by the same character name. On the biochemical level the similarity may be exact without necessarily being phylogenetically homologous. In such a case, the term **structural homologue** may be appropriate.

Example. Both *Amia calva* and higher teleosts have cycloid scales covering the body. Such scales are termed "cycloid" because they display a certain similar structural arrangement. Nevertheless, they are not homologous, as cycloid scales (Patterson, 1973), and would not be considered the same phylogenetic character.

2. Functional characters. These are characters that are similar in basic function in such a way that the parts may be compared in terms of that function. Such characters may also be structural and/or phylogenetic characters. Many such functional characters are given the same name.

Example 1. Birds and butterflies have wings. They are termed wings because they have the same function. However, they are not comparable in terms of structure nor are they phylogenetically homologous.

Example 2. Many sarcopterygians (rhipidistians and tetrapods) as well as euselachian sharks, gars, *Amia calva*, and higher teleosts have vertebral centra which have the same basic function, and many (the actinopts) have highly similar structure. Yet the name "vertebral centrum" applies to attributes that are in some cases only functional (euselachian sharks vs. the others), only structural (*Amia* vs. teleosts), or fully functional, structural, and phylogenetic (all tetrapods).

Example 3. The rectus arcus branchial muscles of *Amia*, gars, and euteleosts function to retract the upper gill arches. In each group the basic structure is similar (a pair of muscles connecting the upper gill arches with the vertebral column—details differ). The name refers to functional and structural attributes that are not the same phylogenetic character.

3. Phylogenetic character. A phylogenetic character comprises the attributes of two or more organisms that are hypothesized to be homologues. As such, a phylogenetic character shared by two or more organisms implies a phylogenetic relationship between the organisms displaying the character. Phylogenetic characters are expected to be similar from organism to organism at a level of similarity set by the investigator, but because of evolutionary divergence we do not always expect phylogenetic characters to exhibit detailed similarity.

Example. The lower jaws of gnathostomes show a wide range of variation in their basic makeup, shape, and dimensions. Sharks have only a Meckel's cartilage, mammals only a dentary, and teleosts no less than four or five separate bones. The presence of the lower jaw itself is a character, regardless of the relative dissimilarity of the jaws in various groups. This is a legitimate character because *all* lower jaws of gnathostomes are homologous.

We shall see in later sections of the chapter how the various components of similarity, structure, and function interplay in our interpretations of phylogenetic characters. We shall also see that only certain phylogenetic characters are useful in assessing phylogenetic relationship. But first, we must consider the relationships between characters and the individuals that display these characters.

Holomorphology and Semaphoronts

Individual organisms display a changing variety of characters during their life span. This may involve a certain number of rather discrete life stages, as in individuals that undergo metamorphosis. Or, it may involve a continuous growth pattern with changes in morphology, physiology, and behavior grading into one another, as in individuals displaying direct development. Hennig (1966) termed the totality of an individual's characteristics from fertilization to death that organism's "total form" or **holomorphology**. An organism at a particular stage in life history was termed a **semaphoront**.

A species also has a holomorphology. The holomorphology of a species, in a theoretical sense, is the sum holomorphology of all individuals belonging to the species. In practice, the holomorphology of a species is estimated (or sampled) by examining specimens representing various life stages, sexes, and so on. The species may have a relatively simple holomorphology. Examples are species that are asexual, or bisexual species without complicating factors. Other species may have complex holomorphologies. For example, many lower plants display alternation of generations where gametophytes and sporophytes occur every other generation. The gametophyte is usually different from the sporophyte, both morphologically and ecologically. Both have different arrays of semaphoronts. The holomorphology of a species of, say, a moss is the totality of the characteristics of both gametophyte and sporophyte. Other complex holomorphologies include many coelenterates (alternate polyp and medusa generations frequently complicated by colony structures) and certain hymenopteran insects (the simple bisexual reproductive strategy of the species is overlain by a caste system of various individuals which do not reproduce).

The most basic step in systematics involves comparison of individual organisms to assess whether their characters are similar or different. The examination should involve **comparable semaphoronts**, that is, two individual specimens at similar stages in their life history (Hennig, 1966). For example, if we draw samples of fishes from two different geographic regions and compare them for similarities and differences, we will make comparisons of adult males vs. adult males, juveniles vs. juveniles, or fry vs. fry. Our analysis may show that adult males differ, but juveniles do not. Or, it may show no discernible differences. What would happen if we had only juveniles from one sampling site and a full range of semaphoronts from our other sampling site? We would be limited to comparing only the juveniles because contrasting adults with juveniles involves noncomparable semaphoronts. In the discussions and examples throughout the remaining part of this chapter, we will assume that comparable specimens are always utilized unless the discussion explicitly involves comparisons of individuals at different stages of their life cycle.

HOMOLOGY AND NONHOMOLOGY

Owen (1848) felt the need to have a set of terms for comparing different parts of the anatomy of different groups of animals. One term would indicate valid comparisons of identity between structures while the other term would indicate invalid comparisons of identity between structures. Owen coined the word "homologue" for the same parts of two different organisms regardless of particular differences in form or function. A homology was correspondence of a part or organ of one species with a part or organ of another species as evidenced by relative position and connection with other parts of the body. For example, the forelimbs of bats and mice are homologous because, in spite of the fact that they differ in overall form and function, their parts are composed of the same bones which have the same general relationship to each other and to other bones in the body. In contrast to "homologue," Owen (1848) used the existing term "analogue" to denote different parts of organs in different organisms which perform the same function. The wing of the bat and the wing of the butterfly are analogues (analogous) because they are composed of very different parts which perform the same function (flight). Owen coined "homologue" to serve the needs of comparative anatomy, it had no evolutionary connotation. But with the rise of a general acceptance of the fact of evolution, the connotation of homologue became different. It changed from general and specific correspondence to similarity due to common descent. Homologues became structures derived from descent from a common ancestor while analogues became structures performing the same function but derived from very different ancestors.

From evolutionist's thinking about Owen's original terms it became apparent that there were several other kinds of similarities that were neither homologues nor analogues. Evolutionists needed terms to describe characters that were similar in composition and position but were nevertheless not derived from characters present in the common ancestor of the forms showing the similarity. Evolutionists began to speak of **convergent similarities**—similarities in both general and special composition but derived from different ancestry. Convergent similarities are actually very similar in structure, but, because the groups displaying this similarity are hypothesized to have very different ancestry, the similarities could not have been said to be present in the common ancestor of the two (or more) forms showing the similarity. Therefore, the similarities can not be homologies. The development of similar adaptations in the African Euphorbiaceae and the Cactaceae of North America is a good example of convergence (Fig. 5.1). Evolutionists also began speaking of **parallelisms**, structural similarities which corresponded so closely that if the ancestor was not known (i.e., the phylogenetic position was not known) the structures would certainly be considered homologies. The usual way that evolutionists distinguished between convergence and parallelism was to consider that convergences were similarities brought about by different genes producing similar structures while parallelisms were similarities brought about by the same genes producing the similarities (Simpson, 1961). Actually, we do not know whether the same genes are involved in the vast majority of cases. And the differences between convergence and parallelism are best thought of as differences in degree of relationship. Nonhomologous similarity in closely related species (such as within a single genus) is usually thought of as parallelism (Bigelow, 1958; see Hennig, 1966: 120–121 for additional discussion). The important point is not whether the terms can be readily distinguished by some objective criterion. Rather, the importance lies in recognizing that both of these types of similarities are nonhomologies and thus both are irrelevant to justifying natural taxa.

In Chapter 1 I discussed briefly the various types of homologous and homoplasous (nonhomologous) characters and gave hypothetical examples of each. We may now consider more formal definitions of homology and homoplasy and discuss examples of the various types of homologous characters, and their relationships to each other.

Homology.   A character of two or more taxa is homologous if this character is found in the common ancestor of these taxa, or, two characters

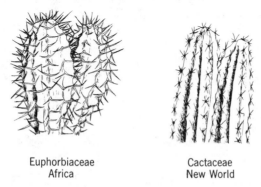

Euphorbiaceae
Africa

Cactaceae
New World

Figure 5.1   Convergence in gross similarity between African Euphorbiaceae and New World Cactaceae (From Ehrlich and Holm, 1963. Copyright McGraw-Hill Book Company, Inc. Used with permission).

(or a linear sequence of characters) are homologues if one is directly (or sequentially) derived from the other(s).[1]

Homoplasy (nonhomology). A character found in two or more species is homoplasous (nonhomologous) if the common ancestor of these species did not have the character in question, or if one character was not the precursor of the other.

Analogous structures rarely or never produce errors in phylogeny reconstruction but other nonhomologies frequently do produce error. I will discuss these sources for error later in the section. For now, I will concentrate only on homologous characters and their recognition criteria.

Evolutionary divergence in morphology (be it structural or biochemical) does not proceed in a smooth and steady fashion. Rather, some parts of the organism change very little or not at all while other parts of the organism may depart from the ancestral condition during or after the origin of a species. As a result of this, some homologues will have their origins in very recent common ancestors while other homologues will have their origins in more ancient common ancestors. For example, the jaws of Recent teleosts and mammals are very ancient homologues derived in the common ancestor of all gnathostomes. The hair of horses and whales is of more recent origin, being derived in the common ancestor of all mammals. The subocular teardrop of the species of the *Fundulus nottii* species group of topminnows (Fig. 5.2) is a very recent development hypothesized to have originated in the immediate common ancestor of the five species (Wiley, 1977b). Hennig (1950, 1966, and other works) stressed the importance of distinguishing between ancient and more recent homologues by terming the former "plesiomorphic" and the latter "apomorphic." These terms are defined and two examples are presented:

Apomorphic character. Of a homologous pair of characters, the apomorphic character is the character evolved directly from its preexisting homologue.[2]

Plesiomorphic character. Of a pair of homologues, the plesiomorphic character is the character that arose earlier in time and gave rise to the later, apomorphic, character.[3]

---

[1] Similar definitions are given by most evolutionists; see Simpson, 1961; Hennig, 1966; Mayr, 1969; and others.

[2] Apomorphic character is frequently shortened to "apomorphy." Synonyms include: apotypic character, apotypy, apomorphous character, derived character, advanced character, and specialized character.

[3] Plesiomorphic character is frequently shortened to "plesiomorphy." Synonyms include: plesiotypic character, plesiotypy, plesiomorphous character, primitive character, ancestral character, and generalized character.

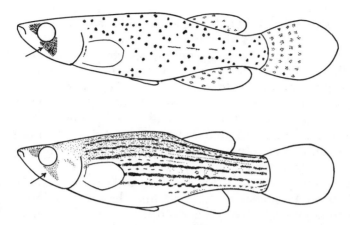

Figure 5.2 Male (upper) and female (lower) *Fundulus blairae*, a member of the *F. nottii* species group, with the subocular teardrops indicated by an arrow. (From Wiley and Hall, 1975. Copyright American Museum of Natural History. Used with permission.)

Example 1. Prokaryotes have a single strand of DNA relatively unbound by proteins whereas eukaryotes have two or more strands of DNA associated with proteins to form chromosomes. The single strand condition of prokaryotes is hypothesized to be the more ancient condition from which chromosomes were evolved once during organismic descent. If so, then the prokaryotic condition is the plesiomorphic character and the eukaryotic condition is the apomorphic character.

Example 2. The protonephridial "excretory" system of acoelomate and pseudocoelomate animals is considered homologous with the nephridial system (and its derivatives) of coelomate animals. Of the two systems, the nephridial system is hypothesized to have arisen from the protonephridial system. If so, the protonephridial condition is the plesiomorphic character and the nephridial system is the apomorphic character.

Apomorphic characters may either be shared between two or more species or they may be unique to a single species. Plesiomorphic characters are usually shared, although there is a possibility that in an analysis only a single species might retain the plesiomorphic character because of extinction. A shared plesiomorphic character is termed a **symplesiomorphic character**. Shared and unique apomorphic characters are defined:

Synapomorphic character (synapomorphy). A homologous character found in two or more taxa that is hypothesized to have arisen in the ancestral species of these taxa and in no earlier ancestor.

Autapomorphic character (autapomorphy). A character evolved from its plesiomorphic homologue in a single species.

I should stress that although synapomorphic characters are passed on from an ancestral species to its decendants in an unmodified form, the descendants may undergo later evolutionary change to modify the character further. Thus, when we speak of the presence of a lower jaw as synapomorphic for gnathostomes or a vascular system synapomorphic for tracheophytes, we do not mean that all lower jaws of gnathostomes and all vascular systems of tracheophytes look the same. Much modification in both character complexes has occurred since their origins. But this modification is hypothesized to have occurred *after* the origin of the first descendants of each group.

This inhibiting array of terms is meant to facilitate exact communication between phylogeneticists regarding character evolution and organismic relationships. To understand the relationships between the terms themselves, we must understand that characters, like taxa, have particular times of origin and thus characters originate at particular levels of phylogenetic descent. To illustrate this point we may examine Fig. 5.3 which shows the evolution of a group of species (A–F) and the evolution of a series of homologous characters (0, 1, 1′, 1″). Each speciation event in the history of this group may be looked on as a level of evolution, with earlier events being described as occurring at a higher level and later events being described as occurring at a lower level. Level 1 in Fig. 5.3 is considered the highest level in this example because the history of descent from the common ancestor that was living at this particular time period involves more descendants than the history of any other ancestral species in the diagram. Taking a term from Popper (1968a), Wiley (1975) described these levels as **levels of universality**. Higher levels of universality represent more inclusive groups than do lower levels of universality. There is an exact analogy between levels of universality and levels of taxonomic rank in a fully ranked phylogenetic classification. Kingdom Eukaryota exists at a higher level of universality than phylum Chordata and Chordata exists at a higher level than class Osteichthyes, and so on.

Turning to our example in Fig. 5.3, we note that character evolution may be correlated with levels of universality. For example, between levels 1 and 2 the character 1 has originated and replaced its plesiomorphic homologue 0. Note that this has occurred in a single species and thus 1 can be described as originating as an *autapomorphy* in the ancestral species W. Also note that during the time period represented by the interval between levels 1 and 2 species W was the only member of the group. At level 2 the ancestral species W speciated, and this process produced the descendant species A and X. Character 1 was retained in both descendant species and became, at this level of universality, a *synapomorphy*. At level 3 the ancestral species X speciated to produce descendant species M and Y. Both M and Y initially retain character 1, but during the evolution of species Y a new modification arises, character

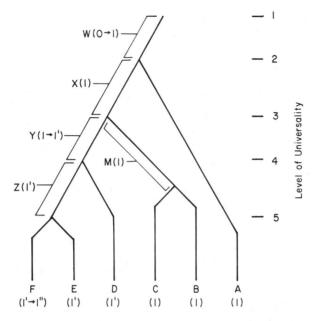

Figure 5.3 Levels of universality. See text for discussion.

1′. With the fixation of character 1′ in the species Y, the character 1 becomes a *plesiomorphy* relative to 1′. Likewise, further modification produced character 1″ in species F. Thus, 1′ has become plesiomorphic relative to 1″. Every homologue may go through three phases depending on the amount of lineage splitting and the amount of character evolution that occurs. These phases are summarized as follows:

Phase 1. A character originates as an evolutionary novelty from a preexisting character and exists as a polymorphism in the ancestral lineage until it becomes fixed as an autapomorphy. (However, it is also possible that it will remain polymorphic through several lineage splits and this complicates phylogenetic analysis. Here I am presenting only the simple case.)

Phase 2. A species with an autapomorphy speciates and passes its autapomorphy to its descendants as a synapomorphy.

Phase 3. Subsequent species may speciate and pass on their synapomorphy to their own descendants. Any subsequent modification of the character results in a new evolutionary novelty and in a change in status of the original character from an apomorphy to a plesiomorphy. This includes reversals to the structurally primitive condition (which may be frequent on the molecular level).

The importance of making these distinctions was pointed out in Chapter 3 where we saw that natural groups may only be justified by characters used at the level of universality where they are hypothesized to be synapomorphic characters. It is the synapomorphic character which gives evidence that two or more species shared a common ancestor among themselves that no other species shared. In contrast, a symplesiomorphic character is one that is hypothesized to have arisen earlier in evolution and to justify a monophyletic group that is larger than the group in which the character is observed and which would include this group within it.

LEVELS OF UNIVERSALITY

A phylogenetic hypothesis incorporating all organisms that have ever existed on earth would occupy the highest level of universality in organismic evolution because it would include all groups of organisms. Let us consider the unlikely prospect of having such a phylogenetic hypothesis before us with all of the characters shared by organisms placed on the tree. Under these conditions it is important to note that all of these characters would be placed on the tree at the point where they arose as evolutionary novelties; that is, they would be placed on the tree as apomorphies. No character would be placed on the tree where it exists as a plesiomorphy. When we work at levels of universality lower than this ultimate level we must understand that symplesiomorphies are not relevant to reconstructing the phylogenetic relationships at this restricted level because they have already been used to reconstruct phylogenetic relationships at higher levels. In other words, it is not a matter of disregarding symplesiomorphies so much as it is a matter of understanding that symplesiomorphies have already been employed to elucidate certain phylogenetic relationships at a higher level of universality where they exist as synapomorphies. Because the methods of phylogenetic systematics are designed to search for characters that unite organisms and taxa by unique common ancestry, we may disregard any character thought to have originated in an ancestor that is older than any involved with our restricted problem. For example, at the level of universality of Mammalia the presence of hair would be a synapomorphy because the ancestor in which this homology is hypothesized to have evolved is included in the phylogenetic hypothesis (albeit as a hypothetical entity). The presence of a pair of frontal bones in mammals, however, is a plesiomorphy at this level of universality because the ancestor which evolved this homology is *not included in the phylogenetic hypothesis*. Rather, this ancestor is included in a higher level phylogenetic hypothesis. This relationship points out a clear difference between synapomorphy and symplesiomorphy. The ancestral mammal had both hair and frontal bones. But only hair is hypothesized to have developed in that particular ancestor and to have been retained in all subsequent ancestors up to and including the ancestral mammal. But, what does the presence of hair become within the Mammalia? It becomes a symplesiomorphy. At the level of universality that considers the interrelationships of the horses (family Equidae), the presence of five toes of the hind foot is a symplesiomorphy because the ancestor in which the character is hypothesized to have arisen (the ancestral tetrapod) is not included in the hypothesis. However, we note that the ancestral mammal, the ancestral horse, and all ancestors in between had four or five toes on the hind foot (Fig. 5.4a). Having three or fewer toes on the hind foot (Fig. 5.4b–d) would be a synapomorphy for a subgroup of horses because the equid ancestor (hypothetical or actual) is included in the hypothesis.

Another concept has recently been introduced by Nelson (1978a)— the concept of **generality**. Whereas level of universality pertains to a phylogenetic hypothesis, level of generality may be construed to apply to a particular character. We may say that *the higher the level of universality that a homology exists as an apomorphy, the more general the character is, relative to other characters.*

In other words, characters of high generality belong to phylogenies of high universality as apomorphic characters and as plesiomorphic characters at any lower or more restricted level of universality. This distinction has merit because it permits a phrase to be used for the phylogenies themselves and another phrase to be used for the characters associated with these phylogenies.

Apomorphies, plesiomorphies, and nonhomologies are all features of organisms that enable these organisms to live and cope with the world.

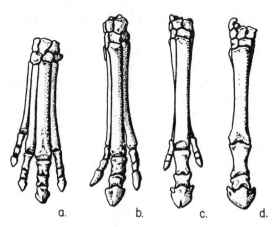

Figure 5.4 Forelimbs of horses: (*a*) *Hyracotherium* ("*Eohippus*"); (*b*) *Miohippus*; (*c*) *Merychippus*; and (*d*) *Equus*. (From *Vertebrate Paleontology* by A. S. Romer. Copyright 1966 by the University of Chicago Press. Used with permission.)

The fact that a particular feature happens to be plesiomorphic does not mean that it is irrelevant to the biology of the organism. The presence of jaws in mammals may be plesiomorphic but no one would question that this feature is essential for survival. Phylogenetic systematics is not concerned with the total biology of an organism *per se*, but how that total biology evolved. To serve this goal, phylogenetic hypotheses must be as precise as possible, given the evidence available. The most precise statements which can now commonly be made are statements about immediate common ancestry.[4] Thus, if we wish to be precise, we will use only synapomorphous characters to test phylogenetic hypotheses because they are the only characters capable of justifying or rejecting statements of immediate common ancestry (sister group statements). We might say that only synapomorphies can be used to justify natural groupings of organisms because only synapomorphies can corroborate or refute the monophyletic (natural) status of these groupings.

If we had a phylogenetic hypothesis of the highest level of universality, one including all of life, it would include all common ancestors within it, and all of the homologies placed with this phylogenetic hypothesis would be synapomorphies and would be placed at the level at which they arose. These synapomorphies may be termed **proper components** of the phylogenetic hypothesis because their place on the hypothesis is logically justified by the axioms of descent with modification. Symplesiomorphic homologies would not be proper components of this hypothesis because all homologies would have already been incorporated at other levels of universality as synapomorphies (the level at which they arose). This is quite apparent in our highest level hypothesis because all features of all organisms would be placed on the tree and we would see immediately that one or more had been used twice, once at the level of universality where it belongs as a proper component and justifies a hypothesis of monophyly and one or more times where it is an improper component and does not justify a hypothesis of monophyly.[5] Because a plesiomorphy incorporated into such a hypothesis serves no useful function and because it represents the same homologous character used twice in the same analysis, it is excluded. This may not be so apparent when only a part of the phylogeny of all life is investigated. Why, for example, can we not use a plesiomorphy to justify a grouping at lower levels of universality? It certainly fits the basic requirements of being present in two and only two of the three groups dealt with, as outlined in Chapter 4. There are two reasons. First, such usage violates the principle of

---

[4] The most precise statements would be those of immediate ancestry, not immediate common ancestry. Statements of immediate ancestry are possible in some groups (hybrid species with known parental species, some invertebrate fossils) but not for most.

[5] One might wonder who would bother to place the same character twice on the same phylogeny. Evolutionary taxonomists do this each time they erect a paraphyletic group and justify it on the basis of plesiomorphic characters.

overall parsimony. Second, it leads to an infinite regression for which no logical solution is possible. For example, if one homology can be used twice, then there is no logical reason why it cannot be used three, four, or an infinite number of times. If this is allowed, then there is no reason why all homologies cannot be used as many times as suits the whim of the particular investigator. Because there would be no logical way to set a limit on the number of times a particular character can be used, nor a limit on the numbers and kinds of characters that can be used two or more times, the system would degenerate into an authoritarian system that could not be evaluated objectively. Or, it would call for a system of overall similarity identical to the phenetic system. The first result is simply beyond the scope of science unless we accept authoritarian systems as scientific. The second must be rejected because overall similarity cannot justify natural groupings of organisms.

If all systematists worked only at the highest level of universality there would be no problems with plesiomorphic characters. But, no systematist works at this level. Rather, only subsets of the highest level phylogeny are evaluated. At any level, other than the highest, both plesiomorphies and apomorphies must be dealt with because the investigator must sort out those characters that demonstrate immediate common ancestry from those characters that do not. Wiley (1976: 9) stated, "This may be framed as a question: which characters are uniquely derived in ancestors included in the hypothesis and which are uniquely derived in ancestors not included in the hypothesis but retained by one or more ancestors included in the hypothesis?"

The synapomorphies of a phylogenetic hypothesis are those homologies that demonstrate (justify or corroborate) immediate common ancestry (monophyly) at the level of universality under consideration. Symplesiomorphies are those homologies that do not demonstrate monophyly at the level of universality of the problem under consideration but that do demonstrate monophyly at a higher level of universality than the phylogeny under consideration. In other words, synapomorphies are relevant to the hypothesis at hand whereas symplesiomorphies are irrelevant to the problem at hand but relevant to a larger phylogenetic hypothesis that includes the problem at hand (and where these plesiomorphies exist as synapomorphies).

I will illustrate this point with an example from Wiley (1975) concerning the Tetrapoda. The character "tetrapod limb present" belongs, as a synapomorphy/proper component, to the level of universality of the sarcopterygian amphibians and their sister group, the amniotes (reptiles, birds, and mammals). This character is hypothesized to justify a monophyletic Tetrapoda because it is hypothesized to have been uniquely evolved in the common ancestral species of all tetrapods. Within the Tetrapoda, at a lower level of universality, we have another monophyletic group, Lepidosauria, composed of rhynchocephalians (the tuatara of New Zealand), snakes, and lizards. Snakes, of course, lack the fully developed tetrapod limb as adults. We may ask, does the presence of a fully developed tetrapod limb in lizards and rhynchocephalians corroborate a hypothesis that these two groups shared a common ancestor not shared by snakes? No, because the character "tetrapod limb present" has already been used to corroborate a monophyletic Tetrapoda—to corroborate the hypothesis that amphibians and amniotes share an immediate common ancestral species not shared by the fishlike sarcopterygians (the dipnoans and rhipidistians). The use of this character to test a snake-lizard-rhynchocephalian hypothesis is an example of the use of a plesiomorphy to test at the incorrect level of universality.

If phylogeneticists had some absolute criterion for identifying synapomorphous homologies, phylogenetic analyses would be error free. Phylogeneticists, like all other scientists, do not have an insight to such absolute truth and can only present their hypotheses as best estimates of the truth. Errors in the identification of a homology or errors in assigning a homology to the correct level of universality result in mistakes in phylogenetic hypotheses. Such errors are discussed in a later section.

RECOGNITION CRITERIA OF HOMOLOGY

Before it was generally accepted that testing homologies was a two-step process involving both the similarity criterion and the criterion of phylogenetic position relative to other characters, there existed what is known as the "problem of homology." The problem, stated simply, was that the theoretical definition of homology was based on descent from a common ancestor whereas the actual recognition criteria were provided by phenetic similarity. Homology was defined in one way but tested in another way. Hennig (1950, 1966) solved the problem by pointing out that the most critical test of homology was a congruence between hypotheses of synapomorphy and phylogeny. In other words, homologies, as synapomorphies, and the phylogenies they are associated with as proper components are self-illuminating systems (Hennig, 1966). As more information is gained through the application of many hypotheses of homology to phylogenetic hypotheses we gain a better understanding about character evolution. Because the testing of hypotheses of homology is a two-step process, I have divided this section into two parts. The first is a discussion of morphological testing, the second is a discussion of phylogenetic testing.

Morphological Criteria of Homology

Phylogeneticists must develop strategies for discovering possible synapomorphic characters and for evaluating characters said by previous investigators to be synapomorphies. One strategy that may be applied is to use various morphological criteria to see if the characters in question fit these criteria for being homologues. Morphological criteria do not provide an infallible guide for identifying two shared characters as homologues, but they do provide an array of evidence that can be brought to bear on the problem of evaluating characters. In some cases simple morphological observation may refute a hypothesis that was based on a mistaken observation. In other cases a close inspection of morphology may provide evidence that a synapomorphy that seemingly refutes a phylogenetic hypothesis may actually be two nonhomologous characters. But most importantly, the criteria discussed below should be viewed as criteria for finding characters that are likely to be of interest in a phylogenetic analysis and not as criteria for absolutely identifying homologues. As we shall see in subsequent sections, two characters that pass all the morphological tests for homology may turn out to be homoplasies. And we shall see that two characters that may flunk some of the morphological criteria may, indeed, be homologues.

Probably the most extensive and philosophically sound discussion of the morphological criteria of homology is that of Remane (1956). Taking the view that homologues are the product of descent from a common ancestor, Remane suggested certain criteria for identifying any two or more characters are homologues: (1) the criterion of position, (2) the criterion of quality of resemblance, and (3) the criterion of continuance of similarity through intermediate species. I will discuss these criteria in some detail.

Similarity of Position

This criterion considers three basic characteristics, topographic position, geometric position, and position in relation to other parts of the body. These characteristics are self-reinforcing and not easily separated. For example, topographic position is frequently determined by relationship of one part to another (examples are shown in Figs. 5.5 and 5.6). Topographic position may also be assessed by the use of a Cartesian coordinate system, suitably distorted to cover comparable parts of the organism, as has been done extensively by Thompson (1942) to demonstrate similar topographic position in very differently shaped animals. Such analyses have also been done on various fossil skulls (Fig. 5.7). Similarity in geometric position refers to identifying corresponding parts in organisms whose body sizes differ considerably. If two species differ in total body size but their parts correspond to a simple geometry, these parts may be homologues. Likewise, if there is allometry in growth, a formula of this allometry may permit the recognition of corresponding parts that might otherwise have been considered as different and nonhomologous.

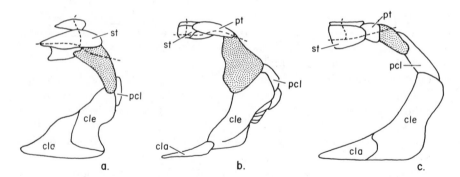

Figure 5.5 Remane's criterion of topographic position, I. Dermal shoulder girdles of (*a*) a sturgeon (*Acipenser*); (*b*) a birchir (*Polypterus*); and (*c*) an osteolepiform "crossoptery-gian" (*Eustenopteron*). The supracleithrum (stippled) has a similar topographic position to the cleithrum (cl) and postcleithrum (pcl) in (*a*) and (*b*) and a similar topographic position in relation to the posttemporal in (*b*) and (*c*). The sturgeon differs dorsally in having lost the posttemporal (pt) and enlarged supratemporal (st) while *Eustenopteron* differs in having an enlarged postcleithrum, displacing the supracleithrum from its association with the cleithrum. (Redrawn and modified from Jollie, 1973.)

The Criterion of Special Similarity

This criterion may either reinforce or falsify homologues based on positional criteria. Reinforcement is rather uninteresting to those who insist
on critical tests of falsification. That is, given that two structures correspond by positional criteria, it is not at all unexpected that finer structure
or similar ontogeny should be demonstrated for these structures. But,
when the criterion of position is refuted by the criterion of special qual-

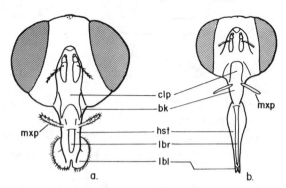

Figure 5.6 Remane's criterion of topographic position, II. Frontal views of (*a*) a sponging fly (*Musca*) and (*b*) a piercing fly (*Stomoxys*). The labled mouth parts have the same relative topographic positions although they differ in shape and, in some cases, function. Abbre-viations: bk, rostrum; clp, clypeus; hst, haustellum; lbl, labellum; lbr, labrum; mxp, max-illary palp. (Modified from Borror et al., 1976.)

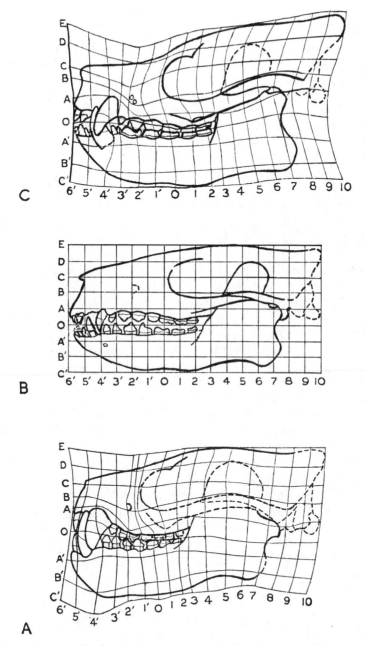

Figure 5.7 Cartesian coordinate diagram of the skulls of three genera of taeniodonts (an extinct group of mammals): (a) *Wortmania*, (b) *Onychodectes*, and (c) *Conocytes*. [From Bryan Patterson, "Rates of Evolution in Taeniodonts," in *Genetics, Paleontology* and *Evolution*, ed. by Glenn L. Jepsen et al. (Copyright 1949 ©1977 by Princeton University Press): Fig. 5. p. 261. Reprinted by permission of Princeton University Press.]

ity, then the analysis becomes interesting. I will give two examples of conflicts between the two criteria.

Gegenbaur (1873) proposed that the endoskeletal shoulder girdle of gnathostomes was the serial homologue of the gill arches. On the basis of topology alone this was a reasonable assertion because the endoskeletal shoulder girdle has approximately the same topographic relationships to the gill arches posteriorly as do the jaws anteriorly (Fig. 5.8a). Gegenbaur's postulated derivation of the shoulder girdle from a gill arch to the condition found in one primitive shark is shown in Fig. 5.8b–f. The problem with this hypothesis is that the endoskeletal shoulder girdle of gnathostomes is derived from lateral plate mesoderm while the gill arches are derived from neural crest cells, indicating that they have quite different embryological origins and thus are not homologous (see Balinsky, 1970: 411, 433, for a discussion of development and Zangerl and Case, 1976, for a discussion of Gegenbaur's thesis).

As another example, the vertebrae of bowfins and teleosts have the same positional relationships to other parts of the body such as the neural and haemel arches and the body myomeres (Fig. 5.9). Yet each has different embryological development and their special qualities are different as a result of this different embryological development (see Schaeffer, 1967, for discussion).

Perhaps even more interesting is the situation where rather dissimilar structures can be homologues because of their special qualities. For example, the transverse ventralis muscles of dipnoans are different in insertion from the oblique ventralis muscles of actinopterygians. And, in some lungfishes, these transverse ventralis muscles may even differ in topographic relationships to other muscles because of the more posterior position of their gill arches in relation to the pectoral girdle and other skeletal features. Yet, embryologically the actinopterygian oblique ventrales go through a transverse stage of development and thus can be identified as homologues of the transverse ventrales of lungfishes (Wiley, 1979d). As another example, the inner ear bones of mammals are very different adult structures from the hyomandibular, quadrate, and articular bones of reptiles yet are recognized as homologues because of (among other criteria) similar embryological development.

The Criterion of Continuance Through Intermediate Forms

Like the other criteria, this one may be used to either justify or refute a hypothesis of homology based on the other criteria. A frequently cited example of the use of this criterion is the modification of the angular, quadrate, and hyomandibular bones from the condition in fishes to the ear ossicles of mammals (Fig. 5.10). An example of refutation is the case of the scales of *Amia* (bowfins) and Recent teleosts. The scales of both groups are thin and (primitively) cycloid. They have a rather similar embryological development. Yet known taxa of both groups (on the basis

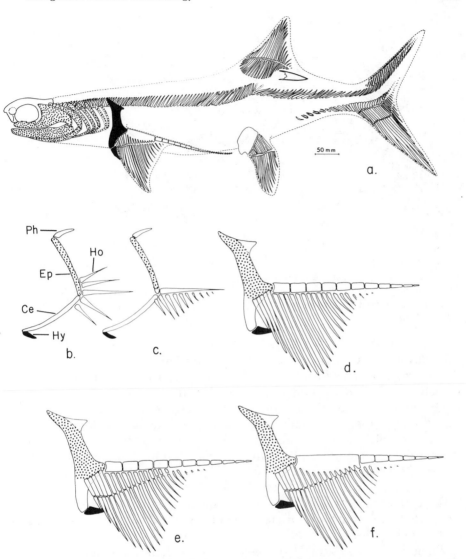

Figure 5.8 *Cobelodus aculeatus* (*a*) and Gengenbaur's hypothesis of the origin of the gnathostome shoulder girdle (*b–f*). In (*a*) the girdle is shaded black and the jaws and gill arches are stippled. In (*b*) the normal parts of the endoskeletal visceral arches are labled, and subsequent modification to produce the shoulder girdle of *C. aculeatus* is shown in subsequent drawings (Modified from Zangerl and Case, 1976.) Abbreviations: Ce, ceratobranchial; Ep, epibranchial; Ho, holobranch, Hy, hypobranchial; Ph, pharyngobranchial.

of other synapomorphies) have the primitive rhomboid scales of primitive actinopterygians.

Hanson (1977: 80–81) has suggested that the first two criteria discussed above form one major criterion, the criterion of position and special similarity. The third criterion (which Hanson terms the criterion of serial

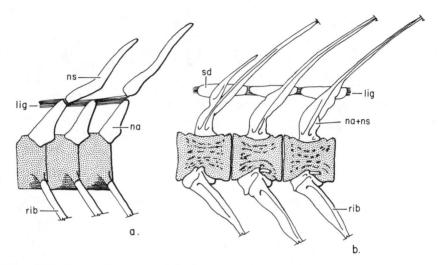

Figure 5.9 Series of vertebrae of the halecomorph *Amia* (*a*) and the teleost *Salmo* (*b*). Although the centra of both genera have the same topographic relationship to other ossifications, the centra develop quite differently and they are not homologues. (Modified from Jollie, 1973.)

relationship) is an extension of the major criterion. Riedl (1978) suggests that all criteria are subsumed by the first two major criteria. I see no objection to looking at Remane's criteria in these ways if it leads to a better understanding of the morphological testing of homology. What I find valuable about Hanson's discussion of Remane's criteria is his reformulation of Remane's "subsidiary criteria." These criteria deal with species not sufficiently complex to allow for a full implementation of his major criteria (sponges and coelenterates for example). Hanson redefines these criteria thus.

Homologues are probable between characters of two organisms which share *other characters* of sufficient complexity to be judged homologues by the major criteria.

When the same relatively simple character is found in large numbers of species it is probably homologous in all of the species.

Sets of characters distributed in the same way among the same species are probably sets of homologues. The more sets of characters shared, the more likely the homologies exist (this is similar to Hennig's "correlation of transformation series" criterion).

These subsidiary criteria do not, of course, provide as critical a series of tests as the major criteria. For example, it is not clear that one is justified in assuming that two simple characters are homologues simply because their distributions are correlated with more complex characters. We must therefore be satisfied with such characters if they are congruent with our best estimate of phylogenetic relationship. Their usefulness in reconstructing these relationships compared to more complex characters is relatively low because they can only be postulated homologues based on the distributions of other characters rather than any intrinsic merits of their own.

Most other similarity criteria discussed by other authors reflect Remane's criteria in whole or in part. Some authors stress one criterion while other authors stress other criteria. This depends in part on the overall point of view of the author. For example, Simpson (1961), a paleontologist, naturally stresses the criterion of intermediate forms. Jardine (1967, 1969) stresses similarity in topographic position and special composition (see also Jardine and Jardine, 1969). Bock (1969, 1974) stresses similarities of all kinds. Sneath and Sokal (1973) term "operational homology" as similarity "in general and in particular." Actually any of these criteria are adequate to arrive at a *morphological hypothesis*

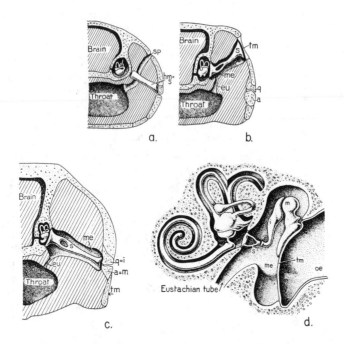

Figure 5.10 Remane's criterion of intermediate forms, the evolution of the mammalian ear: Cross sections through hypothetical skulls of (*a*) a fish, (*b*) an amphibian, (*c*) a reptile, and (*d*) a mammal. Abbreviations: a, articular; eu, eustachian tube (homologue of part of the spiracle); hm, hyomandibular (homologue of stapes); i, incus (homologue of quadrate); m, malleus (homologue of articular), me, middle ear cavity (homologue of part of the spiracle); q, quadrate; sp, spiracle; tm, tympanic membrane. (From *Vertebrate Paleontology* by A. S. Romer. Copyright 1966 by the University of Chicago Press. Used with permission.)

of homology (Wiley, 1975). But, all are subject to testing by what we may call the phylogenetic testing of homology. Thus, although it is important to test in as many ways as possible by the various criteria provided by phenotypic similarity, ontogeny, and so on, the critical test is the congruence or incongruence of a particular hypothesis of synapomorphy with other hypotheses of synapomorphy in an open system of the testing of competing phylogenetic hypotheses. This will be the bulk of my discussion in the following section.

Phylogenetic Criteria of Homology

The "problem of homology," the supposed incongruence between a theoretical definition and its practical application, has led many workers to abandon an evolutionary definition of homology (see Sneath,and Sokal, 1973; Hanson, 1977). Hanson (1977: 82) for example, defines homology thus:

> Homology is direct (positional and compositional relationships) or derived (serial relationship) similarity of structural and/or functional aspects of different organisms, especially as they are members of different species.

Those who will not abandon the theoretical definition are frequently led to admit that the theoretical definition has low resolving power to separate homologues from nonhomologues (see Bock, 1969, 1974). The problem with such an admission is that unless some other criterion can be brought to bear, the whole question of the ability of an investigator to reconstruct phylogenies becomes open. If homologues must be identified before they can be used for reconstructing phylogenies, and if such identity is dependent on criteria of low resolving power, then how can the resultant analysis be anything but very weak?

I feel that this dilemma is the direct result of the exclusive use of induction in framing hypotheses of homology and hypotheses of phylogenetic relationships. Bock (1974) most clearly expresses the dilemma of induction—similarities are recorded and, via induction, identified as homologies. These homologies can then induce hypotheses of relationships. The fact of homology is needed to induce the phylogenetic hypothesis.

The "problem of homology" does not exist, however, if we admit phylogenetic relationships based on other characters as a major criterion in the hypotheticodeductive mode of hypothesis testing. Hennig (1966) correctly stressed that any similarity criterion is secondary to the major criterion of phylogenetic position. Of course, the absolute truth about any hypothesis of homology will never be demonstrated because the one true phylogeny for any given set of organisms will never be known. But the "problem of homology" is broken by simply realizing that homologies can be treated as hypotheses which are tested by other hypotheses of homology and their associated phylogenetic hypotheses (Wiley, 1975, 1976). Such an outlook reflects the idea that truth is approached asymptotically, that is, by testing and retesting in a system of reciprocal illumination. Perhaps more important, it is a direct appeal to the evolutionary process.

The Criterion of Phylogenetic Position

The major criterion for determining whether a particular character or its alternate homologue is the apomorphic character is the criterion of phylogenetic position. Testing is accomplished by **out-group comparison**. The process may be expressed most simply by the use of what we shall term the "out-group rule."

The Out-group Rule. *Given two characters that are homologues and found within a single monophyletic group, the character that is also found in the sister group is the plesiomorphic character whereas the character found only within the monophyletic group is the apomorphic character.*

Because out-group comparison requires an out-group (or in many circumstances several possible out-groups), the investigator is required to use the open system of phylogenetic hypothesis testing to determine which of a particular pair of homologues is the apomorphous character (as discussed in Chapter 4).

Hennig (1966) termed the process of hypothesizing which of two characters was the apomorphous expression as the **argumentation scheme** of the holomorphological method. As used here, argumentation is a series of logical deductions based on the phylogenetic definitions of apomorphous and plesiomorphous homologies.

Argumentation begins by postulating, based on similarity criteria, that structures (for example) in two different species are homologues and further that they are synapomorphic. If they are considered homologues but plesiomorphic homologues, then we must raise the level of universality of the problem to include all taxa showing the character if we wish to test for homology because the only way that we can reject these plesiomorphic similarities as homologues is to demonstrate that they were not present in the ancestor of the taxa having the similarities. And the only way that we can do this is to include that hypothetical ancestor in the phylogenetic hypothesis, thus rendering the character synapomorphous. We may say that *plesiomorphous homologies are accepted as homologues by convention gained by assumption or previous experience* (Wiley, 1975; Gaffney, 1979). There is nothing wrong with this, not every homologue must be tested in every hypothesis, only the hypothesized synapomorphies. It is only common sense to accept that the hair of horses and foxes are plesiomorphic homologues, common sense gained from previous experience about the phylogenetic relationships of Mammalia.

Given that we have two characters that are hypothesized as synapomorphous homologies, we may apply deductive logic to the hypothesis. *If* character 1 in taxon X and character 1 in taxon Y are synapomorphous at the level of universality represented by X, Y, and Z, *then* X and Y share an immediate common ancestral species not shared by Z (Fig. 5.11*a*). We argue from a premise to a conclusion that must be true (logically) given that the premise is true.

We then go to another character (2) thought because of the similarity criterion to be homologous. We look at the distribution of this character. Is it found only in X and Y and not in Z (Fig. 5.11*b*)? If so, then the character might be another synapomorphy and additional corroboration for the monophyletic status of X and Y (Fig. 5.11*b*). If, however, we find that this second character is found in X and Z but not in Y (Fig. 5.11*c*), then we have a potential conflict. The conflict can be summarized by a series of statements:

1. *If* character 1 found in X and Y is a true synapomorphy, *then* X and Y shared a common ancestral species not shared with another taxon, including Z (Fig. 5.11*a*).

2. *If* character 2 shared by X and Z is a true synapomorphy, *then* X and Z shared a common ancestral species not shared by any other taxon, including Y (Fig. 5.11*c*).

3. Both statements cannot be true, because there is only one true phylogeny of X, Y, and Z. *Therefore,* we may conclude that one (or both) of the characters is not a true synapomorphy.

We may then go to a third character in the analysis, character 3. Character 3 is found only in X and Y, it is not found in Z or any other taxon (Fig. 5.11*d*). Further, using similarity criteria, we have no reason to reject the hypothesis that 3 is a homology. We now have reached the point where we may consider the rejection of character 2 as a synapomorphy on the basis of parsimony. Have we then rejected 2 as a homology? No, because there are two possibilities for this character; it may be a nonhomology or it may be a plesiomorphy. We test for plesiomorphy vs. nonhomology by looking outside the group X, Y, and Z to see what the distribution of character 2 might be. If we find 2 in close relatives of X, Y, and Z (i.e., if we find them in out-groups), then we may conclude that 2 is the plesiomorphic homologue in the character transformation series $2 \rightarrow 2'$, and that 2 is, indeed, homologous in taxa X and Z (Fig. 5.11*e*). If, however, we find 2' in the close relatives of X, Y, and Z, then we cannot dismiss 2 as a plesiomorphy. Rather, we are faced with two possibilities. Either 2 is a shared homoplasy (Fig. 5.11*f*), or 2 is a synapo-

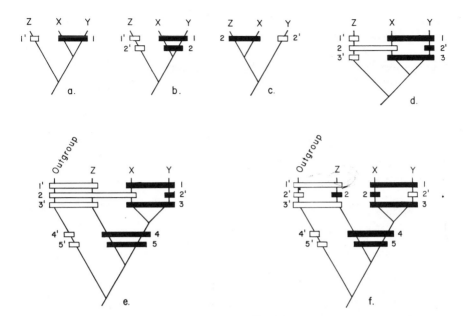

Figure 5.11 Phylogenetic argumentation I. Black bars connecting taxa are hypothesized apomorphies whereas open bars are plesiomorphies. See text for discussion.

morphy and 1 and 3 are shared homoplasies. We may attempt to resolve this conflict in several ways. Additional characters can be analyzed to see which support 2 as a synapomorphy and which support 1 and 3 as synapomorphies. If other characters are congruent with the hypothesis that 1 and 3 are synapomorphies, then the hypothesis that 2 is a homoplasy is reinforced. We may also reexamine the morphology and ontogeny of 2 in X and Y to see if they are really very similar or if they display differences that might lead us to suspect that they are homoplasic similarities. If, for example, we found that 2 in X had a different ontogeny than 2 in Z, we would reinforce our hypothesis that 2 is a shared homoplasy. We might also consider the quality of the characters. For example, if 2 is a relatively simple character whereas 1 and 3 are complex characters, it would be reasonable to postulate that 2 evolved twice whereas 1 and 3 evolved only once. In actual practice, the phylogeneticist should pursue all three approaches.

In this example we have assumed that the group XYZ is monophyletic. In all phylogenetic analyses the beginning premise is that the group analyzed is, indeed, monophyletic. The amount of information available to the investigator before analysis determines how reasonable this premise may be. Let us say that we are interested in the relationships among species in a genus. If a previous analysis has shown that the members of this genus share one or more synapomorphies within a demonstrated monophyletic family, then the investigator may have confidence that the premise is reasonable. If such an analysis has not been performed, then the premise may not be reasonable. In other words, if we simply accept the latest taxonomy of the family and if that taxonomy is not strictly phylogenetic, then we run the risk of analyzing a paraphyletic or even polyphyletic taxon. Analyzing such a group can lead to problems. To illustrate this point, let us consider the following taxonomy.

Family Aidae
   Subfamily Ainae
      Genus A
      Genus B
      Genus C
      Genus D
   Subfamily Einae
      Genus E
Family Fidae
   Genus F

The problem the investigator is interested in is the relationships among the genera of Ainae. Accepting the monophyletic nature of Ainae, genus E would naturally be selected as the out-group for comparison. The distribution of characters in five-character complexes are examined (Table 5.1, forget for the moment the distribution of characters in Fidae). Applying the out-group criterion by examining genus E, and assuming that Aniae is monophyletic, we come to the conclusions shown in Fig.

Table 5.1 The distributions of five character complexes among six hypothetical taxa; characters justifying the monophyly of Aidae plus Fidae have been omitted for simplicity

Taxon	Character complex				
	1	2	3	4	5
Aidae					
Ainae					
A	$1'$	2	$3'$	$4'$	$5'$
B	1	2	3	$4'$	5
C	$1'$	$2'$	$3'$	$4'$	$5'$
D	1	$2'$	3	$4'$	5
Einae					
E	1	$2'$	3	4	5
Fidae					
F	1	2	$3'$	$4'$	$5'$

5.12a and summarized thus: (a) Ainae is monophyletic by virtue of the synapomorphy 4'; (b) genera A and C form a monophyletic group by virtue of three synapomorphies (1', 3', and 5'); (c) genera B and D form an unresolved trichotomy with the group $A + C$; and (d) the characters coded 2 in genera A and B are homoplasies. These conclusions would be reasonable and acceptable *if* it were true that Ainae is a monophyletic group. However, if we examine the character distributions of genus F (family Fidae, Table 5.1), we arrive at a quite different conclusion, shown in Fig. 5.12b. Rather than a monophyletic Ainae defined by the synapomorphy 4' we have a monophyletic group composed of genera B and D plus the supposed out-group, genus E, and defined by the synapomorphies 3 and 5 (which were considered plesiomorphies in the analysis presented in Fig. 5.12a). Inspection of Fig. 5.12b shows other differences. In fact, the only part of the original analysis to survive is the monophyletic group $A + C$ and the synapomorphy 1'. This simple example shows that the conclusions drawn from analyzing a paraphyletic group as if it were a monophyletic group may lead to incorrect results. It points out to the beginning investigator that background analysis must be done before beginning work on a taxon to insure that this taxon can be reasonably considered as monophyletic. And, unless there is substantial evidence for both the monophyletic nature of the taxon and for possible sister groups to use as out-groups, the investigator should study the relationships of the taxon to other taxa before attempting to analyze the relationships within that taxon. It also suggests that utilizing more than one out-group is a good strategy.

The two examples used above have dealt with relatively few characters. In reality, argumentation using out-group comparison should extend to as many characters as possible and as many out-groups as necessary. In the first example (summarized in Fig. 5.11), inspection and testing of additional characters would help resolve the question of whether character 2 was a homoplasy. In the second example (Fig. 5.12), the addition of a second out-group completely changed the analysis, making the supposed out-group, E, an in-group.

The phylogenetic method of argumentation is not directed simply toward solving some aspect of the relationships among organisms. It is also directed toward explaining character evolution. When we have a conflict between two or more characters we must attempt to provide an evolutionary explanation as to why one set of characters is not relevant to the problem we are attempting to solve. In other words, we must give reasons why observed similarities exhibited between organisms are not evidence for a close relationship between these organisms. In this regard, plesiomorphies are relatively easy to explain because they are accepted as homologues, but homologues at a level of universality higher than the immediate problem at hand. Homoplasies are more difficult to explain because we must seek to demonstrate that there is reason to

144

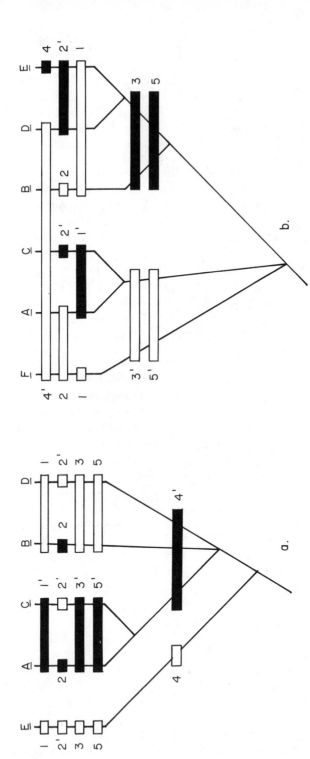

Figure 5.12 Phylogenetic argumentation II. Based on character distributions shown in Table 5.1. Black bars are inferred apomorphies, open bars are inferred plesiomorphies. See text for discussion.

believe that the similarities we observe have evolved independently of each other, or in failing to do so, we must have enough evidence (in the form of other characters) that they have evolved independently to accept these similarities as homoplasies on the basis of parsimony arguments. Frequently there is no strong morphological or ontogenetic evidence that can be forwarded to demonstrate that two similar characters are nonhomologous. Instead, they are rejected as synapomorphies on the basis of a parsimony argument. For example, the course of the spinal nerves in bowfins (*Amia calva*) and gar fishes (Lepisosteidae) are very similar. Jessen (1973) called attention to the observation that only these two taxa, among all actinopterygian fishes, had spinal nerves that penetrate the body musculature ventral to the transversely oriented pleural ribs (Fig. 5.13b). All other gnathostomes (including all other actinopterygians) have spinal nerves that follow the inner side of the body musculature medial to the pleural ribs (Fig. 5.13a). Jessen (1973) hypothesized the course of the spinal nerves in bowfins and gars as a synapomorphy uniting these two groups. The logical deduction of his hypothesis may be stated: *If* the spinal nerve pattern of gars and bowfins is synapomorphic, *then* gars and bowfins share a common ancestral species not shared by any other actinopterygian. This hypothesis may be expressed as the phylogenetic hypothesis shown in Fig. 5.14b.

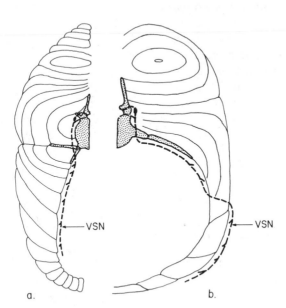

Figure 5.13 Similar features interpreted as homoplasies based on phylogenetic position: cross sections through the bodies of (a) *Salmo* and (b) *Amia* illustrating the course of the spinal nerves (vsn). Ossifications are stippled. (Modified from Jessen, 1973. © Linnean Society of London, 1973. Used with permission.)

Jessen's (1973) hypothesis was tested by Patterson (1973). He was able to show that most other characters that could be hypothesized as synapomorphic at the level of universality of actinopterygian interrelationships (by out-group comparison) were incompatible with the hypothesis that the pattern of spinal nerves is a synapomorphy of gars and bowfins. This alternate hypothesis (including other characters cited by Wiley, 1976) is shown in Fig. 5.14a. Because the spinal nerve pattern found in gars and bowfins is not found in any other group of gnathostomes, then we must conclude, on the basis of parsimony, that the character in the two groups is nonhomologous, or that the nerve pattern of teleosts is a reversal to the primitive structural condition. Our logical deduction takes the form of: *If* the synapomorphies shown in Fig. 5.14a are true synapomorphies, *then* either the similarity in the course of the spinal nerves of gars and bowfins a nonhomologous similarity, or the course of the spinal nerves of teleosts is nonhomologous with that of lower actinopterygians.

The important point of this example is that no criterion of similarity has yet been applied which would refute the supposed synapomorphic homology of the spinal nerve patterns of gars and bowfins. Rather, the supposed synapomorphy has been refuted (at least for the present) on the basis of character distributions that make it unparsimonious to conclude that the course of the spinal nerves of gars and bowfins is a synapomorphic character.

In summary, the strength of a method of testing homology that takes advantage of the criterion of phylogenetic position as shown by other hypotheses of homology rests on the fact that more than similarity can be applied to the problem. In admitting this phylogenetic criterion in addition to a number of similarity criteria we are at the same time admitting the phylogenetic definition of homology into the realm of testable science.

Auxiliary Criteria of Phylogenetic Apomorphy

In addition to the major criterion of phylogenetic position as evidenced by out-group comparisons, Hennig (1966: 95–99) recognized four auxiliary criteria for distinguishing between apomorphic and plesiomorphic characters. Two of these criteria, geologic character precedence and chronological progression, are little more than rules of thumb and so completely dependent on the major criterion that they are useful only in special circumstances. The third criterion, correlation of character transformations, is also dependent on the criterion of phylogenetic position. The fourth criterion, ontogenetic character precedence, is a major criterion which is complementary to out-group comparisons and thus is a powerful phylogenetic tool. I shall briefly discuss each of these aux-

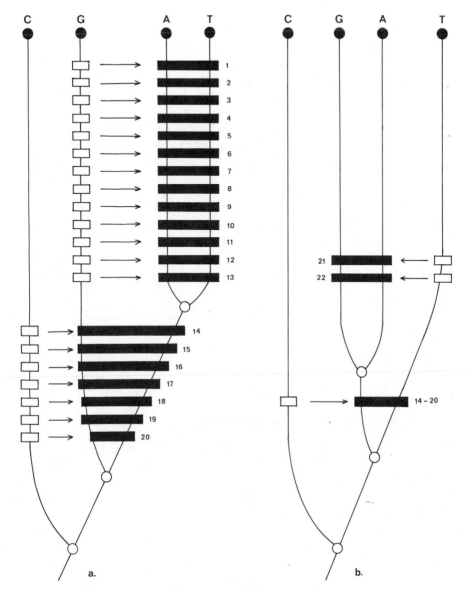

Figure 5.14 Two alternate hypotheses of the relationships of Recent actinopterygian fishes. Taxon abbreviations are: C, Chondrostei; G, Ginglymodi; A, *Amia*; and T, Teleostei. Character justifications are given in Wiley (1976).

147

iliary criteria:

1. *The criterion of geologic character precedence.* This criterion rests on the rule that when one character is found entirely in the geologically older members of a monophyletic group while the alternate homology is found only in the geologically younger members of the same monophyletic group, then the older homologue is the plesiomorphic character. Note that some members of the more recent (younger) monophyletic group can retain the plesiomorphic character, but the apomorphic character cannot be found in the older members of the monophyletic group. The validity of this criterion has been criticized by Schaeffer et al. (1972), who argue that stratigraphic position should not be considered as *a priori* evidence for character evaluation. Rather, these authors suggest that all organisms should be treated by examining characters without regard, initially, to stratigraphic position. Only after the phylogenetic hypothesis has been put forward would the investigator examine the correlation between stratigraphic position and relative primitiveness.

Nelson (1978a) has also criticized this criterion (which he calls the "paleontological argument"). It is his opinion that it is infallible in the usual sense that paleontologists employ it because the ad hoc proposition of an incomplete fossil record is introduced to save the criterion from supposed refutation.

Hennig would probably have agreed with Schaeffer et al. (1974) and with Nelson (1978a) to this extent: he recognized that the geological precedence rule is not a major phylogenetic criterion in the sense that it is the first criterion applied to character analysis. Hennig (1966) suggested the geologic precedence rule for situations where out-group comparison, ontogeny, and correlated transformation series do not help in determining relative apomorphy among two or more homologous characters. The rule of geologic precedence is, in fact, theoretically valid because the plesiomorphic character of a pair of homologous characters must have originated before the more apomorphous character. In other words, in a character transformation series of a → a', character a must be older than character a'. This is a logical deduction based on the way characters evolve. The general application of this rule leads to the circular argument that what is older must be primitive and what is primitive must be older. However, just because a taxon is older and occupies a more primitive phylogenetic position we cannot then assume, *a priori*, that its characteristics are all plesiomorphic. As a practical example, gars may occupy a primitive phylogenetic position, but their characters (or at least 26 of them: Wiley, 1976) are generally more derived than those of teleosts. Another problem is missing taxa. It is quite possible that whole groups of organisms are not found in the fossil record and yet retain more primitive characters than some groups with a long fossil record. For example, there is no doubt that amoebae are more primitive than brachiopods, yet brachiopods are older in the fossil record. Such gaps in knowledge make the geological precedence rule suspect and useful only in certain circumstances. Additionally, the geological precedence rule only occasionally can be separated from the criterion of phylogenetic position.

Let us consider a hypothetical example. In taxon 1 we have character a. In taxon 2 we have members with character a'. Because taxon 1 is the older taxon on the basis of stratigraphic occurrence, can we then conclude that character a is the plesiomorphic character of the homologous pair? No, because first we have to demonstrate that taxa 1 and 2 belong in the same monophyletic group—this requires out-group comparisons. Once the sister group relationships of 1 and 2 are determined by out-group comparisons, then we have the sister group of 1 and 2 and can easily examine this sister group to see whether its members have character a or a'. If the sister group of 1 and 2 has character a, then on the basis of the criterion of phylogenetic position, a is the primitive character. If the sister group has character a', can we still conclude that a is primitive within 1 and 2 because it is older? No, because this calls for some ad hoc hypothesis of missing taxa between 1, 2, and the sister group or 1 + 2. Thus, such conflicts can only be worked out by out-group comparison and in cases such as this the geologic precedence rule collapses into the criterion of phylogenetic position. For example, are four toes primitive for the Equidae because *Hyracotherium* is the geologically oldest horse or because other perissodactyls have five toes? I would argue the latter. We can see, however, that the rule of geologic precedence is useful in certain circumstances. For example, Hennig (1966) discusses the problem of determining which wing venation pattern in bibionomorph flies (Diptera) is primitive or derived. Recourse to Lias fossils was necessary to resolve this conflict because the possible outgroup do not have a comparable wing venation system to determine the pleisiomorphic pattern within Diptera (Fig. 5.15). I would suspect that this situation has been brought about by the extinction of relevant intermediate sister groups. The point, however, is that the geologic precedence rule can be of some benefit in certain situations.

It cannot be denied that geologic age and plesiomorphy are highly correlated. Fossils come into their own by filling in the morphological gaps encountered between many Recent sister groups because, as Patterson and Rosen (1977) have stated, the inclusion of increasingly older fossil species in a phylogenetic analysis makes it more and more likely that primitive morphologies will be analyzed. This is of direct value in two practical circumstances. First, the most meaningful comparisons between two sister groups is a comparison of the hypothesized ancestral morphotypes of these sister groups. The greater number of primitive groups included in the analysis of the two sister groups, the better the

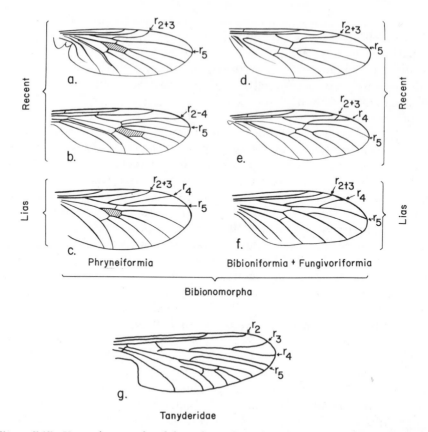

Figure 5.15 Hennig's example of the paleontological method, radial wing venation in two groups of bibionomorph flies (Diptera): (*a–c*), two Recent (*a, b*) and one fossil (*c*) species of Phryneiformia; (*d–f*), two Recent (*d, e*) and one fossil (*f*) species of Bibioniforma + Fungivoriformia; (*g*) , wing venation in a crane fly illustrating a more plesiomorphic condition of the radial veins. Outside Bibionomorpha there are usually four radial veins (*g*, r_2–r_5). In more apomorphic members of each subgroup of Bibionomorpha (cf. *a, d*), there are two veins (r_{2+3} and r_5). More primitive Recent members of each subgroups (*b, e*) show intermediate vein reductions; however, tracing the reduction series back to other Diptera is problematical. Examination of Lias fossils of each subgroup indicates that the first step in reduction involved the fusion of r_2 and r_3. Note that although there are considerable differences in wing venation between the two groups of Recent bibionomorphs, the Lias fossils bridge the morphological gap between them. Also note that the similarities exhibited by (*a*) and (*d*) were gained independently. (*a–f* from *Phylogenetic Systematics* by Willi Hennig. Copyright the University of Illinois Press, 1966. Used with permission. g modified from Borror et al., 1976.)

estimate of the hypothetical morphotypes and the more critical the comparisons between these sister groups in terms of the actual morphological gap separating them. That is, a better estimate of the actual divergence of the two ancestral species can be achieved. Thus, fossils can have great affects on testing various models of evolutionary divergence such as saltation and anagenetic gradualism. Second, the incorporation of increasingly primitive groups into the analysis has the potential of testing hypothesized synapomorphies. For example, *Amia* and teleosts both lack suborbital bones in the skull (Fig. 5.16*b,c*). Without reference to older and more primitive fossils of these two groups, we would conclude that the absence of suborbital bones was a synapomorphy uniting *Amia* and teleosts. This can be deduced from the observation that gars (Fig. 5.16*a*) and more primitive actinopterygians have suborbitals. Yet, suborbitals are found both in fossil relatives of *Amia* (parasemionotids, Fig. 5.16*d*, and caturids) and in fossil teleosts (pholidophorids and leptolepids, Fig. 5.16*e*). Thus, the supposed synapomorphy of *Amia* and teleosts is actually a nonhomology.

2. *The criterion of chronological progression.* Hennig (1966) suggested that two factors might be used to determine which of a homologous pair of characters might be the apomorphic character: ecology and biogeography. This criterion, also termed the "progression rule," pos-

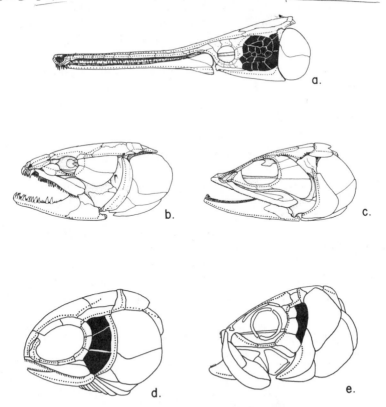

Figure 5.16 The distribution of subopercular bones (shaded black) in neopterygian fishes: (*a*) a gar; (*b*) *Amia*; (*c*) *Elops*, a teleost; (*d*) *Paracentrophorus*, a fossil relative of *Amia*; and (*e*) *Leptolepis coryphaenoides*, a fossil teleost. (*a–c* modified from Patterson, 1973; *d* redrawn and modified from Gardiner, 1960; *e* redrawn and modified from Rayner, 1948.)

tulates that ancestral species, or the descendants of that ancestral species that retain the less apomorphic epiphenotype, are found in the center of origin of the group. Dispersal out of the center of origin coupled with sequential speciation produced descendant groups of more apomorphic epiphenotypes because the ecological requirements of the new environment were different than the ecological requirements of the original ancestral range. Thus, if differentiation did occur, the descendants occurring outside the original range of the ancestral species would tend to have apomorphic characters in direct relationship to the difference of the environment and the extent of the dispersal. This is tested by demonstrating that various morphological transformation series are correlated with the geographic distributions of the taxa involved.

The problem with the progression rule is the initial assumption of a center of origin and the auxiliary assumption that dispersal is accompanied by speciation. Nelson (1975) has demonstrated, on logical grounds, that some examples of the progression rule (linear dispersal and speciation from a center of origin) can also be formulated as examples of fragmentation (vicariance), i.e., the dispersal of the ancestral species followed by later vicariance so that the original range of the ancestral species *at the time of the first speciation event* is equal to the sum of the ranges of its descendants. I will discuss this issue in the chapter on biogeographic analysis (Chapter 8). For now, we shall say that biogeography and phylogenetic relationships are, indeed, highly correlated, but not always in the way that Hennig (1966) formulated the progression rule.

3. *Criterion of the correlation of transformation series.*   Two character transformation series always found in the same taxa are said to be correlated. For example, in the three taxa (A, B, and C) shown in Table 5.2, there are two transformation series that are correlated, the series 1 → 1' → 1", and the series 2 → 2' → 2". The criterion of transformation series states that *if one of the correlated series can be determined to be a transformation from plesiomorphic to apomorphic to fully apomorphic, then* the other transformation series which it is correlated with can also be deduced as a series from plesiomorphic to apomorphic to fully apomorphic. In our example, if, by some other criterion, we can determine

Table 5.2 Two character transformation series in three taxa.

Transformation Series	Taxa		
	A	B	C
1	1	1'	1"
2	2	2'	2"

that character 1 is plesiomorphic, 1' is apomorphic, and 1" is fully apomorphic, then we can logically conclude that 2 is plesiomorphic, 2' is
apomorphic, and 2" is fully apomorphic. Note that this criterion demands
transformation series of three or more characters. In other words, if there
are only two characters in the transformation series, such as the distribution of character transformation series 1 and 2 in Table 5.2, the fact
that 2' is correlated with 1', and the fact that 2' is shown apomorphic by
some other criterion do not lead to the logical conclusion that 1' is
apomorphic because we have found that 2' is apomorphic. So far as I am
aware, this criterion has not been extensively used by English-speaking
phylogeneticists who use out-group comparison to resolve the problem
of apomorphy without resorting to the correlation criterion. In reference
to Table 5.2, rather than hypothesizing that character 2 is plesiomorphic
because it is correlated with character 1 in A, B, and C, we simply look
at the out-group to see if it has character 2 rather than 2' or 2". If there
is a conflict, the actual correlation cannot resolve this conflict (although
it may very well create the conflict). In other words, if there are two
correlated transformation series which run in opposite directions, then
there is obviously something wrong in the analysis. This may very well
lead to a rejection of the sister group relationship between the group
and the out-group. Thus, like all other phylogenetic criteria, the correlation criterion is most interesting if it points to a conflicting result in
independent hypotheses of transformation series.

   4.   *The criterion of "common equals primitive".* This criterion was
not discussed by Hennig (1966), but has been advocated by several
workers and is best dealt with here. In one form, this criterion states that
if a character is common in the close relatives of a group analyzed, then
the character is probably plesiomorphic, whereas its alternate homologue is apomorphic (Wagner, 1961). *This is the out-group criterion.* In
another form, the criterion states that if a character is widely distributed
within a taxon, then the character is probably primitive (Estabrook,
1977). *In this form, the criterion is ad hoc* because it assumes that the
evolutionary process tends to conserve plesiomorphic characters or primitive taxa. Although this may be true for some groups, it is not true in
others. Thus, we have no way of evaluating the assumption because we
cannot predict that the assumption holds in all or even a majority of
cases. In every case where an apomorphic character arises early in the
evolution of a group, and thus is observed in the majority of the taxa
comprising the group, the application of this criterion will lead to mistaken conclusion regarding character evolution.

Ontogeny and Phylogeny

Hennig (1966) listed another auxiliary criterion—ontogenetic precedence. Hennig ascribed its origin to Naef (1931). After reading the anal-

yses of Løvtrup (1974, 1978) and Nelson (1978a), I doubt that this criterion is an auxiliary criterion of the same quality as the four just discussed. Rather, it deserves to be thought of as complementary to the criterion of out-group comparison.

The ontogenetic criterion assumes that ontogenetic transformation toward a particular character reflects the phylogenetic development of that ontogeny. Apomorphic characters of a clade will go through ontogenetic stages of development recognizable as plesiomorphic or embryonic characters of that clade's more primitive relatives (primitive in respect to that character only). This sounds rather similar to the biogenetic law, the law of recapitulation, and it is. We shall see, however that the modern interpretation of this law (i.e., von Baer's, 1828, 1837, interpretation) is biologically sound and a powerful phylogenetic tool.

Organisms are the product of an interplay between their genomes, their cytoplasm, the spatial relationships of their cells, and the external environment. The production of an oak tree or an amphibian from a fertilized egg entails a number of ontogenetic events which must be carried out in a specified order if the organism is to continue to grow. Lineage splitting (speciation) makes possible the independence of lineages and thus makes possible the modification or elaboration of a primitive ontogeny in one or both of these lineages. It is this change in ontogeny that is the mechanical cause for biological diversity (Garstang, 1922; Naef, 1931; Holmes, 1944; Waddington, 1957; Løvtrup, 1974, 1977, 1978) among multicellular organisms. Obviously, the base cause for some of this is genetic. Just as obviously, other causes are not primarily genetic (Waddington, 1957; Løvtrup, 1974). The elaboration of the interplay between the factors affecting ontogeny and the biological diversity it produces when coupled with speciation is one of the great mysteries of science and I would be writing another kind of book if I had the answers to the mystery. We do, however, understand the basic principles well enough to utilize ontogenetic information to deduce patterns of evolutionary descent.

The concept of recapitulation "implies a certain parallelism between the changes which are thought to have occurred in the course of phylogenetic descent and those observable in various embryos" (Løvtrup, 1978: 348). Løvtrup (1978) distinguishes between two types of characters, epigenetic and nonepigenetic. Epigenetic characters are causally interrelated and the modification of one depends on some change in an earlier stage of ontogeny. Nonepigenetic characters are not so readily explained. Løvtrup (1978) makes the distinction on the basis of whether the character is essential for normal morphogenesis. Epigenetic characters are seen as essential for normal morphogenesis whereas nonepigenetic characters are seen as not essential. Thus, a notochord would be an epigenetic character because normal vertebrate morphogenesis cannot proceed without it, whereas the presence of hair is nonepigenetic because it is possible to have a viable mouse without hair (Løvtrup, 1978). As an additional example, secondary sex characteristics are seen as nonepigenetic—thus neoteny itself involves nonepigenetic characters.

I find this distinction rather arbitrary. I would prefer to speak of heritable and nonheritable characters. Heritable morphological characters, whether a product of genetic or cytogenetic or positional factors would correspond to the epigenetic characters of Løvtrup (1978). Many of the so-called nonepigenetic characters discussed by him actually are ontogenetically controlled and heritable—the hair of mammals or skin pigmentation. Some types of neoteny are heritable and thus epigenetic; others such as the suppression of the adult characteristics of the salamander *Eurecia nana* are due to the lack of certain trace elements in the water (in this case iodine) which inhibit certain characters but not others and are therefore not heritable.

Løvtrup (1978) goes on to distinguish between two additional concepts, both of which are invaluable in our understanding of the interplay of ontogenetic information and phylogenetic patterns:

1. **Terminal character.**   a character that is the last in a sequence of ontogenetic characters.
2. **Nonterminal character.**   a character that occurs in the ontogenetic sequence before a terminal character.

The key to understanding the relationship between ontogeny and phylogeny lies in the affects of changes in ontogenetic sequences involving these two types of characters. A character added onto the end of an ontogenetic sequence is a terminal character. This character is also epigenetic and thus heritable. Now, if the immediately preceding ontogenetic character is deleted from development, the terminal character will not develop. We can state this in the form:

$$X^1 \rightarrow X^2$$

$$X^2 \rightarrow X^3$$

where $X^1 \rightarrow X^2$ represent the nonterminal characters in the ontogenetic sequence and X^3 represents the terminal character. Now, if X^3 occurred as an evolutionary novelty in clade A, we would expect the sister group, clade B, to have the ontogenetic sequence:

$$X^1 \rightarrow X^2.$$

Thus the terminal character of clade B is X^2, and clade A goes through an ontogenetic stage that is present in the *adults* of its sister group, clade B. This type of recapitulation is **Haeckelian recapitulation** (Løvtrup, 1978).

Not all changes in ontogeny are the result of the addition of new terminal characters. Indeed, many major changes in adult morphotypes (such as, say, from primitive archosaur to bird) have involved changes in nonterminal characters (as evidenced by the observation that the wings of embryonic birds are never covered with the scales of crocodiles). As a hypothetical example we may consider the following ontogenetic sequence:

$$Y^1 \rightarrow Y^2 \rightarrow Y^3 \rightarrow Y^4$$

Rather than add Y^5 as a new terminal character, let us say that some gene, or gene complex, which transcribes to produce Y^3 mutates in an ancestral species. This may result in an inviable embryo or it might result in the sequence:

$$Y^1 \rightarrow Y^2 \rightarrow Y^7 \rightarrow Y^8$$

Now, if that ontogenetic sequence is fixed in that ancestral species, the result is two developmental sequences:

Clade C: $Y^1 \rightarrow Y^2 \rightarrow Y^3 \rightarrow Y^4$

Clade D: $Y^1 \rightarrow Y^2 \rightarrow Y^7 \rightarrow Y^8$

We would never expect the ancestor of clade D to have an *adult* terminal character Y^4. We do note, however, that the sister groups C and D do have embryonic stages which are similar (Y^1, Y^2). This type of recapitulation is termed **von Baerian recapitulation**. As Garstang (1922: 98) states, "The phylogenetic succession of adults is the product of successive ontogenies." And Løvtrup (1978: 351) characterizes this phenomena thus: "Von Baer generalized his empirical observations in his laws of development which imply that in the course of ontogeny there is a gradual change from the general to the special." Coupled with the tenets of phylogenetic systematics, we may state:

In the course of their ontogeny, the members of two sister groups will follow the same course of recapitulation up to the stage of their divergence into separate taxa. (modified slightly from Løvtrup, 1978: 352).

This "theorem" (as Løvtrup terms it) applies to both recapitulation phenomena. Equally applicable is what may be termed the rule of correspondence between phylogenetic and ontogenetic character modification:

Given an ontogenetic character sequence which goes from a character found in the outgroup to a character found only within the group considered, the character found only in the group considered is the derived character and the character found in the out-group is primitive. (combined and modified from Nelson,1978a; and Løvtrup, 1978).

In other words, given the sister groups and ontogenetic patterns:

$$\text{Clade A:} \quad Y \rightarrow Y^1 \rightarrow Y^2$$
$$\text{Clade B:} \quad Y \rightarrow Y^1$$

Character Y^2 is apomorphic relative to Y^1. And given (1) the following sequence:

$$\text{Clade A:} \quad Y^1 \rightarrow Y^2 \rightarrow Y^3 \rightarrow Y^4 \rightarrow Y^5$$
$$\text{Clade B:} \quad Y^1 \rightarrow Y^2 \rightarrow Y^7 \rightarrow Y^8 \rightarrow Y^9$$

(2) that A and B are sister groups and C is the sister group of A + B, and (3) that C's ontogeny is

$$Y^1 \rightarrow Y^2 \rightarrow Y^3 \rightarrow Y^4 \rightarrow Y^5$$

we may conclude that the terminal character Y^9 is apomorphic relative to Y^5.

At this point a practical example is in order. The teleost killifishes (Order Cyprinodontoidea) are small fresh and brackish water inhabitants of streams and near-shore habitats. The group exhibits two basic conditions in their lateral line sensory system. In many genera the sensory neuromasts of the supraorbital line sit in an open groove or a series of open grooves on top of the head (see Fig. 5.17a). In other genera, these neuromasts are enclosed in tubes of epithelial tissue (see Fig. 5.17b).

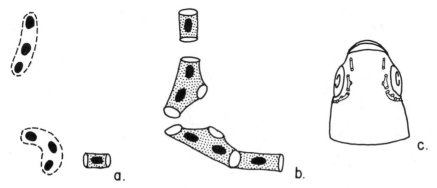

Figure 5.17 Supraorbital sensory canals in two genera of killifishes: (*a*) the open groove neuromast system of *Aplocheilichthys*; (*b*) the pore-canal system of *Fundulus*; and (*c*) the location of the pore-canal system in *Fundulus*. The neuromasts are shaded black, grooves are shown by dashes, canals are stippled, and pores are unshaded.

The juveniles of those genera with the tubular canal system go through an ontogenetic stage characterized by the open groove and a polymorphism of tubular and grooved lines is characteristic of many of these juveniles. The development of the groove is a necessary ontogenetic precursor to the tubular system and thus a tubular canal system must be derived rather than primitive.

The most interesting point of this example is that most of the closest relatives of killifishes, the flying fishes and silversides have a tubular canal system. On the basis of out-group comparison we would have assumed that the tubular canal system was primitive. This would have quite an effect on the final phylogenetic hypothesis because (1) killifishes with tubular canal systems have apomorphic caudal skeletons and gill arches whereas (2) killifishes with open grooves have plesiomorphic caudal skeletons and gill arches. Thus, accepting open grooves as a synapomorphy would result in the erection of two major hypothesized monophyletic groups when only one major group can be justified on the basis of the sensory line data at hand.

The killifish sensory system is an example of Haeckelian recapitulation (terminal addition). As an example of von Baerian recapitulation, I can cite the phenomena of gastrulation. The formation of a gastrula is apomorphic for Eumetazoa (given that Mesozoa are not degenerate flatworms). Although Haeckel postulated an adult "gastrulata" capable of leaving offspring, it is very doubtful that such an organism ever really existed. Indeed, it is possible that the ancestor of eumetazoans used the gastraea as a temporary feeding structure, like the living placozoan *Trichoplax adhaerens*. Gastrulation is an embryonic phenomena which no doubt involved the modification of nonterminal characters in a Precambrian species that was the ancestral eumetazoan. And this was probably the result of a cytoplasmic reorganization in the egg. Thus, the basic body plan of eumetazoans is the result of nonterminal changes and the comparative ontogeny is probably a phenomenon of von Baerian recapitulation.

In summary, the basic strength of the ontogenetic criterion is that it can be utilized independently of the out-group criterion when terminal additions are involved. Because of this, the two can be used to check each other. Ontogenetic criteria come into their own in situations where there are many possible out-groups and ontogeny may be used to do intragroup analysis. This, in turn, might lead to characters that might help in determining which of the possible out-groups is the sister group. It is my opinion that this little used criterion is a powerful phylogenetic tool.

Errors in Character Analysis

There is an array of error in character analysis as well as in tree analysis. Errors in testing homologies fall into two basic classes which are analogous to the Types I and II errors discussed in Chapter 4. Each of these types has two subdivisions:

Type I Errors. Accepting a false hypothesis and rejecting a true hypothesis.

Type Ia. Accepting a nonhomologous similarity as a synapomorphy.

Type Ib. Accepting a plesiomorphic homology as a synapormorphy.

Type II Error. Rejecting a true hypothesis in favor of a false hypothesis.

Type IIa. Accepting a synapomorphy as a nonhomologous similarity.

Type IIb. Accepting a synapomorphy as a plesiomorphic homology.

CHARACTER ANALYSIS—SOME EXAMPLES

The first half of this chapter is devoted to the theory of character analysis and some examples of ways of determining relative apomorphy. In this section I would like to integrate these theoretical considerations with the principles of tree testing (Chapter 4) by showing examples of actual analyses that phylogeneticists have performed. In each example I will state the general background information which leads to the analysis. Selected characters will be considered individually. The criterion or criteria for considering one particular character as apomorphic rather than its alternate will be stated. Each character is then applied to the various alternate hypotheses to be tested. Finally, tables and/or trees with all characters are presented for completeness.

Example 1: The Phylogenetic Relationships of *Neocteniza*

Neocteniza is a genus of mygalomorph spiders of the family Actinopodidae (Fig. 5.18). They are a Central American and northern South American group of about seven species. Platnick and Shadab (1976) became interested in the genus when they received a male collected in Amazonia. Prior to this no male was known. Platnick and Shadab decided to investigate the phylogenetic relationships of the genus.

Background Information

Neocteniza, Actinopus, and *Missulena* comprise three monophyletic genera of the family Actinopodidae. That actinopodids are a monophyletic group is indicated by the excavated depressions in the posterior sigilla of females. This character is not found in any other mygalomorph family and thus may be considered a unique evolutionary innovation (synapomorphy by out-group comparison). Actinopodids are also characterized by having eyes which are spread across the carapace (Fig. 5.18)

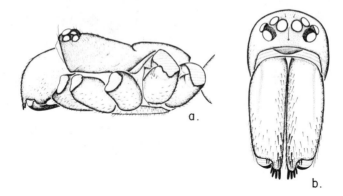

Figure 5.18   The spider *Neocteniza fantastica*: (*a*) lateral view of body; (*b*) frontal view of the head, showing the arrangement of the eyes. (from Platnick and Shadab, 1976. Copyright by the American Museum of Natural History, Used with permission.)

rather than eyes localized and closely clustered. This spread of eyes across the carapace is also found in the family Migidae. But it is not found in other mygalomorphs nor is it found in more primitive spiders such as liphistiids. Platnick and Shadab (1976) interpreted the condition of the eyes in actinopodids and migids as synapomorphic (synapomorphy by out-group comparison). Finally, that no genus within Migidae is more closely related to the three genera of actinopodids than to other migids is indicated by the presence in all migids of vertically inclined chelicerae, a character not found in actinopodids or other mygalomorphs (synapomorphy by out-group comparison).

At this point, Platnick and Shadab have established three points:

1.   Actinopodidae is monophyletic.
2.   Migidae is the sister group of Actinopodidae.
3.   Migidae is monophyletic.

The assumptions on which this is based are:

1.   That other mygalomorph spiders are more closely related to actinopodids and migids than to any other taxon.
2.   Other mygalomorph spiders may therefore serve as out-groups for comparison.

Given this higher level set of relationships, Platnick and Shadab were then faced with a standard problem of the phylogenetic relationships of the three genera using another family for out-group comparison. The four possibilities are shown in Fig. 5.19. Platnick and Shadab (1976) then proceeded to analyze 19 characters. In each case there were five possibilities:

1.   The character is found in all three genera and the out-group Migidae. *Conclusion:* not relevant to the resolution of the problem.
2.   The character is present in one of the three genera and in Migidae. *Conclusion:* the character is plesiomorphic at that level of analysis, but the alternate character is probably synapomorphic.
3.   The character is present in two of the three genera but the alternate is present in one genus and the out-group. *Conclusion:* the character is synapomorphic for the two genera and the alternate is plesiomorphic.
4.   The character is found in all three genera but not the out-group. *Conclusion:* the character is synapomorphic for the family but irrelevant for analyzing relationships among the genera.

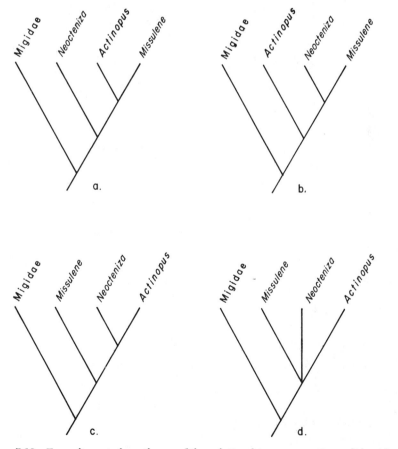

Figure 5.19   Four alternate hypotheses of the relationship among actinopodid spiders.

5. The character is found in one of the genera and is absent in the out–group. *Conclusion*: the character is synapomorphic for the genus.

Character Analysis

The 19 characters used to analyze the relationships of *Neocteniza* to *Actinopus* and *Missulena* are shown in Table 5.3. These fall into two classes, those represented by alternates 2, 3, and 5 above. The reason no characters fall into categories 1 and 4 is that Platnick and Shadab had already analyzed the higher relationships of the family to other families, thus characters fitting 1 or 4 were properly disregarded. I shall discuss three characters. The logic for determining relative apomorphy for the remaining characters is exactly the same as one or the other of the three discussed.

Character 1, Coloration of the Legs of Males. In *Neocteniza* the only male known (the holotype of *N. fantastica*) has a distinct color pattern on the legs. Males of *Actinopus*, *Missulena*, migids, and most other mygalomorphs have legs of uniform coloration. Note that Platnick and Shadab have no idea whether the males of other *Neocteniza* actually have a color pattern on the legs. They have preferred to propose a bold hypothesis—that the color pattern is synapomorphic for the genus rather than autapomorphic for *N. fantastica*. This hypothesis will be tested with the very next male of the genus collected (as will characters 3 and 4).

Character 2, the Presence of Cusps of Femur IV. This character is found only in *Neocteniza*. Cusps are absent in migids, *Actinopus*, and *Missulena*. By the criterion of out-group comparison, the presence of cusps is a synapomorphy uniting the seven known species of *Neocteniza* into a monophyletic group.

Character 10, the Thoracic Groove. The thoracic groove of *Neocteniza* is recurved. The same condition is found in migids. *Actinopus* and *Missulena* have procurved grooves. Procurved thoracic grooves may be considered synapomorphic relative to recurved grooves. If we assumed that recurved grooves were synapomorphic, then we would have to conclude that *Neocteniza* was more closely related to migids than to *Actinopus* and *Missulena*. This is, of course, a possibility if we reject our original higher level hypothesis. However, if most mygalomorphs have recurved thoracic grooves, then this is additional evidence that the procurved condition is the apomorphic condition.

The 19 characters used by Platnick and Shadab (1976) are organized into Fig. 5.20 as a summary hypothesis of relationship. In terms of our original four possibilities (Fig. 5.19a–d) we see that Fig. 5.19a is more justifiable than the other three.

Before leaving this example I should like to point out one additional assumption that Platnick and Shadab (1976) have made implicitly. The

Table 5.3 Data table used by Platnick and Shadab (1976) to analyze the phylogenetic relationships of three genera of actinopodid spiders; apomorphies are in italics

Character	Outgroup (Migidae)	*Neocteniza*	*Actinopus*	*Missulena*
1. Coloration, legs of males	Uniform	*Patterned*	Uniform	Uniform.
2. Cusps on femur IV	Absent	*Present*	Absent	Absent
3. Ant. metatarsi of males	Long apical spine absent	*Long apical spine present*	Long apical spine absent	Long apical spine absent
4. Palpae tibia of males	Elongate	*Incrassate*	Elongate	Elongate
5. Embolus	Short	*Long*	Short	Short
6. Labium of females	With spinules	*No spinules*	With spinules	With spinules
7. Sclerotized bursae copulatrix	Absent	*Present*	Absent	Absent
8. Spermathecae	Short and straight	*Long and sinuous*	Short and straight	Short and straight
9. Pars cephalica of males	Flattened	Flattened	*Greatly elevated*	*Greatly elevated*
10. Thoracic groove	Recurved	Recurved	*Procurved*	*Procurved*
11. Labium	Suture lines	Suture lines	*No suture lines*	*No suture lines*
12. Post. sigilla of males	Flattened plates	Flattened plates	*Excavate depression*	*Excavate depression*
13. Post. tarsi of males	Not scopulate	Not scopulate	*Scopulate*	Not scopulate
14. Tibia III	Apical comb absent	Apical comb absent	*Apical comb present*	*Apical comb present*
15. Post. tarsi	Unarmed	Unarmed	*Armed*	*Armed*
16. Maxillae of males	Armed	Armed	*Unarmed*	Armed
17. Paired tarsal claws of males	Multidentate	Multidentate	*Unidentate*	Multidentate
18. Tibia IV	Spines only	Spines only	Spines only	*Cusps*
19. Carapace margin of males	Flat	Flat	Flat	*Reflexed*

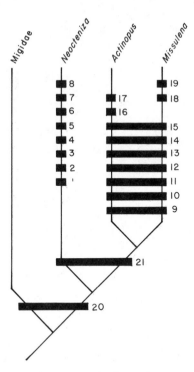

Figure 5.20 A phylogenetic hypothesis of relationships among actinodpodid spiders. Black bars are hypothesized synapomorphies. Character transformation series 1–19 are shown in Table 5.2. Character 20 is the condition of the eyes and Character 21 is the presence of excavated depressions on the posterior sigilla of females. (Adapted from Platnick and Shadab, 1976.)

seven species of *Neocteniza* are represented by only eight specimens, the only material available to the authors. Thus, they assume that the characters examined are not subject to variation of the type that would render them useless for analysis. Because Platnick and Shadab are trained arachnologists who are aware of this problem, I assume that the characters were picked for this reason. Similar problems face vertebrate paleontologists who may have only a single specimen well enough preserved to perform character analyses.

Example 2: Recent Neopterygian Fishes

Statement of the Problem

There are three major groups of Recent neopterygian fishes: gars (Ginglymodi), the bowfin (*Amia calva*, Halecomorphi), and teleosts (Teleostei). All recent workers (e.g., Nelson, 1969; Patterson, 1973; and Wiley, 1976) except Jessen (1972, 1973) conclude that neoterygians are a monophyletic group. This is evidenced by at least seven synapomorphies shared among neopterygians and no other group. Neopterygians together with bichirs, the paddlefish, and sturgeons form a monophyletic group (Actinopterygii) related to lungfishes and tetrapods (Sarcopterygii).

There are four possible phylogenies for the three groups of neopterygians (Fig. 5.21a–d). Neither 5.21b nor 5.21d have been supported by any worker and apparently there is no evidence for their support. The other hypotheses (Fig. 5.21a,c) have been justified by various workers and thus warrant serious consideration. These alternatives may be stated as two hypotheses:

Hypothesis 1. *Amia* and teleosts form a monophyletic group (Halecostomi) that is the sister group of gars (Ginglymodi): Fig. 5.21a.

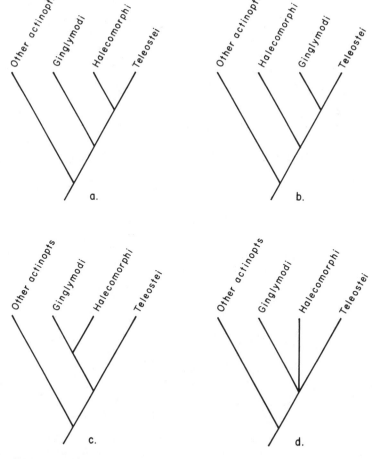

Figure 5.21 Four alternate hypotheses of the relationships among neopterygian fishes.

General Predictions of Hypothesis 1. If hypothesis 1 is preferred we should find characters in *Amia* and teleosts not found in gars, paddle-fishes, sturgeons, or polypterids. If characters are found in *Amia* and gars and not in teleosts or the various other actinopterygian groups, they must be rare compared to the synapomorphies found in *Amia* and teleosts.

Hypothesis 2. *Amia* and gars form a monophyletic group (Holostei) that is the sister group of teleosts: Fig. 5.21*c*.

General Predictions of Hypothesis 2. We would predict the opposite results of hypothesis 1.

Character Analysis

General Approach. This example will concentrate on the analyses of Nelson (1969), Patterson (1973), and Wiley (1976). All three workers used the comparative method and analyzed their characters phyloge-netically. Nelson (1969) concentrated on gill arch morphology, Patterson (1973) on skull morphology, and Wiley (1976) on a general review of the problem. The characters that will be reviewed in detail are selected for their clarity to nonichthyologists. The totality of evidence for and against the two alternate hypotheses (some of which is rather esoteric) is pre-sented at the end of the discussion.

Character 1, the Interoperculum. The interoperculum is a small der-mal skull bone found in front of the suboperculum in *Amia* and teleosts but not in gars (Fig. 5.22). Further, there is no interoperculum in lower actinopterygians nor in the sarcopterygians. This character meets the conditions of topographic homology, and ontogenetic development. It

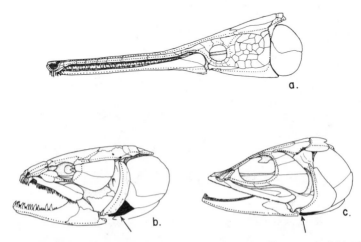

Figure 5.22 Lateral views of the skulls of (*a*) *Lepisosteus*, (*b*) *Amia*, and (*c*) *Elops* illus-trating the distribution of the interoperculum (indicated by an arrow and shaded black in *b* and *c*). (Adapted from Patterson, 1973.)

also meets the criterion of being present in only two of the three groups. Out-group comparison supports the hypothesis that the presence of an interoperculum is a synapomorphy uniting *Amia* and teleosts (Patterson, 1973).

Character 2, the Supratemporals. Supratemporals are dermal bones that are found on the top of the skull between the parietals, dermopterotics, and the posttemporals in actinopterygians and primitive sarcopterygians (Fig. 5.23). In gars there are two or more supratemporals on each side of the skull. This is also true of more primitive actinopterygians and in those sacropterygians with supratemporals. Only *Amia* (and its fossil relatives) and teleosts have a single supratemporal per side. The presence of supratemporals is a plesiomorphic similarity shared between gars, lower actinopterygians, and sarcopterygians. The reduction (or fusion) of numerous supratemporals in *Amia* and teleosts was interpreted by Wiley (1976) as a synapomorphy of halecostomes (by out-group comparison).

Character 3, the Rectus Communis Muscle. The rectus communis is a muscle found on the ventral gill arches of *Amia* and teleosts (Fig. 5.24). It is thought to be a derivative of the fourth obliquus ventralis because in *Amia* these two muscles share fibers and because the rectus communis is innervated by the same nerve as the obliquus ventralis (Nelson, 1967). The fourth gill arch muscle of gars is a simple obliquus ventralis as are similar muscles in sturgeons and paddlefishes (Wiley, 1976; from earlier authors). The rectus communis of teleosts varies in both origin and insertion, but no author has suggested that the muscle is nonhomologous to that of *Amia*. The rectus communis satisfies both the morphological criterion of homology and the out-group criterion for considering this homology as a synapomorphy of halecostomes.

Character 4, the Infrapharyngobranchials. Infrapharyngobranchials are dorsal gill arch elements that articulate with epibranchials and/or suprapharyngobranchials and the neurocranium (Fig. 5.25). In gars and *Amia* there are three infrapharyngobranchials (Fig. 5.25*b,c*). In teleosts there are four (Fig. 5.25*d*). Nelson (1969) suggested that the primitive condition for gnathostomes (jawed vertebrates) was for each gill arch to have an infrapharyngobranchial. Thus if there are five gill arches, there should be, primitively, five pharyngobranchials. (Such is the case in sharks.) Based on this, Nelson considered the four infrapharyngobranchials of teleosts to be plesiomorphic compared to the three infrapharyngobranchials in *Amia* and gars. Thus, Nelson (1969) considered the similarity of gars and *Amia* as synapomorphic and evidence of a monophyletic Holostei.

Patterson (1973) reanalyzed this character. He pointed out that chondrosteans also lacked a fourth infrapharynobranchial (Fig. 5.25*a*). He concluded that three infrapharynobranchials were primitive for actinopterygians and thus not evidence for a monophyletic Holostei. In terms

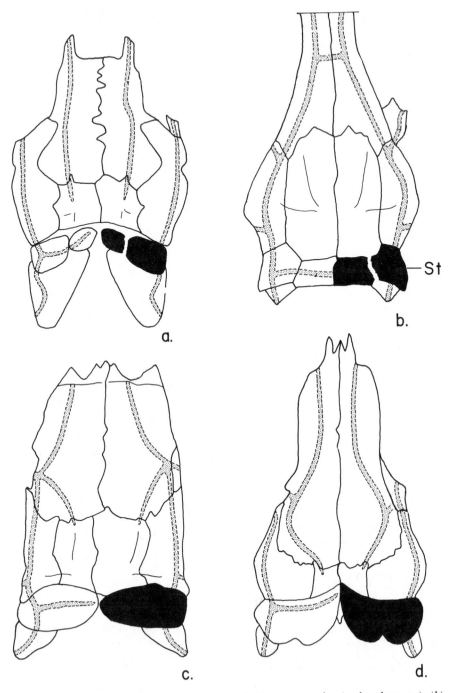

Figure 5.23 Dorsal views of the skull roofs of (*a*) *Pteronisculus* (a chondrostean), (*b*) *Lepisosteus*, (*c*) *Amia*, and (*d*) *Elops*, Illustrating the number of supratemporals (ST, shaded black). (Adapted from several sources by Wiley, 1976.)

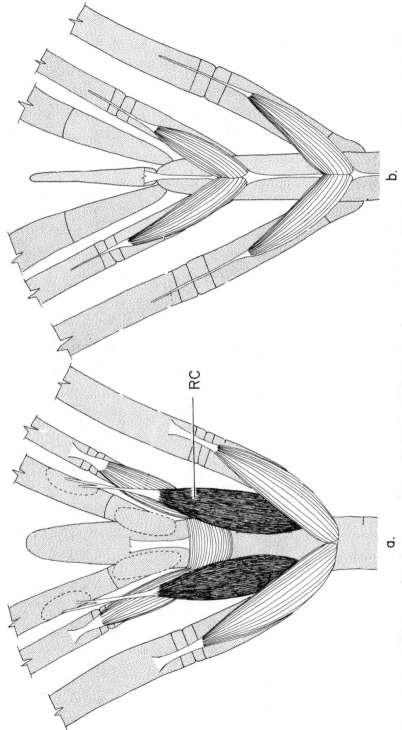

Figure 5.24 Ventral views of the posterior gill arches and associated muscles of *Amia* (left) and *Atractosteus* (right, a gar), illustrating the rectus communis muscle (RC) of halecostomes. (From Wiley, 1976.)

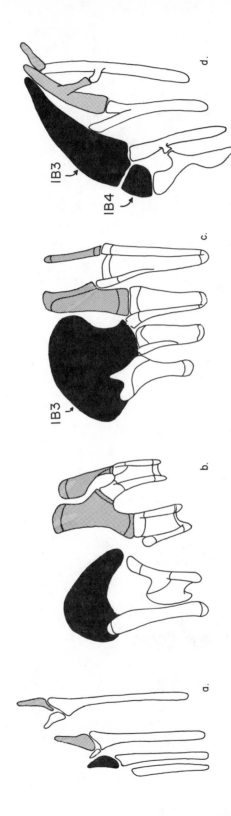

Figure 5.25 Dorsal views of the upper gill arch ossifications in (*a*) a "Gogo palaeoniscoid" (Chondrostei), (*b*) *Atractosteus*, (*c*) *Amia*, and (*d*) *Alopus* (a teleost), illustrating the infrapharyngobranchials. In all drawings the first two infrapharyngobranchials are stippled. In (*a–c*) the third infrapharyngobranchial is shaded black. In (*d*) the third and fourth infrapharyngobranchials (IB3, IB4) are shaded black. (*a* adapted from Gardiner, 1973; *b* and *c* from Wiley, 1976; *d* from Rosen, 1973.)

of errors, Patterson (1973) was suggesting that Nelson (1969) committed a Type I error. This is worth discussion because it points out some of the logic involved in phylogenetic argument. The following points can be made.

If Nelson (1969) is correct that three infrapharyngobranchials is a synapomorphy, then two conclusions may be drawn:

1. The Neopterygii is not a monophyletic group because *Amia* and gars must be more closely related to chondrosteans than they are to teleosts.

2. Even if three infrapharyngobranchials are synapomorphic, this does not justify a monophyletic Holostei because three infrapharyngobranchials must be synapomorphic for the group composed of sturgeons, paddlefishes, gars, and *Amia* because all four taxa have the character.

There are many other characters justifying Patterson's position. Given that neopterygians are a monophyletic group, out-group comparison forces the conclusion that three infrapharyngobranchials are primitive within Actinopterygii. This means that a reduction in infrapharyngobranchials has occurred somewhere between elasmobranchs and actinopterygians and thus the character has potential for defining a monophyletic group at some higher level of universality.

Other Characters.   For completeness, the totality of evidence supporting both hypotheses (Fig. 5.21*a,b*) is shown in Fig. 5.14*a,b*. Note that two similarities remain in support of a monophyletic Holostei despite efforts to refute them on morphological grounds. Both were rejected by Patterson (1973) and Wiley (1976) as homologies because that interpretation is inconsistent with other characters. If we are correct, their interpretation as synapomorphies (Jessen, 1972, 1973) represent Type I errors.

Example 3: The Phylogenetic Relationships of *Leysera*

Leysera L. is a small group of composite shrublets. Three species are found in southern Africa (*L. gnaphalodes*, *L. tenella*, and *L. longipes*) and one is found in the Mediterranean region (*L. leyseroides*). As a continuation of previous work on closely related genera, Bremer (1978a) decided to study the interrelationships of these four species.

Background Information

Leysera (Fig. 5.26) is a member of the Compositae, tribe Inuleae. Along with 23 other genera, *Leysera* was placed by Merxmüller et al. (1977) in the subtribe Athrixiinae. They grouped the 24 genera of Athrixiinae

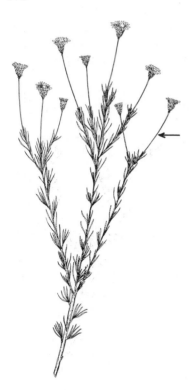

Figure 5.26 The composite plant *Leysera gna-phalodes*, illustrating long penducle typical of the genus (arrow). (From Bremer, 1978a. Used with permission of the editor, *Botanica Notiser*, and the Lund Botanical Society.)

into seven informal generic groups. *Leysera* belongs to the *Athrixia* genus group.

Bremer (1976a) justified the monophyly of four members of the *Athrixia* genus group on the basis of leaf and involucre characters. *Leysera*, *Antithrixia*, *Relhania*, and *Rosenia* have ventrally furrowed and pubescent leaves and wide, yellowish brown and scarious involucral bracts. These characters are uncommon in the Athrixiinae and are considered by Bremer (1978a) to be a synapomorphy linking the four genera (outgroup comparison).

Bremer (1976a,b) also analyzed the relationships of the five genera. In other members of the *Athrixia* genus group and within Inuleae generally, the pappus consist of several barbellate bristles. *Antithrixia* has the same condition (Fig. 5.27*a*). *Leysera*, *Rosenia*, and *Relhania* have scales and show a reduction of bristles on the disc-floret pappus (Fig. 5.27*b*) and a complete loss of bristles on the ray-floret pappus. These conditions were interpreted as synapomorphic and Bremer (1978a) concluded that *Leysera*, *Rosenia*, and *Relhania* form a monophyletic group whose sister group is *Antithrixia* (Fig. 5.28). Finally, Bremer (1978a) pointed out that only species of *Leysera* have a solitary capitula on a long peduncle (see Fig. 5.26) whereas the other three genera have sessile

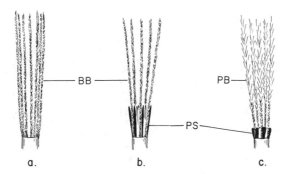

Figure 5.27 Features of the disc-floret in (*a*) *Antithrixia,* (*b*) *Leysera longipes,* and (*c*) *Leysera tenella.* Abbreviations: BB, barbellate bristles, PB, plumose bristles; PS, pappus scales. Transformation series in both characters proceed left to right. (From original drawings by Kåre Bremmer included in Bremer, 1978a. Used by courtesy of Kåre Bremer.)

capitulas (even though some *Relhania* have developed solitary capitulas independently).

Bremer (1978a) has established three points:

1. *Leysera* is monophyletic.
2. *Leysera* is most closely related to *Relhania* and *Rosenia.*
3. *Antithrixia* is the sister group of *Leysera* + *Relhania* + *Rosenia.*

The assumption is: the *Athrixia* genus group established by Merxmüller et al. (1977) is monophyletic. This assumption is necessary for the analysis of the morphology of the four genera.

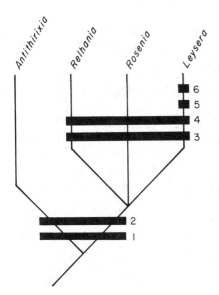

Figure 5.28 The phylogenetic relationships of *Leysera* and closely related genera. Synapomorphies are: (1) leaves ventrally furrowed and pubescent; (2) involucral bracts wide, yellowish brown and scarious; (3) ray-floret pappus with scales but no bristles; (4) disc-floret pappus with reduced bristles and no scales; (5) solitary capitula on long peduncle; (6) chromosomes 2*N* = 8. (Adapted from Bremer, 1978a.)

Character Analysis

Unlike the other two examples discussed above, Bremer's (1978a) analysis is a four-taxon analysis involving two possible sister groups (*Relhania* and *Rosenia*). Further, there was the problem of *"Leysera" montana*, a species with a capitula on a long peduncle and the pappus characters of a primitive Athrixiinae. I would say that this type of problem is more common than "clean phylogenetic analyses". Bremer (1978a) took one step to simplify the problem: he removed *"Leysera" montana* from consideration.[6] This leaves a four-taxon + two-sister-group problem of which there are 11 possible phylogenetic trees (given that no analyzed species of *Leysera* is ancestral to the other species). Because this is a four-taxon problem, there are two levels of universality for which synapomorphic characters must be applied if complete dichotomy is achieved:

1.   One level demonstrating a monophyletic subgroup composed of three of the four species (in this example, *L. gnaphalodes*, *L. tenella* and *L. leyseroides*).
2.   And another level demonstrating the sister group relationship of two of the three species forming the monophyletic subgroup within the genus (in the example, *L. leyseroides* and *L. tenella*).

One character from each level will be discussed. The complete list of characters is shown in Table 5.4 and they are placed on the phylogenetic hypothesis in Fig. 5.29. The characters specifically discussed are Character 1, Table 5.4 (the receptacle) and Character 5, Table 5.4 (the pappus scales).

Character 1, the Receptacle.   Within *Leysera* there are two conditions of the receptacle. In *L. longipes* the receptacle is more or less smooth, without scalelike outgrowths. In the remaining three species the receptacle is rough, and this roughness is caused by scalelike outgrowths. Receptacles with scalelike outgrowths are not known in any of the two closely related genera, *Relhania* and *Rosenia*, nor in other members of the *Athrixia* genus group (a similar condition is known in less closely related genera of Compositae). By the criterion of out-group comparison, Bremer concluded that the presence of scalelike projections on the receptacle was evidence for the hypothesis that *L. gnaphalodes*, *L. tenella*, and *L. leyseroides* were more closely related to each other than any one was to *L. longipes*.

---

[6] Bremer (1978b) subsequently placed *"L." montana* in the new genus *Oreoleysera*.

Table 5.4 Characters used by Bremer (1978a) to analyze the phylogenetic relationships of *Leysera*. Autapomorphies are not listed. Characters are shown in the hypothesis presented in Fig. 5.29. All determinations were made by out-group comparison

Character	Plesiomorphic	Apomorphic
1. Receptacle	Smooth	With scalelike growths
2. Floret tubules	Glands present	Hairs present
3. Pappus	Barbellate	Plumose
4. Achenes surface	Smooth	Cells imbricated
5. Pappus scales	Subulate	Wide and flat
6. life cycle	Perennial	Annual

Character 5, Pappus Scales. There are two conditions of the pappus scales in *Leysera*, subulate, wide and flat. In *L. tenella* and *L. leyseroides* the pappus scales are wide and flat. In *L. gnaphalodes* and *L. longipes* the scales are subulate. Given that the characters that demonstrate the monophyly of the *L. gnaphalodes–L. tenella–L. leyseroides* group are true synapomorphies, an out-group comparison is furnished

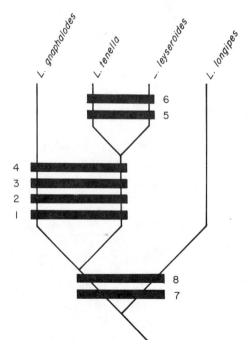

Figure 5.29 Bremer's (1978a) hypothesis of the relationships among species of *Leysera*. Black bars are synapomorphies. Character numbers 1 to 6 correspond to characters shown in Table 5.4, the two synapomorphies (7, 8) for the genus are shown in Fig. 5.29 as characters 1 and 2. Autapomorphies for each species are not shown.

by *L. longipes*. In other words, within *Leysera*, the out-group for the three-taxon problem is their sister group, *L. longipes*, just as the out-groups for the four-taxon problem were the two probable sister groups *Relhania* and *Rosenia*. Because subulate pappus scales are found in the out-group *L. longipes*, Bremer concluded that subulate pappus scales were plesiomorphic relative to wide, flat pappus scales. This is justification for the hypothesis that *L. tenella* and *L. leyseroides* are more closely related to each other than either is to *L. gnaphalodes*.

OTHER METHODS OF ARGUMENTATION

Thus far we have considered the construction of phylogenetic trees from what might be called the classical perspective. There are other methods of arriving at the same conclusions, methods that may seem different from Hennig's argumentation but that are similar in that they are deductive and group only on the basis of synapomorphy. Further, like Hennig's method, they arbitrate between conflicting hypotheses using parsimony. Given the same data base, all phylogenetic methods should give the same result. This is true of the Wagner groundplan divergence method and the latest versions of the Wagner tree programs of J. S. Farris. It may also be true of the computer algorithms of Fitch (1971) and Dayhoff (1969). I would like to discuss Wagner trees in this section and demonstrate that they are phylogenetic sensu Hennig given that the decisions concerning relative apomorphy are determined by an appropriate criterion. Because the algorithms of Fitch (1971) and Dayhoff (1969) are usually applied to amino acid sequence data, I will defer discussion on them until Chapter 9.

Wagner's Groundplan Divergence Analysis

In 1948 Dr. H. W. Wagner (University of Michigan) developed a technique for determining the phylogenetic relationships among organisms that he hoped would replace intuition with analysis. The method was based on determining the apomorphic (termed advanced) characters present within a taxon and then linking the subtaxa based on relative degrees of apomorphy. It is important to note that only apomorphic characters were used. Wagner's methods for determining relative apomorphy were not as critically formulated as Hennig's. However, given Hennig's methods for determining relative apomorphy, and given that no taxon will occupy an ancestral position, Wagner's method of clustering produces the same result as Hennig's. Thus with appropriate constraints, Wagner's groundplan divergence analysis is phylogenetic sensu Hennig.

Wagner's method received (so far as I can tell) no attention from zoologists until Kluge and Farris (1969) used Wagner's basic method to produce a phylogenetic computer algorithm and analyzed frog phylogeny. However, many botanical studies have used the method with varying degrees of success based on the extent to which a critical criterion for determining relative apomorphy was employed. The method proceeds in the following manner:

1. Determining which of the various characters in a series of character transformations are the apomorphic characters.

2. Assign the plesiomorphic character in each transformation series the score of 0 and each apomorphic character the score of 1. If the transformation series contains more than two homologues then Wagner (pers. comm.) recommends that these "intermediate apomorphies" be scaled between 0 and 1. That is, a transformation series involving four different characters might be scaled 0, 0.25, 0.50, 1.

3. Construct a table of taxa and coded characters. This has been done in Table 5.5 for the *Neocteniza* data discussed earlier.

4. To determine divergence, total the values for each taxon for all transformation series ("divergence index" in Table 5.5).

5. Plot the taxa on a graph, placing each taxon on a concentric semicircle that equals its divergence index. The lines (radii) connecting taxa are determined by shared synapomorphies. Ancestral species are indicated by open circles and each is placed on the semicircle which denotes the number of synapomorphies shared by its descendants. Actual taxa are placed on the semicircle which is the total of their divergence indices. This has been done on the *Neocteniza* data in Fig. 5.30. Note that the hypothetical common ancestor of *Actinopus* and *Missulena* is placed on the semicircle labeled "6" because these two genera share six synapomorphies.

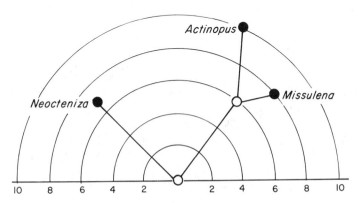

Figure 5.30 A phylogenetic analysis of *Neocteniza*, *Actinopus*, and *Missulena* using the Wagner groundplan divergence method. Concentric semicircles represent anagenetic divergence based on the number of synapomorphies and autapomorphies of each taxon. See text and Table 5.5 for additional discussion and characters.

Table 5.5 Wagner groundplan divergence analysis of
Platnick and Shadab's *Neocteniza* data. Transformation
series are coded as 0 (plesiomorphy) or 1 (apomorphy).
Order of the transformation series follows Table 5.3

Transformation Series	Taxon		
	Neocteniza	*Actinopus*	*Missulena*
1	1	0	0
2	1	0	0
3	1	0	0
4	1	0	0
5	1	0	0
6	1	0	0
7	1	0	0
8	0	1	1
9	0	1	1
10	0	1	1
11	0	1	1
12	0	1	1
13	0	1	0
14	0	1	0
15	0	1	1
16	0	1	0
17	0	1	0
18	0	0	1
19	0	0	1
Divergence index	7	10	8

Groundplan divergence analysis in its original form does allow for
some peculiar features not found in other phylogenetic analyses. A taxon
can be placed at an ancestral node. This could be the result of not
incorporating the diagnostic features of the taxon, or if that taxon is a
species it is always possible that it is an ancestor. However, the analysis
itself is not sufficient to demontrate this.

Computerized Wagner Algorithms

Kluge and Farris (1969) suggested an algorithm for computing Wagner
trees which could be implemented using a computer. Subsequent papers
dealt with strategies for computation (Farris, 1970) and the application

of Wagner analysis to distance data such as that derived from immunological studies (Farris, 1972). Kluge and Farris' algorithm is based on applying the parsimony criterion to derive an estimated tree of minimum evolutionary steps (character transformation steps) and thus is similar to several other computer algorithms discussed in Chapter 10.

The purpose of this section is to present briefly and simply the Wagner algorithm so that the reader will understand how it works. The actual internal workings of these programs, such as Farris's "Wagner 78" or "Willi Hennig Memorial Program" are likely to be incomprehensible to all but the most knowledgeable computer person. Yet the basic algorithm that these programs implement is relatively simple and straightforward, and they are user-oriented. Another point is in order. It has taken many years to derive the programs which fully implement Hennig's methods. These efforts have left programs strewn around the country. Some are better than others. Some are worse than nothing. Most are defective in one respect or another. I would caution any who wish to use a computer algorithm to survey the literature carefully and use only those specific programs that are reliable.

Features of the Wagner Algorithm

The Wagner algorithm operates on the working assumption that the best estimate of the phylogenetic relationships among members of a monophyletic group is that estimate that requires the smallest number of character transformations. That is, it works by applying the principle of simplicity or parsimony. To accomplish the task of finding the simplist tree for the data at hand, the computer program that implements the algorithm (such as "Wagner 78") computes a tree of minimum length, with length directly related to the number of character transformations.

Although the tree may be "rooted" (i.e., an out-group or ancestral bodyplan specified) or "unrooted" (no out-group or ancestral bodyplan specified), we shall be interested only in "rooted" trees. To produce such a tree by Wagner computer programs the investigator must first determine the best estimates concerning which characters in each transformation series are plesiomorphic and which are apomorphic. This is accomplished by specifying the character conditions in the out-group or an ancestral bodyplan (i.e., a set of characters for the hypothetical ancestral species of the group analyzed). We shall see that the out-group does not have to be entirely plesiomorphic. The character matrix, therefore, will be coded in such a way that all presumed apomorphies and plesiomorphies are coded consistently. The important point is that the algorithm requires all of the background study normally associated with phylogenetic analysis. The choice of using a computer program, then, rests more with the nature of the data than with any difference in the

workings of phylogenetic analysis. Small data sets with little or no homoplasy can be done by hand (this includes data sets with many characters). Large data sets, where it is possible that homoplasy is present but
not detected by character coding or where homoplasy may be unde-
tectable because of the nature of the data (amino acid sequences, for
example), are easier to analyze by an appropriate computer program.

One nice feature of phylogenetic computer programs is that although
the various characters may be discrete data (0 = plesiomorphic; 1 =
apomorphic, etc.), the programs may also be used for variable data (means
or modes) and for distance data (immunological or even phenetic). Further, weights may be given to the data. For example, we might wish to
give lower weight to characters which show reversal on the first run of
the program, or we might wish to give lower weight to "losses" and
higher weight to "gains." If such loss and gain characters conflict, the
program will weight the characters as specified by the investigator. If,
however, they do not conflict, then the weights are not used (*a priori*
weighting is not a feature of the system, weights only count if there is
a conflict in two mutually exclusive hypotheses of relationship in the
phylogenetic system, see discussion by Wiley, 1975).

Wagner Algorithm Definitions

The following list of definitions is necessary to understand the algorithm
of Kluge and Farris (1969). These definitions will be used in the subsequent examples of the implementation of the algorithm.

  1.   A particular character for a particular taxon is designated by:

$$X(A, i)$$

where: A = the taxon A
         i = the ith character of taxon A

  2.   The array of characters displayed for a taxon in a particular analysis
is:

$$\sum_i X(A, i)$$

  3.   The **difference, D, between two taxa** is defined as the sum of the
absolute differences between their characters:

$$D(A, B) = \sum_i |X(A, i) - X(B, i)|$$

For example,

Taxon	Character Matrix
A	0 0 1 1 1
B	1 1 0 0 0

$$D(A, B) = \sum_i |X(A, i) - X(B, i)|$$

$$= |0 - 1| + |0 - 1| + |1 - 0| + |1 - 0| + |1 - 0|$$

$$= 5$$

4. The length of a line between a taxon and its hypothetical ancestral species is termed the **interval of the taxon, INT** (taxon); see Fig. 5.31a.

$$INT(B) = D[B, ANC(B)]$$

Where: INT(B) = the interval of taxon B
          ANC(B) = the hypothetical ancestral species of taxon B
     D[B, ANC(B)] = the distance of B to its ancestor

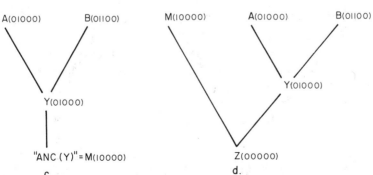

Figure 5.31 Steps in computing Wagner trees using the algorithm of Kluge and Farris, I: (a) the concept of "interval"; (b–d) steps in computing a tree based on data presented in the text.

For example,

Taxon	Character Matrix
ANC(A, B)	0 0 0 0 0
A	0 1 1 1 0
B	1 1 0 0 0

$$D[B, ANC(B)] = \sum_i | X(B, i) - X\{[ANC(B)], i\} |$$

$$= | 1 - 0 | + | 1 - 0 | + | 0 - 0 |$$
$$+ | 0 - 0 | + | 0 - 0 |$$

$$D[B, ANC(B)] = 2$$

$$INT(B) = 2$$

The Wagner Algorithm

Given a matrix of characters, we may proceed in the following series of steps (from Kulge and Farris, 1969):

1. Specify an ancestor or a sister group.
2. Find among the taxa to be analyzed that taxon which shows the least amount of difference from the designated ancestor or sister group. To do this, we compute D for each taxon from the ancestor or sister group and pick that taxon where D is smallest. We then connect this taxon with the ancestor/sister group, creating an interval for that taxon.
3. Find the next taxon that shows the least difference from the ancestor/sister group. We may do this by inspection of the previous computations.
4. Find the interval that shows the least difference from this selected taxon. To find this we compute the difference between the selected taxon and the interval of each taxon that is already connected to the tree. For example, if we have a taxon, A, connected to an ancestor and our next taxon is B, then we compute the quantity $D[B, INT(A)]$, which is defined thus:

$$D[B, INT(A)] = \frac{D(B, A) + D[B, ANC(A)] - D[A, ANC(A)]}{2}$$

5. Attach this taxon to the interval with which it differs least. To accomplish this we construct a hypothetical common ancestor for the two taxa such that its characters are the median of the characters of the first taxon, its original ancestor, and the new taxon. In the example

equation, this would be the median computed for $X(A, i)$, $X(B, i)$, and $X[ANC(A), i]$.

6. If any taxa remain, go to "3" and begin again. Otherwise, stop.

Wagner Calculations—First Example

For simplicity, we will examine a case where there are only two taxa and their specified sister group, as shown in the following matrix:

Taxon	Character Matrix
M	1 0 0 0 0
A	0 1 0 0 0
B	0 1 1 0 0

Step 1. Calculate D for each taxon from the sister group.

$$D(A, M) = \sum_i |X(A, i) - X(M, i)|$$

$$= |0 - 1| + |1 - 0| + |0 - 0| + |0 - 0| + |0 - 0|$$

$$= 2$$

$$D(B, M) = \sum_i |X(B, i) - X(M, i)|$$

$$= |0 - 1| + |1 - 0| + |1 - 0| + |0 - 0| + |0 - 0|$$

$$= 3$$

Step 2. Because taxon A shows the least distance from taxon M, we connect A to M to form the interval, INT(A), as shown in Fig. 5.31b.

Step 3. The only other taxon is B. We compute the difference between B and INT(A) by the formula given in the algorithm step 4 described earlier.

$$D[B, INT(A)] = \frac{D(B, A) + D[B, ANC(A)] - D[A, ANC(A)]}{2}$$

Because the "ancestor" of taxon A is M, the formula becomes:

$$D[B, INT(A)] = \frac{D(B, A) + D(B, M) - D(A, M)}{2}$$

$$= \frac{1 + 3 - 2}{2}$$

$$= 1$$

Step 4.   We connect B to INT(A) by constructing a hypothetical ancestor (say, Y) whose characters are computed as the median of $X(A, i)$, $X(B, i)$, and $X(M, i)$, as shown in the following data matrix:

Taxon		Character Matrix				
M		1	0	0	0	0
A		0	1	0	0	0
B		0	1	1	0	0
Y	(median)	0	1	0	0	0

The position of Y and its characters are shown in Fig. 5.31c. Because there are no other taxa, we stop.

In this example, I would like to call attention to two points. First, although the sister group M was used to root the tree it was not considered plesiomorphic for all characters. The character of the first transformation series in M was a synapomorphy shared by the members of M. Note that the actual hypothetical ancestor of A and B (that is, Y) is correctly constructed such that it has the synapomorphy for the group AB. We may "clean up" the analysis by constructing another ancestor for M and Y, as shown in Fig. 5.31d, by the construction of ancestor Z. Second, note that $D(A, Y) = 0$. This is a by-product of the fact that I did not specify any unique characters for A and does not imply that A is Y.

Wagner Calculations—Second Example

This example uses four terminal taxa (A–D) and a specified hypothetical ancestor. Thus it requires two passes through the algorithm and a comparison of more than one interval. The hypothetical ancestor was constructed by out-group comparison. Thus, the ancestor is a vector of zero values and each 1 is considered a possible synapomorphy within the taxon ABCD. The data matrix is:

Taxon	Character Matrix				
ANC(A–D)	0	0	0	0	0
A	0	0	1	1	0
B	0	0	0	1	0
C	0	1	1	1	1
D	0	0	1	1	1

Step 1.   Calculate D for each taxon from the ancestor.

$$D(A, ANC) = \sum_i | X(A, i) - X(ANC, i) |$$

$$= |0 - 0| + |0 - 0| + |1 - 0|$$
$$+ |1 - 0| + |0 - 0|$$

$$D(A, ANC) = 2$$

$$D(B, ANC) = \sum_i | X(B, i) - X(ANC, i) |$$

$$= |0 - 0| + |0 - 0| + |0 - 0|$$
$$+ |1 - 0| + |0 - 0|$$

$$D(B, ANC) = 1$$

Further calculations show that:

$$D(C, ANC) = 4$$
$$D(D, ANC) = 3$$

Thus, $D(B, ANC)$ is the smallest distance and taxon B is connected to ANC to form the interval INT(B), as shown in Fig. 5.32a. The length of this interval is equal to the distance of B to ANC and thus is INT(B) = 1.

Step 3.   We then find the taxon which differs least from the ancestor, ANC. From our calculations in step 1 we see that taxon A has the next lowest value [i.e., $D(A, ANC) = 2$].

Step 4.   Because there is only one available interval [INT(B)], taxon A must be connected to this interval. Thus, calculations for finding the interval are not necessary. We may attach taxon A to INT(B) by constructing the new common ancestor of A and B, which we shall term Y, by taking the median of each transformation series of A, B, and ANC.

Taxon		Character Matrix
ANC		0 0 0 0 0
A		0 0 1 1 0
B		0 0 0 1 0
Y	(median)	0 0 0 1 0

We connect A to INT(B) by connecting it to Y, as shown in Fig. 5.32b.

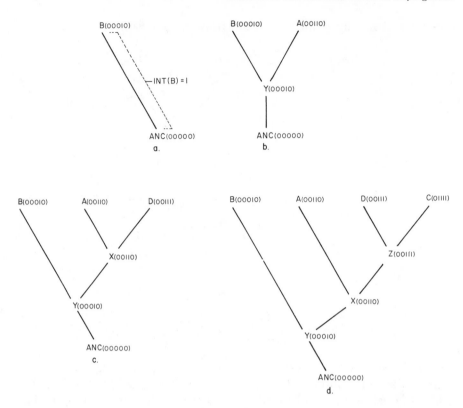

Figure 5.32 Steps in computing Wagner trees using the algorithm of Kluge and Farris II: see text for data computations and discussion.

Step 5. Since there are taxa left we proceed up the algorithm and find the next taxon that is the least different from ANC, taxon D.

Step 6. There are now three intervals in the tree (Fig. 5.32b), INT(Y), INT(A), and INT(B). We must now determine which of these intervals to connect taxon D onto. Thus, we must calculate the distances of taxon D to each of these intervals.

$$D[D, INT(Y)] = \frac{D(D, Y) + D(D, ANC) - D(Y, ANC)}{2}$$

$$= \frac{2 + 3 - 1}{2}$$

$$= 2$$

$$D[\text{D, INT(B)}] = \frac{D(\text{D, B}) + D(\text{D, Y}) - D(\text{B, Y})}{2}$$

$$= \frac{2 + 2 - 0}{2}$$

$$= 2$$

$$D[\text{D, INT(A)}] = \frac{D(\text{D, A}) + D(\text{D, Y}) - D(\text{A, Y})}{2}$$

$$= \frac{1 + 2 - 1}{2}$$

$$= 1$$

The distance $D[\text{D, INT(A)}]$ is minimal, thus taxon D will be connected to INT(A). We connect D to INT(A) by constructing another hypothetical common ancestor, designated X, in such a way that its characters are the median of the characters of Y, D, and A:

Taxon		Character Matrix				
Y		0	0	0	1	0
A		0	0	1	1	0
D		0	0	1	1	1
X	(median)	0	0	1	1	0

The placement of D and X are shown in Fig. 5.32c.

Step 7. There is only one taxon left, C. We proceed directly to interval calculations. There are now five intervals, so we must compare INT(B), INT(Y), INT(X), INT(A), and INT(D) (see Fig. 5.32c).

$$D[\text{C, INT(B)}] = \frac{D(\text{C, B}) + D(\text{C, Y}) - D(\text{B, Y})}{2}$$

$$= \frac{3 + 3 - 0}{2}$$

$$= 3$$

$$D[\text{C, INT(Y)}] = \frac{D(\text{C, Y}) + D(\text{C, ANC}) - D(\text{Y, ANC})}{2}$$

$$= \frac{3 + 4 - 1}{2}$$

$$= 3$$

$$D(C, INT(X)] = \frac{D(C, X) + D(C, Y) - D(X, Y)}{2}$$

$$= \frac{2 + 3 - 1}{2}$$

$$= 2$$

$$D[C, INT(A)] = \frac{D(C, A) + D(C, X) - D(A, X)}{2}$$

$$= \frac{2 + 2 - 0}{2}$$

$$= 2$$

$$D[C, INT(D)] = \frac{D(C, D) + D(C, X) - D(D, X)}{2}$$

$$= \frac{1 + 2 - 1}{2}$$

$$= 1$$

Thus, taxon C is nearest INT(D).

Step 8. We connect C to INT(D) by constructing a common ancestor, Z, in the same manner as above by taking the median of characters of C, D, and X.

Taxon		Character Matrix
X		0 0 1 1 0
C		0 1 1 1 1
D		0 0 1 1 1
Z	(median)	0 0 1 1 1

The connection of taxon C is shown in Fig. 5.32*d*. Because this is the last taxon, we stop.

Wagner Analysis—Neocteniza *and Related Taxa*

The only remaining point that I would like to make is that Hennig's argumentation scheme, Wagner's groundplan divergence analysis, and current computer algorithms using the algorithm of Kluge and Farris (1969) result in the same phylogenetic tree. I have already given an example of Wagner's groundplan analysis using the data from Platnick

Table 5.6 Table of Platnick and Shadab's *Neocteniza* data used to compute a Wagner tree according to the algorthm of Kluge and Farris (1969). Transformation series contain characters coded as 0 (plesiomorphic) and 1 (apomorphic). Series 1–19 follow Table 5.3 Series 20 contains the synapomorphy of Actinopodidae (1 = excavated depressions in the posterior sigilla of females. Series 20 contains the synapomorphy of Migidae (1 = inclided chelicerae).

Transformation Series	Characters			
	Migidae	*Neocteniza*	*Actinopus*	*Missulena*
1	0	1	0	0
2	0	1	0	0
3	0	1	0	0
4	0	1	0	0
5	0	1	0	0
6	0	1	0	0
7	0	1	0	0
8	0	0	1	1
9	0	0	1	1
10	0	0	1	1
11	0	0	1	1
12	0	0	1	1
13	0	0	1	0
14	0	0	1	0
15	0	0	1	1
16	0	0	1	0
17	0	0	1	0
18	0	0	0	1
19	0	0	0	1
20	0	1	1	1
21	1	0	0	0

and Schadab (1976) on *Neocteniza*. That data table (Table 5.5) is expanded in Table 5.6 to include Migidae as the outgroup. Steps in analysis are:

Step 1. Migidae is selected as the out-group.

Step 2. We calculate *D* for each taxon. For simplicity I shall use the following abbreviations: NEO = *Neocteniza*, ACT = *Actinopus*, MISS = *Missulena*, and MIG = Migidae.

$$D(\text{NEO}, \text{MIG}) = \sum_i X(\text{NEO}, i) - X(\text{MIG}, i) \mid = 9$$

$$D(\text{ACT}, \text{MIG}) = \sum_i X(\text{ACT}, i) - X(\text{MIG}, i) \mid = 12$$

$$D(\text{MISS}, \text{MIG}) = \sum_i X(\text{MISS}, i) - X(\text{MIG}, i) \mid = 10$$

Step 3. *Neocteniza* shows the least difference from Migidae. Thus, we form INT(NEO) as shown in Fig. 5.33*a*.

Step 4. *Missulena* is less distant from Migidae than is *Actinopus*. Thus we shall connect MISS to INT(NEO) by constructing a hypothetical common ancestor, X, by taking the median of MIG, NEO, and MISS.

Taxon	Character Matrix
MIG	0 0 0 0 0 0 0 0 0 0 0 0 0 0 0 0 0 0 0 0 0 1
NEO	1 1 1 1 1 1 1 0 0 0 0 0 0 0 0 0 0 0 0 0 1 0
MISS	0 0 0 0 0 0 0 1 1 1 1 1 0 0 1 0 0 1 1 1 0
X (median)	0 0 0 0 0 0 0 0 0 0 0 0 0 0 0 0 0 0 0 1 0

The tree at this point is shown in Fig. 5.33*b*.

Step 5. Only a single taxon remains, *Actinopus*. We compare it to the three intervals present at this stage of analysis, INT(NEO), INT(MISS), and INT(X).

$$D[\text{ACT}, \text{INT(NEO)}] = \frac{D(\text{ACT}, \text{NEO}) + D(\text{ACT}, X) - D(\text{NEO}, \text{ACT})}{2}$$

$$= \frac{17 + 10 - 7}{2}$$

$$= 5$$

$$D[\text{ACT}, \text{INT(MISS)}] = \frac{D(\text{ACT}, \text{MISS}) + D(\text{ACT}, X) - D(\text{MISS}, X)}{2}$$

$$= \frac{6 + 10 - 8}{2}$$

$$= 4$$

$$D[\text{ACT}, \text{INT(X)}] = \frac{D(\text{ACT}, X) + D(\text{ACT}, \text{MIG}) - D(X, \text{MIG})}{2}$$

$$= \frac{10 + 12 - 2}{2}$$

$$= 10$$

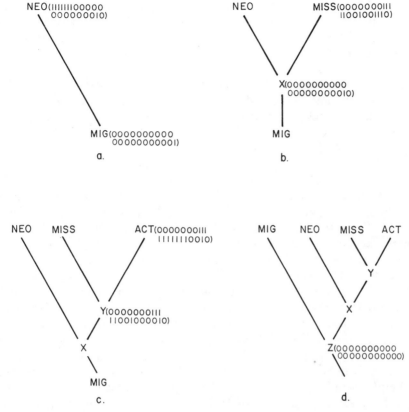

Figure 5.33 Steps in computing Wagner trees using the algorithm of Kluge and Farris III: the relationships among *Neocteniza* (NEO), *Actinopus* (ACT), *Missulena* (MISS), and Migidae (MIG). See text for data and discussion.

Actinopus is least distant from INT(MISS). We connect ACT to INT(MISS) by constructing another hypothetical ancestral species by taking the median of the characters of ACT, MISS, and X:

Taxon		Character Matrix
MISS		0 0 0 0 0 0 0 1 1 1 1 1 0 0 1 0 0 1 1 1 0
X		0 0 0 0 0 0 0 0 0 0 0 0 0 0 0 0 0 0 0 1 0
ACT		0 0 0 0 0 0 0 1 1 1 1 1 1 1 1 1 1 0 0 1 0
Y	(median)	0 0 0 0 0 0 0 1 1 1 1 1 0 0 1 0 0 0 0 1 0

The tree has now been built to the point shown in Fig. 5.33c.

Step 6. We may now normalize the tree by constructing a hypothetical common ancestor for taxa MIG and X, by assigning this ancestor (Z) any apomorphic characters shared by both taxa or a plesiomorphic 0 for each shared plesiomorphy or any situation where one has an apomorphic and the other a plesiomorphic character. In this case, Z is a vector of zero values (see Fig. 5.33d).

I picked this example of real data to show that the algorithm could be carried out by hand. Obviously, when there are many characters or when there is a data matrix with suspected homoplasy, a computer program comes into its own.

Compatibility Algorithms

Compatibility analysis works on the assumption that the distribution of the largest collection of mutually compatible characters (or character states as the case may be) best estimates the phylogenetic relationships of the taxa displaying these characters (LeQuesne, 1972; Estabrook et al., 1977; Estabrook, 1979; Gardner and La Duke, 1979). It is also termed clique analysis because it defines "cliques" of characters shared by a certain array of taxa. Compatibility analysis is an approach to estimating phylogenetic trees. However, I cannot agree with Estabrook (1979) that it is either superior to the parsimony approach or that it is Hennig's method. If there is complete congruence between synapomorphies and if these synapomorphies are correctly identified by outgroup comparison, then a clique analysis should give the same result as either manual or computerized parsimony methods. In other words, given a "clean data set," compatibility analysis should be equal in veracity to parsimony analysis, whether the parsimony analysis was done manually or by a computer algorithm. If, however, plesiomorphies and apomorphies are determined by some internal criterion such as "common equals primitive," then an internally parsimonious but possibly externally unparsimonious result is possible. The other problem I see is that even if we can correctly code the "true synapomorphies" as synapomorphies, compatibility analysis may disregard other perfectly valid synapomorphies if there is homoplasy in the data matrix (Kluge and Farris 1979). As homoplasy becomes increasingly frequent, compatibility analysis explains less of the original data. Thus, in terms of veracity (Hennig's prime criterion), I cannot see that compatibility analysis is superior to parsimony analysis. In terms of the claim that compatibility analysis is Hennig's method, I must disagree. Hennig was interested in (1) explaining all characters (character states) and (2) classifying only strictly monophyletic groups. The compatibility theorists who advocate "convex groups" in classification are clearly not following Hennig because some convex groups are paraphyletic and Hennig's phylogenetic principles do not allow paraphylectic groups in classification.

Chapter 6

Phylogenetic Classification

Classifications are systems of words. Biological classifications are systems of words which are used to organize the diversity of life and/or to reflect man's estimate of nature's own organization of life. Phylogenetic classifications are biological classifications which accurately reflect hypotheses concerning the genealogical descent of organisms and are usually accomplished on the species level or above. Put simply, phylogenetic classifications reflect the best estimate of the evolutionary history of organisms (Brundin, 1966).

More-or-less phylogenetic classifications have been around ever since taxonomists embraced evolution. Further, classifications containing natural taxa have been around long before Linnaeus much less Darwin. This indicates that the pattern of evolutionary descent in at least some groups is clear enough that their pattern of descent can be recovered whether one believes in an orderly creator or in evolution (cf., Agassiz vs. Darwin). As Patterson (1977) has stressed, it was as much the nature of these very distinctive groups that made them apparent as it was the methods of taxonomists. Thus, it was left to Hennig (1950, 1960) to develop a system of reconstructing phylogeny which could be applied across all groups and a system of classification whose general philosophy was based on the insistance that taxonomic classifications should consistantly reflect estimates of the branching sequence (the speciation sequence) observed in nature. It is because of this attitude and not the particular ways Hennig suggested to accomplish his goals that he should be credited for establishing the phylogenetic system of classification. Hennig accomplished the task of developing the genealogical system called for by Darwin (1859).

In this chapter I would like to discuss the general nature of classifications. From this, I will then discuss three ways to produce phylogenetic classifications and the reasons why I prefer a modified Linnaean system. This will be followed by a detailed discussion of the annotated Linnaean hierarchy (Wiley, 1979c). The merits of this system are discussed and examples given. Finally, I will discuss the differences in classification produced in those circumstances where we are working with phylogenetic dendrograms and not phylogenetic trees. The reasons for preferring the phylogenetic system of classification over the competing systems of evolutionary taxonomy and phenetics are reserved for the next chapter (7).

CLASSIFICATIONS—SOME GENERAL TYPES

The process of classifying is the activity of grouping entities or phenomena and giving names to the resulting groups. The placing of some things into one group to the exclusion of other things implies that the members of the group share some type of relationship not shared by the nongroup things.

There are many ways to classify classifications themselves. One could, for example, recognize a basic dichotomy between hierarchical and non-hierarchical classifications. Or, one might like to discuss natural vs. nonnatural classifications. A more basic difference can be utilized to distinguish between three major types of classifications: (1) natural classes, (2) historical groups and individuals, and (3) convenience classes.

1. *Classifications of natural classes.* As discussed in Chapter 3, classes are spatiotemporally unrestricted constructs that have definitions and that contain individual entities or historical groups that fit the class definition. A natural class is one that contains individual entities that fit the definition and whose origins and behavior are governed by natural processes. That is, their origins and actions can be predicted by what we term "natural law" or hypotheses concerning natural processes. The similarities displayed by members of a particular class are the products of the workings of particular laws, the entities are not necessarily similar because they share a similar history. In fact, the entities placed in a natural class may be quite unrelated in a historical sense. Let me illustrate this with some examples.

The periodic table is an array of natural classes which classifies the entities we call atoms. Each class (Hydrogen, Helium, etc.) contains entities fitting the definition of the class and the reactions of these entities to natural processes can be predicted by both the general laws governing the behavior of atoms and the particular qualities of the atoms themselves.

Astronomers classify stars based on their size, luminosity, and temperature into such groupings as white dwarfs, main sequence, and super giants. The pattern on which this classification is derived is the Hertzsprung-Russell (H-R) Diagram shown in introductory astronomy texts (Fig. 6.1). This classification is not based on historical connections. The stars classed as main sequence did not share an ancestral star in common. In fact, most of these stars are quite unrelated historically except for the fact that they are all products of the cosmic egg. The classification is based on certain physical laws which predict the "ontogeny" of any particular star based on (among other things) its initial mass. The stars in the main sequence classified as B0 are so classified not because they are descendants of a common ancestral B0 star, but because their initial mass was equal to about 16 suns.

Note that "similarity" in this context does not necessarily connote physical similarity. Rather, it denotes similarity in ontogeny given initial conditions of star formation. Main sequence stars may be quite different in spectral class and mass. Some main sequence stars may have masses equal to some giants while others may have masses equal to white dwarfs. It is the process behind the pattern and not necessarily physical similarity which provides the basis for the classification.

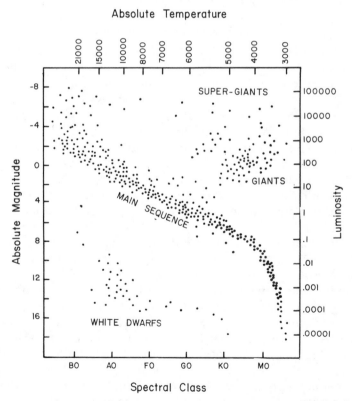

Figure 6.1 A Hertzsprung-Russell classification of stars based on mass and temperature as reflected by absolute magnitude and spectral class. This is an example of a natural nonhierarchical classification. (Compiled from various sources.)

As another example, we might examine Dana's system of mineral classification. Diamonds are classified as diamonds because of their particular physical properties which were brought about by certain physical processes and not because of any historical connection shared by all diamonds. Interestingly, preevolutionary biological classifications are also of this type. A creator, of course, provided the necessary process.

Classifications of natural classes, such as the periodic table, are inherently nonhierarchical. That is, they are not structured in such a way that there is a natural group and groups-within-groups arrangement. Just as astronomers may produce classifications based on, for example, spectral types, chemists have organized elements into groups and series. However, such arrangements do not produce a hierarchy like that found in historical classifications. For example, the Group O elements include those elements which are largely inert (Helium, Neon, etc.). Such a grouping, read directly from the periodic table, is analogous to the grouping BO stars in that the elements grouped display certain characteristics. However, we should not forget that the reasons for these similar characteristics are the result of the arrangements of electrons in different orbitals. Without pushing an analogy too far, this is analogous to classifying organisms by convergence.

2. *Classifications of individuals and historical groups.* These classifications are based upon inferred historical connections between the entities classified. The inferred relationship is unique simply because it is part of history. There is, of course, process behind the historical pattern. For example, we might classify the present chunks of continent according to their hypothesized history of being chunks of Gondwana or Laurasia. These in turn could be classified as parts of Pangea. Interestingly, the actual processes behind these patterns do not need to be demonstrated before we can perceive the pattern. Indeed, getting at the process would be impossible because we would not know that we needed an explanation before we have something to explain. Phylogenetic classifications are another example of such historical classifications.

Classifications of individuals and historical groups differ from classifications of natural classes in that classifications of individuals and historical groups are inherently hierarchical. The hierarchy may be explained in one of two ways. First, it might have been produced by the logical mind of a creater. Linnaeus and Agassiz perceived the basis of the hierarchy in this manner. Second, it might have been produced by the fact that all organisms (or languages, or continents) are historically connected in the genealogical sense. Lamarck, Buffon, Darwin, and other evolutionists perceived the basis of the hierarchy in this manner. The former perception is an adequate basis for searching for order among living organisms (surely this must be correct or nonevolutionists would not have done such a good job of classifying). However, it is an inadequate philosophy because it stops short of attempted explanations of the hierarchy. Everything is ascribed to the hand of God. The latter (evolutionist) perception is also an adequate basis for searching for order. It also provides an adequate philosophy in that it provides a basis for attempting natural explanations of the origin of the hierarchy. One runs the risk, however, of setting up a causal process that is not based on pattern and then trying to fit the organisms on the basis of this causal process. When one is searching for process, one should deduce process based on pattern and not attempt to force the pattern into a particular process. In other words, if we begin with a process and then attempt to fit the organisms to that process, we run the risk of producing an ad hoc system. However, this is not a weakness of thinking about evolution; rather, it is a weakness of individual scientists.

3. *Classification of convenience classes.* These classifications have no particular basis in terms of process or history, or, they may purport to reflect either process or history and fail to do so. The former I call **convenience classifications**. The latter I call **ad hoc classifications**.

The vast majority of our classifications are convenience classifications. They are a by-product of our desire to bring order into language. The purpose of convenience classifications is to ease communication so that every object or phenomena in the universe does not have to be described in order to communicate about it. For example, we might classify automobiles into size categories (standard, compact, etc.) or style categories (tudor, convertable, etc.) The term "tudor subcompact sedan" carries quite a bit of information and is therefore quite useful in our efforts to communicate. We could be more specific and say "1966 Mini-Cooper 1263S." The Dewey Decimal and Library of Congress systems of classifying books are additional examples. Who said that "500" must denote a science book?

Convenience classifications, like metaphysics, are neither irrational nor useless simply because they are artificial. At the same time, convenience classifications can hardly be termed scientific since they cannot be justified in terms of natural process or pattern.

BIOLOGICAL CLASSIFICATION

Biological classifications may fall into any of the three types outlined above, or they may be combinations of these types. Further, biological classifications may be hierarchical or nonhierarchical. An example of a hierarchical classification combining both historical and nonhistorical factors would be a classification of cell types. The justifications for recognizing a cell as a nerve cell and further as a brain cell has a historical component in the ontogeny of the cell from an original fertilized egg cell, and it has a historical component in the history of descent of the ontogenetic pathway itself.

Taxonomic biological classifications are hierarchical. The structure of the hierarchy (however conceived) does not provide a descriptive commentary on the organisms classified. To know from a group name, such as Aves, the special attributes of the members of the group requires special knowledge. The classification cannot tell you that birds have feathers. However, biological classifications do give a relative idea about the relationships one wishes to express by indicating group membership. All we have to do is to decide on which type of relationship we wish to express. For the purpose of this chapter, I will assume that we wish to express the genealogical relationships among species and natural supraspecific taxa. Following Hennig (1966), I shall treat phylogenetic classifications as systems of words used to communicate knowledge about the inferred pattern of descent with modification (evolution). This summary knowledge can be used in one of three ways. First, it may be used for communication. Second, it may be used by other investigators as a basis for criticizing the original investigator's ideas about the pattern of evolution. Third, it may be used by other investigators to study process or make further comparisons based on the inferred pattern as reflected by the classification.

Components of Phylogenetic Classifications

The components of phylogenetic classifications are taxa. A **taxon** is a grouping of individual organisms that is given a formal name. Phylogenetic classifications contain taxa hypothesized to be monophyletic sensu Hennig (1966) and only these taxa have evolutionary connotations. Taxa that are of unknown status (i.e., there is no evidence of monophyly) either do not appear in a phylogenetic classification or they are clearly specified as groups of unknown status. When a nonmonophyletic taxon is placed in a phylogenetic classification it is considered to have no critical evolutionary connotations and serves only as an artificial reservoir for organisms whose genealogical relationships are unspecified.

There are two basic kinds of natural taxa. The first is the species and it is considered monophyletic by virtue of its individual nature. (Individuals are neither monophyletic, paraphyletic, or polyphyletic.) Species are the largest groups of organisms subject, as a group, to the sequence of processes involving cladogenesis, anagenesis, and stasis (Chapter 2; Wiley, 1978). The second, the supraspecific taxon (monophyletic taxon, historical group), is a collection of species linked by unique genealogical descent from a single common ancestral species (Chapter 3). This common ancestral species was, from the historical point of view, the sole member of the supraspecific taxon at its time of origin (Hennig, 1966). The rationale for grouping species into supraspecific taxa is entirely historical, and is based on the discovery of characters thought to reflect genealogy (i.e., synapomorphies). The rationale for placing individual organisms in a single species is not purely historical and calls for evidence of a lineage continuum in the absence of historical speciation events. Subspecific grouping may also be recognized and given formal taxon names but the term "monophyletic" sensu Hennig is not relevant because the genealogical and nongenealogical connotations of subspecific taxa imply a different set of processes than those between species.

For convenience I shall distinguish between two types of taxon names. **Formal taxon names** are names associated with particular taxonomic ranks and formulated (when applicable) according to an applicable nomenclatural code. The vast majority of taxa are given formal names. Examples include family Ericaceae, genus *Homo*, and superphylum Deutrostomata. **Informal taxon names** are formed by coupling a formal taxon name, the rank of that taxon, and the noun "group." An informal taxon name can be used to group two or more taxa of the same rank together without the need for a formal taxon name of the next highest rank. Although informal taxon names might be used at any level in the hierarchy of a classification, I find them most useful when applied to taxa within families or within genera. For example, rather than name a new subgenus (or infragenus) for *Fundulus nottii* and its four relatives, Wiley (1977) simply designated the taxon as the "*Fundulus nottii* species group." The use of informal taxon names if further illustrated in the classification of the plant genus *Anacyclus* in a later section.

Grouping and Ranking Taxa

As Hennig (1975) stated, the actual classification of a particular phylogenetic hypothesis is a relatively straightforward procedure accomplished by applying whatever conventions the investigator wishes to adopt. Any classification that exactly reflects the phylogenetic relationships of the taxa classified is a **phylogenetic classification**. If two phylogeneticists produce different classifications of the same phylogeny, then the difference lies in the conventions applied. For example, one investigator may produce a completely subordinated classification in which each branch point is named (McKenna, 1975, for example) while another investigator may use a listing convention in conjunction with incomplete subordination (Nelson, 1974; Cracraft, 1974a; Schuh, 1976). Both classifications are logically correct if both reflect the phylogeny on which they are based. However, one may be more convenient than the other, or simpler in the sense of using fewer taxon names and categorical ranks to reflect the same phylogeny. There are only two necessary rules for phylogenetic classification. I will state these rules (taken from Wiley, 1979c) and then present a series of conventions for producing classifications which can reflect complex phylogenies.

Rule 1. Taxa classified without qualification are monophyletic groups sensu Hennig (1966). Nonmonophyletic groups may be added if they are clearly qualified as such.

Rule 2. The relationships of taxa within the classification must be expressed exactly.

The conventions I will recommend have several advantages which outweigh the disadvantage of having to learn them. First, the convention of the annotated Linnaean hierarchy confers the advantage of using familiar, nonnumerical, categorical names as "tags" for indicating relative position and relative subordination of sister taxa. Second, other conventions result in the relative stability of Recent taxon ranks without impairing the ability of the classification to reflect the exact phylogenetic relationships of both Recent and fossil taxa. Third, redundant names are kept to an absolute minimum consistent with the various rules of nomenclature. Fourth, certain conventions permit the classification of ancestral species and descendant species and of hybrid origin or symbiotic origin without effecting the classification of descendant groups.

The Linnaean Hierarchy and Its Alternatives

The **Linnaean hierarchy** is one of three general conventions for classifying phylogenies. It is a system that tags sets and subsets of taxa with a rank which reflects the relative levels of taxa to each other. Rank position is shown by categories, with the actual name of the category (family, class, etc.) denoting subordination relative to other categories. In the phylogenetic system, taxa of higher categorical rank are hypothesized to have originated (as stem species or ancestral species) earlier than the taxa of low categorical rank contained within them. The exceptions to this statement are redundant taxa which are injected into the system because of various nomenclatural rules.[1] Such redundant taxa add no additional evolutionary information. (They may, of course, be necessary for other reasons discussed later.)

One minor inconvenience of the Linnaean system is that the relative position of one categorical rank must be memorized relative to the others. We must memorize that family is subordinate to order. We must also memorize the fact that since all species have a binomial name, species is not subordinate to genus. With this exception, the Linnaean system is what Nelson (1974a) termed a "group and group-within-groups" subordination scheme.

That the adoption of the Linnaean hierarchy is a convention, and not

[1] A **redundant taxon** is one that contains exactly the same members as the higher taxon that includes it. For example, if an order contains a single family then the family name is redundant because its definition is the same as that of the order.

a biological necessity, is evident from the fact that there are several other ways to classify a phylogeny. One alternate is a numerical prefix scheme. Relative rank is denoted by a numerical prefix which is unique to the taxon and indicates that taxon's position relative to others. This proposal has been discussed by Hull (1966), Hennig (1969), and Griffiths (1974). Using Hennig's (1969) example of the Mecopteroidea whose phylogenetic relationships are shown in Fig. 6.2, I will examine this proposal.

Mecopteroidea is composed of four terminal groups, Diptera, Mecoptera, Lepidoptera, and Trichoptera. A prefix is assigned Mecopteroidea. This prefix reflects the relationships of Mecopteroidea to other insects (and ultimately of Insecta to all other organisms). The group is tagged within Insecta as "2.2.2.2.4.6 Mecopteroidea." Each subgroup within Mecopteroidea is assigned an additional number to the prefix and each subgroup within that group adds yet another prefix. Considering Fig. 6.2, the classification is:

2.2.2.2.4.6 Mecopteroidea
2.2.2.2.4.6.1 Amphiesmenophora
2.2.2.2.4.6.1.1 Trichoptera
2.2.2.2.4.6.1.2 Lepidoptera
2.2.2.2.4.6.2 Antilophora
2.2.2.2.4.6.2.1 Mecoptera
2.2.2.2.4.6.2.2 Diptera

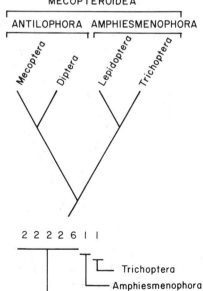

Figure 6.2 A hypothesis of the relationships among mecopteroid insects and the classification of Trichoptera using numrical prefixes. (From Wiley, 1979c, after Griffiths, 1974.)

In spite of its rather unusual nature, this system is completely logical and consistent. And it has certain advantages. Fossil groups can be added or deleted without affecting the hierarchical position of Recent groups and ancestral species can even be added by inserting them wherever they may fit (Griffiths, 1976). However, there are disadvantages that far outweigh these advantages. (1) Numerical prefixes are not the languages of ordinary usage and are foreign to our efforts to communicate. (2) Because each prefix is unique, there will be as many prefixes as taxon names. This means that the prefixes themselves will become too long and therefore too cumbersome. If eight digits are needed for Diptera, how many digits will be needed for the common house fly? Probably enough to cover several lines of text. (3) The scheme has some of the disadvantages of pure indentation without rank tags (discussed later) in that sister groups widely spaced in a classification are not readily apparent as sister groups. The prefix system was advocated by Griffiths (1976) primarily because he felt that the Linnaean system was not fully capable of classifying fossil groups and stem species (ancestral species). We shall see that the Linnaean system is capable of handling these problems with the appropriate conventions outlined in later sections. Therefore, I see no reason to abandon the more traditional way to tag taxa by categorical rank in favor of a system of numerical prefixes which are unique for each taxon. It is not intuitively obvious that "2.2.2.2.4.6" has any advantage over "superorder."

Løvtrup (1977) suggests a modification of Hennig's system by using binary coding. He shortens certain prefixes by summing all 1s which succeed each other. For example, the binary code 1.1.1.0 would be converted to 3.0. The code 1.1.1.1.1.1.1.1.1.1 would translate to 10. Using this system, and assuming that Mecopteroidea has a value of 1, we might classify Fig. 6.2 as:

1 Mecopteroidea
1.0 Amphiesmenophora
1.00 Trichoptera
1.01 Lepidoptera
2 Antilophora (i.e., 1.1 = 2)
2.0 Mecoptera
2.00 Diptera

Løvtrup's system has the definite advantage of cutting down on the number of prefixes needed. However, the system of punctuations must be gotten exactly right if the phylogeny is to be recovered. And it is not clear to me that a really large classification will result in short enough prefixes to be comprehensible. Finally, there is no good way to start. In other words, the system will constantly change as the staring point changes, that is, as we discover more about the relationships of the most primitive organisms. This last problem could be circumvented by, for example, assigning Linnaean ranks or numerical prefixes to each phylum. However, we should have to begin analysis at the highest levels within each phylum in order to assign numerical prefixes to taxa of lower rank.

Subordination by pure indentation is a system in which relative phylogenetic position is indicated by indenting subordinate taxa beneath the higher taxon to which they belong. Sister groups have the same indentation. Such a scheme may use ranked or unranked taxa. When categorical rank is assigned, the rank does not necessarily denote phylogenetic relationship (see for example, Farris, 1976). I reject the proposal that ranked taxa should be used in a system where pure indentation denotes phylogenetic relationship because such a system either makes the categories unnecessary or it produces a mixed system where one must learn by rote where rank denotes phylogenetic relationship and where it does not. Thus, I shall discuss only pure indentation schemes. Two classifications are presented. The first classification is a ranked and indented Linnaean classification, the second classification is a pure indentation scheme with unranked taxon names.

Class Vertebrata
   Subclass Cyclostomata
   Subclass Gnathostomata
      Infraclass Chondrichthyes
      Infraclass Teleostomi

Vertebrata
   Cyclostomata
   Gnathostomata
      Chondrichthyes
      Teleostomi

Both classifications exactly reflect the phylogenetic relationships of three terminal groups (cyclostomes, chondrichthyans, and teleostomes). The pure indentation scheme has an advantage in that if the teleostome's phylogenetic position (for example) was changed by subsequent analysis, it could be moved in the classification without changing the ranks of it or the other taxa. And, one does not have to memorize a relative hierarchy of categories. Unfortunately, pure indentation systems have one very practical difficulty: they demand that the reader be able to line up coordinate taxa to confirm their status as sister groups. We would literally have to measure from the margin or line up the taxa with a ruler. This might be practical for a small classification (such as the one shown above) but it is not practical for a large classification covering several pages of text. What if the sister group is on the next page? Do we measure from the margin? How do we compare a classification published by one author with another author's classification without redrawing the phylogenies? Finally, in large classifications there would be the problem of page space and space wastage. Because of these problems, I reject pure indentation schemes for a general phylogenetic system.

Having rejected numerical prefixes and pure indentation schemes, I will accept categorical subordination as the most convenient way to tag sister groups. The obvious result of this acceptance by earlier workers (of all schools) has been the proliferation of categories as classification increased in content due to an increasing understanding of the world's flora and fauna, and discoveries of new organisms. This is why we have prefixes as well as categories such as "cohort" that were not part of the original Linnaean hierarchy. It would be convenient, however, to minimize the number of categories needed to express exactly the phylogenetic relationships of taxa. It is toward this end that I have examined various conventions recommended by previous workers and have produced the system (Wiley, 1979c) I will now discuss. I will first define the categories of the Linnaean hierarchy and then outline the annotated system.

Definitions of Higher Categories

Five higher categories are demanded by the various codes of nomenclature: genus, family, order, class, and phylum. The genus is a mandatory category to which every species must belong if binomial nomenclature is to be preserved. The other categories are not absolutely necessary for phylogenetic classification, but they are demanded by various codes and they have become essential enough to taxonomic literature, field guides, and faunal works that to abandom them in the name of efficiency would be counterproductive. Therefore, I treat these categories differently from other categories. I allow them to be monotypic (thus redundant) whereas I recommend that all others contain at least two taxa of lower rank (Wiley, 1979c).

A phylogeneticist obviously takes a different attitude toward higher categories than do such workers as Mayr (1969: 92) who defines the genus thus (italics in the original):

> A genus is a taxonomic category containing a single species, or a monophyletic group of species, which is separated from other taxa of the same rank (other genera) by a decided gap.

Because phylogeneticists reject gaps they must also reject definitions based on gaps. I offer the following definitions, taken from Wiley (1979c).

1. Category. A name tag of convenience that expresses relative subordination (rank) in a classification.

2. Genus. A mandatory category to which every species must belong and which contains one species or a monophyletic group of species.

3. Family. A uninomial plural name, with endings set by various codes and which contains a single species or one or more monophyletic genera.

4. Order. A uninomial plural name whose ending varies depending on the conventions of the particular group and which contains a single species or a variable number of monophyletic taxa of lower rank.

5. Class. A uninomial plural name of variable ending depending on tradition which contains a single species or a variable number of monophyletic taxa of lower rank.

6. Phylum. A uninomial plural name of variable ending depending on tradition which contains a single species or a variable number of monophyletic taxa of lower rank.

7. Plesion. A name of variable rank accorded a fossil species or a monophyletic group of fossil species when classified with one or more Recent species or groups of species (modified from the original definition of Patterson and Rosen, 1977). Plesion substitutes for categorical ranks and categorical rank *within* a plesion can be no higher than rank within that plesion's Recent sister group. Plesions in purely fossil classifications take on the categorical rank of their nearest Recent sister group.

CONVENTIONS FOR ANNOTATED LINNAEAN CLASSIFICATIONS

Convention 1. The Linnaean hierarchy will be used, with certain other conventions, to classify organisms.

Convention 2. Minimum taxonomic decisions will always be made to construct a classification or modify an existing classification. This will be accomplished in two ways. First, no empty or redundant categories will be produced unless these categories are necessary taxonomic conventions (i.e., the five mandatory categories may be redundant). Second, natural taxa of essential importance to the group classified will be retained at their traditional ranks whenever possible, consistent with phylogenetic relationships and the taxonomy of the group as a whole.

The convention simply states that classifications should be minimally redundant, minimally novel, and maximally informative. This follows closely the analysis of Farris (1976) to whom I gave credit for the convention (Wiley, 1979c). An example of an application of this convention is provided by my classification of gars (Wiley, 1976):

Division Ginglymodi
    Family Lepisosteidae
        Genus *Atractosteus*
        Genus *Lepisosteus*

The family contains only two genera. There is no need to assign each of the genera to different subfamilies (Atractosteinae and Lepisosteinae respectively). The diagnosis of each subfamily would be exactly the same as the single included genus. Likewise, injecting a single subfamily (Lepisosteinae) between the two genera and the family contributes no new information since the diagnosis of the subfamily is the same as that of the family (i.e., Lepisosteidae = Lepisosteinae). All that injecting one or two subfamilies does is to add one or two names to be learned without adding new information. Note, however, that the classification does have one redundant name because Ginglymodi = Lepisosteidae. We could delete Lepisosteidae and produce a completely nonredundant classification. However, the family, as a category in fish taxonomy, occupies an important place in handbooks, field guides and more formal taxonomic works. This makes it an informative category apart from its place in phylogenetic classifications. Thus, even though it is redundant, it is retained.

We may now consider another recommended convention, the phyletic sequencing convention of Nelson (1972a, 1974; termed the "sequencing convention" herein). Nelson suggested that asymmetrical trees containing a number of monophyletic groups could be placed at the same categorical rank and listed in order of their branching sequence. This would produce an exact reflection of the tree without having to name every branch point. Thus, the number of rank categories and their associated taxon names would be kept to a minimum (Nelson, 1974; Cracraft, 1974a; Schuh, 1976). For example, Schuh's (1976) analysis of the relationships of the hemipteran family Miridae (Fig. 6.3) results in the classification:

Family Miridae
    Subfamily Isometopinae
    Subfamily Psallopinae
    Subfamily Phylinae
    Subfamily Cylapinae
    Subfamily Mirinae
    Subfamily Bryocorinae

Without this convention, Schuh (1976) would have needed 14 taxon names and associated categorical ranks rather than the seven he required. Of the seven extra names, four would be redundant (containing only the terminal taxon, in this case the first four subfamilies). To avoid redundant

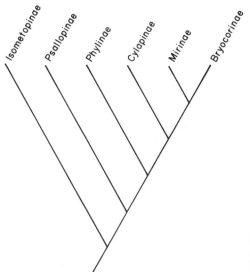

Figure 6.3 Schuh's (1976) hypothesis of relationships among subfamilies of the hemipteran family Miridae. (From Wiley, 1979c, after Schuh, 1976.)

names, Schuh (1976) would have had to place the subfamilies at different categorical ranks or alternately he would have to inject extra categories between family and subfamily. In either case, four extra hierarchical levels would have been needed. These problems were avoided by the use of the sequencing convention.

One important advantage of the sequencing convention is that minimum taxonomic changes may be made to convert a nonphylogenetic classification into a phylogenetic classification. As an example, I will compare four classifications of ratite birds: Storer's (1971) represents a traditional classification similar to previous classifications; Cracraft's (1974b) represents a comparable portion of a phylogenetic classification based on a phylogenetic hypothesis; the third is an alternate phylogenetic hypotheses proposed by Cracraft (1974a) while the fourth was proposed by Wiley (1979c) using conventions 1 through 3 with the express purpose of conserving as many traditional taxon ranks as possible. The common names of the groups are given in parentheses. Extinct taxa are tagged with "daggers" (i.e., "†").

Storer (1971)

Superorder Neognathae (all "higher" birds including ratites)
    Order Tinamiformes (tinamous)
    Order Rheiformes (rheas)
    Order Struthioniformes (ostriches)
    Order Casuariiformes (emus, cassowaries, and Dromornithidae)
    †Order Aepyornithiformes (elephant birds)
    Order Dinornithiformes (moas and kiwis)

Cracraft (1974b)

Order Palaeognathiformes
  Suborder Tinami (tinamous)
  Suborder Ratiti
    Infraorder Apteryges
      Superfamily Apterygoidea (kiwis)
      †Superfamily Dinornithoidea (moas)
    †Infraorder Dromornithes (*incertae sedis*)
    †Infraorder Eremopezithes (*incertae sedies*)
    Infraorder Struthiones
      †Superfamily Aepyornithoidea (elephant birds)
      Superfamily Struthionoidea
        Family Casuariidae (cassowaries and emus)
        Family Struthionidae (ostriches and rheas)

Storer (1971), like Cracraft (1974b), was striving for natural groups. Unlike Cracraft (1974b), Storer (1971) did not attempt to use the hierarchy to classify the various orders into supraordinal groups, but Storer's placement of orders supposedly denotes a loose phylogenetic relationship. There are obvious differences between the two classifications which go beyond the issue of hierarchy or subordination. For example, Storer (1971), following previous workers such as Wetmore (1951, 1960), considered Dromornithidae closely related to Cassowaries and Emus whereas Cracraft (1974b) considered the two groups relatively unrelated. Given such differences and given that we shall use Cracraft's phylogenetic hypothesis, can this hypothesis be turned into a phylogenetic classification which is minimally different from Storer's? Cracraft (1974a) produced several classifications using the sequencing convention that represented minimal modifications of the more traditional classification used by Storer (1974). One such classification, utilizing the sequencing convention is presented (down to suborder). (The extinct families Dromornithidae and Eremopezidae are not included.)

  Superorder Palaeognathae
    Order Tinamiformes (tinamous)
    Order Apterygiformes (= Dinornithiformes, moas, and kiwis)
    †Order Aepyornithiformes (elephant birds)
    Order Struthioniformes
      Suborder Casuari (cassowaries and emus)
      Suborder Struthiones (ostriches and rheas)

It would be possible to sequence rheas and ostriches, thus raising them to ordinal status. Note that while the above classification looks much more like Storer's than Cracraft's (1974b) preferred classification, the implications are quite different. The sequences in Cracraft's classification imply sister group relationships, that is, an exact hypothesis of relationship.

To place the extinct dromornithids and eremopezids in a classification with other ratites and conserve the ordinal rankings of the large extant ratites, one might use a classification similar to that proposed as an exercise by Wiley (1979c).

Subcohort Palaeognathae
   Superorder Tinami
      Order Tinamiformes
   Superorder Ratiti
      Ratiti, *incertae sedis*: Plesion Dromornithidae
                         Plesion Eremopezidae
      Order Dinornithiformes (= Apterygiformes)
      Plesion Aepyornithiformes
      Order Casuariiformes
      Order Rheiformes
      Order Struthioniformes

One might, of course, argue over whether Cassowaries should be a family (Cracraft, 1974b), a suborder (Cracraft, 1974a), or an order (Wiley, 1979c). It does not really matter. The important point is that in each classification the hypothesized relationships of cassowaries is clearly expressed. No doubt a settlement will depend on progress in the phylogenetic classification of Aves as a whole.

I conclude that the sequencing convention of Nelson (1974) is a powerful tool for phylogenetic classification, and I incorporate it as Convention 3 (from Wiley, 1979c).

Convention 3. Taxa forming an asymmetrical part of a phylogenetic tree may be placed at the same categorical rank (or designated "plesion" and indented) and sequenced in phylogenetic order of origin. When an un-annotated list of taxa is encountered in a phylogenetic classification, it is understood that the list forms a completely dichotomous sequence, with the first taxon listed being the sister group of all subsequent taxa and so on down the list.

Note that the convention states "may" rather than "shall" or "will." It is quite possible that an investigator has good reason for not sequencing an asymmetrical tree. The most obvious reason would be biogeographic. Each branch point might represent a vicariance event the investigator wishes to discuss (Wiley, 1979c). Another reason would be systematic. Platnick (1977c) notes that a completely sequenced classification might leave out frequently used taxon names which describe inclusive groupings of interest. In such cases, it might be better to rank rather than sequence.

I shall now consider taxa which form trichotomies or other multiple functions. Such tree topologies may result from true multiple speciation or from the inability of an investigator to justify a dichotomous arrangement (see Chapter 4). The former may be corroborated by a biogeographic argument (see Platnick and Nelson, 1978), the latter is an expression of ignorance.

Regardless of how such multifurcations are viewed, some convention must be adopted to indicate that the taxa involved form, say, a trichotomy rather than an ascending dichotomy which is sequenced. For example, the same classification of amniote tetrapods may be translated into either phylogenetic hypothesis shown in Fig. 6.4*a* and *b*.

Superdivision Amniota
    Division Theropsida
    Division Anapsida
    Division Sauropsida

Put simply a sequencing convention cannot be used unless there is a convention which clearly differentiates a list of taxa forming a series of dichotomies from a list of taxa arranged as a multiple furcation (a polytomy) from the same mode.

One way of tagging multiple furcations would be to list all the involved taxa as *incertae sedis*. This is a logical extension of the term as used by Patterson and Rosen (1977) and one I had considered when Wiley (1979c) was in manuscript. However, *incertae sedis* has traditionally been used to show that a taxon of low rank is doubtfully placed in a higher taxon or that this taxon might be related to one of several high taxa. If a Latin phrase is to be used, it would be more appropriate to indicate "of in-

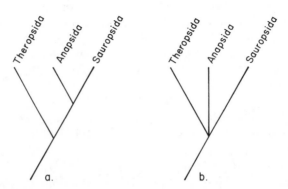

Figure 6.4 Two phylogenetic hypotheses of the relationships of amniote vertebrates (from Wiley, 1979c).

terchangeable taxonomic sequence" rather than "of uncertain taxonomic position." I have suggested that the term *"sedis mutabilis"* (L.—of changeable position) suits this connotation and is phonetically pleasing (see Wiley, 1979c). It neither implies a true multifurcation nor a failure to resolve a dichotomous situation. Rather, it says unambiguously that the order of the list is interchangeable. I incorporate the term in Convention 4 (Wiley, 1979c).

Convention 4.   Monophyletic groups in trichotomous or polytomous interrelationship will be given equivalent ranks and placed *sedis mutabilis* at the level of the hierarchy at which their relationships to other taxa are known.

This convention clearly indicates which lists are dichotomous sequences and which are multiple furcations. For example, the hypothetical phylogeny in Fig. 6.5 is classified using a combination of conventions 3 and 4.

Family XYZ
    Genus *X, sedis mutabilis*
    Genus *Y, sedis mutabilis*
    Genus *Z, sedis mutabilis*
      Z aus
      Z bus
      Z cus
      Z dus

As an alternate, family XYZ could be annotated as containing genera X, Y, and Z whose phylogenetic interrelationships are of trichotomous arrangement.

  Family XYZ (all genera *sedis mutabilis*)
    Genus *X*
    Genus *Y*
    Genus *Z*

Sedis mutabilis has two major advantages. First, it conserves the sequencing convention for use in showing asymmetrical relationships. Second, it calls immediate attention to problems of interrelationship.

The term *incertae sedis* is a valuable tool for indicating unknown or uncertain relationships. Some previous workers who have considered the convention in phylogenetic classifications (e.g., Nelson, 1972a, 1973a; Patterson and Rosen, 1977) have restricted the use of *incertae sedis* to fossil organisms. Others (e.g., McKenna, 1975) have used it for both Recent and fossil taxa. I prefer to use the term for either fossil or Recent taxa, since its restriction to fossils seems rather arbitrary.

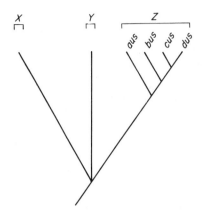

Figure 6.5 Phylogenetic relationships among some members of the hypothetical family XYZ (From Wiley, 1979c.)

Convention 5. Fossil or Recent monophyletic taxa of uncertain relationships will be placed in the hierarchy *incertae sedis* at the level their relationships are best understood.

Such a convention is usually applied to denote ambivalence toward the classification of a taxon of relatively low rank to one or more taxa of higher rank. It may also denote ambivalence toward one taxon's placement when its possible sister groups are known to be dichotomously interrelated. For example, Koponen (1968) analyzed four tribes of bryophytes of the family Mniaceae (Fig. 6.6). Orthomniceae traditionally has been placed in family Mniaceae, but Koponen (1968) was unable to find any synapomorphy uniting the members of this tribe with the other three tribes. Using the sequencing convention plus *incertae sedis*, family Mniaceae can be classified thus (from Wiley, 1979c):

Family Mniaceae
    Mniaceae *incertae sedis*: Tribe Orthomnieae
        Tribe Plagiomnieae
        Tribe Cinclidieae
        Tribe Mnieae

In addition to multifurcations and doubtful relationships, there are two additional problems: (1) the classification of groups whose status (as monophyletic) is unknown, and (2) the problem of known or suspected para- or polyphyletic groups the investigator wishes to add to the classification for completeness. Patterson (1973) indicated such groups by a code put in parentheses beside each taxon's name: M = monophyletic, G = paraphyletic, P = polyphyletic, and IK = unknown status. Such a convention is useful when coupled with *incertae sedis*. I suggest, however, a simpler convention used by Patterson and Rosen (1977), placing the taxon name in shutter quotes (Wiley, 1979c).

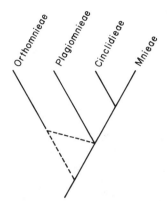

Figure 6.6 A hypothesis of the relationships among mosses of the family Mniaceae (From Wiley, 1979c, after Koponen, 1968.)

Convention 6. A paraphyletic or polyphyletic assemblage, or a group of unknown status may be included in a phylogenetic classification if it is put in shutter quotes that indicate that all of the included taxa are *incertae sedis* at the level in the hierarchy that the group is placed. Such groups will not be given rank (neither formal rank nor plesion status).

This convention was used by Patterson and Rosen (1977) to indicate that relationships of certain paraphyletic or polyphyletic groups of Mesozoic fishes. The "Semionotidae" is a collection of diverse taxa which fits somewhere between the gars and the teleosts in actinopterygian evolution. Some may be closely related to *Amia*, others to teleosts. Or, they may form a paraphyletic assemblage between gars and Recent halecostomes (i.e., *Amia* + teleosts among Recent groups). This status of "Semionotidae" is clearly shown (from Patterson and Rosen, 1977):

    Infraclass Neopterygii
        Division Ginglymodi
        Division Halecostomi
            Halecostomi, *incertae sedis*: "Semionotidae"
                Subdivision Halecomorphi
                Subdivision Teleostei

The conventions presented so far allow the placement of all Recent taxa into a classification that exactly reflects the phylogeny of these groups with two exceptions. These exceptions are (1) Recent species that are ancestral to other Recent species and (2) species of hybrid origin. Conventions for handling these entities will be discussed after a dicussion of combined fossil-Recent classifications because the solutions are similar to the solution for classifying extinct ancestral species.

Combined Fossil and Recent Classifications

Because of the requirements of strict monophyly and exact reflection of phylogenetic position, fossils have been a special problem of phylogenetic classification (Hennig, 1966; Crowson, 1970; and others). There are three distinct problems associated with combined fossil and Recent classifications (summarized in one respect or another by Hennig, 1966; Brundin, 1966, 1968; Crowson, 1970; Nelson, 1972a, 1973a, 1974; Løvtrup, 1974; Cracraft, 1974a; and Patterson and Rosen, 1977). First, fossil organisms are inherently incomplete compared to their Recent relatives. Second, whenever fossil species are classified with Recent species or higher taxa, there is always the chance that one or more of the fossil species is the actual ancestral species (stem species) of one or more of the Recent taxa. Third, as pointed out by Patterson and Rosen (1977), the incorporation of increasingly primitive fossil groups into a fully ranked Linnaean classification calls for more and more categories containing fewer and fewer actual specimens or species.

Patterson and Rosen (1977) reviewed the manner in which phylogeneticists have attempted to deal with these three problems. They viewed the problems encountered by Hennig (1966), Brundin (1966), Crowson (1970), and Griffiths (1974) to be derived from two sources:

1. The impossibility of classifying ancestral species in the same hierarchy with their descendants.
2. The impossibility of ranking ancient fossil species or groups in a Linnaean hierarchy in which absolute age is the ranking criterion.

Patterson and Rosen (1977) concluded that both of these difficulties could be circumvented by considering all fossil species as terminal taxa (also see Farris, 1976). Actually this addresses only one of the problems—how to deal with fossil species that may or may not be ancestors. The second problem is dealt with by simply giving up the idea that categorical rank has anything to do with absolute age (and accepting the idea that it has only to do with relative age). That the Linnaean hierarchy (properly annotated) can incorporate stem species in a satisfactory manner will be dealt with later. For now we are concerned with how investigators who reject ancestors as knowable have handled the other two problems—the inherent instability of a fully subordinated classification incorporating fossil and Recent organisms (because of the inherent incompleteness of fossil specimens) and the problem of the proliferation of categories of increasingly higher rank to handle fewer and fewer fossils (which would be required even if it was assumed that the fossil taxa classified were well enough known to produce a stable classification). Previous workers have handled these two problems in one of three ways, which I have termed proposals 1, 2, and 3 (Wiley, 1979c).

Proposal 1. *Fossils should be classified separately from Recent orga-nisms.* Crowson (1970) proposed that fossils should simply be excluded from classifications of Recent organisms. He suggested that groups of classifications be erected for the fossil species of each geologic period (i.e., one for the Neogene, one for the Upper Cretaceous, etc.). Løvtrup (1973) incorporated Crowson's proposals into his axiomatized classifi-cation scheme. However, Patterson and Rosen (1977: 155) were the first actually to try Crowson's proposal: "Crowson's ideas have been quoted with approval by several writers on cladistics, but so far as we are aware they have not yet been tried out in practice." Patterson and Rosen then produced two classifications, one for Eocene herrings (Family Clupei-dae) and another for Lower Cretaceous herrings. Customary Recent group name endings were used to avoid ambiguity or to avoid changing the meaning of Recent families. Their resultant classifications are:

Eocene classification
   Superorder Teleostei
      Order Elopocephana
         Suborder Clupeocephala
            Superfamily Clupeomorpha
               Family Clupeidae
                  Genus *Diplomystus*
                  Genus *Ornategulum*

Lower Cretaceous classification
   Superorder Teleostei
      Superfamily Elopocephana
         Family Clupeocephala
            Subfamily Clupeomorpha
               Tribe Clupeiformes
                  Genus *Diplomystus*
                  Genus *Spratticeps*

Patterson and Rosen (1977) point out that *Diplomystus*, a genus that ranges from the Lower Cretaceous to the Miocene, would represent a genus in the Lower Cretaceous, a tribe in the Upper Cretaceous, and so on until it represented a family in the Miocene. "This is not simply a problem of nomenclature" (ibid.: 156).

The advantages of Crowson's (1970) proposal are two-fold according to Patterson and Rosen (1977). First, it increases the "tidiness" of Recent classifications. Second, fossil groups that would have been paraphyletic or of uncertain status (*incertae sedis*) in a combined Recent-fossil clas-sification would be monophyletic if considered only with other members of their "horizontal biotas." But, there are also two disadvantages which led Patterson and Rosen to reject Crowson's proposal. First, there is a decrease in information content in this system compared to a combined fossil-Recent system. Second, there would be a needless proliferation of names to accommodate relatively few fossil taxa. To these disadvantages I add the following (Wiley, 1979c): (1) The decision as to where to draw the horizontal divisions are arbitrary and, in fact, not generally useful for fossils of widely different groups as, for example, vertebrates and brachiopods. It would produce an unworkable general system of classification if one wanted to classify Recent groups with large numbers of Recent taxa and few fossil representatives (coelenterates or protozoans for example) along with other groups with relatively few numbers of Recent taxa but large numbers of old fossil taxa (brachiopods or echinoderms for example). (2) The whole system is actually a grade system. How can a taxon be monophyletic in one classification (pure fossil), of uncertain status in another classification (combined Recent and fossil), and still be a natural taxon? (3) In a series of natural classifications, how can a monophyletic group such as the Clupeomorpha be considered a subfamily in one classification and a superfamily in another classification? Such conventions would seem only to create confusion rather than clarity. Because of these reasons I agree with Patterson and Rosen (1977) that Crowson's (1970) proposal cannot serve as a general reference system for biology as a whole (Wiley, 1979c). This also leads to a rejection of Løvtrup's (1974) axioms concerning fossils which rest on acceptance of Crowson's proposal.

Proposal 2.   *Fossil and Recent organisms should be classified together and treated the same.* An excellent example of a completely subordinated (ranked) classification of a large group of fossil and Recent organisms is McKenna's (1975) preliminary classification of Mammalia. However, there are several problems with this classification (which have nothing to do with the correctness of the phylogenetic hypothesis on which it is based). As pointed out by Patterson and Rosen (1977), McKenna did not consider the position of mammals within the context of other vertebrates; he began his classification with the traditional class Mammalia. To place Mammalia within the phylogenetic context of other vertebrates and at the same time retain the rank "class" would involve a prohibitive number of supraclass categories just to get out of the Theropsida, much less to get to phylum Chordata. I do not doubt, however, that such a classification is internally consistent. It would be theoretically possible to produce such a completely ranked system. The very practical problem is that every time a fossil group is reassigned it demands a new subordination of ranks up and down the line. Such a new subordination scheme would be justifiable in the case of a Recent group because Recent groups are of interest to all biologists, not just systematists or evolutionary theorists. Even if most fossil groups were as confidently assignable as their Recent relatives (and many fossil mammals are), it takes only a single fossil taxon of sufficiently ancient origin and controversial relationships to keep a classification in a constant state of flux. To ask all vertebrate comparative biologists to learn a new classification of Osteichthyes every time one expert classified acanthodians as chondrichthyans (Jarvick, 1977) or another reclassified them as teleostomes (Miles, 1977) is simply too much to ask of colleagues who could not study acanthodians even if they desired. I suggest that a general classification system which treats fossils in exactly the same way as Recent organisms will result in such flux as to make the resulting classifications of little use to our nonsystematist colleagues. Therefore, I turn to proposal 3.

Proposal 3.   Fossil and Recent organisms should be classified together but fossils should be treated differently. This is the convention used by Patterson and Rosen (1977), and advocated by both Griffiths (1974) and Nelson (1972a, 1974). This is also the convention I recommend, and for exactly the same reasons—the classifications of those groups with the most characters and of general interest to all biologists should eventually become stable enough to permit rank designations to become stable. Such stability is not necessarily true of the classification of fossil organisms which, because of their inherent incompleteness, may be shifted within the phylogenetic hypothesis as new characters are discovered owing to new techniques or specimens preserved in unique ways. Historically, phylogeneticists have handled combined fossil and Recent classifications in several ways.

Hennig (1969) suggested that all fossil species of a group more primitive than the most primitive Recent species of the same group be referred to as the "stem group," the **stammgruppe**. For example, there are several genera of fossil birds that are thought by many workers to be more primitive than the most primitive Recent bird group, the rattites. These genera, including *Archaeopteryx, Hesperornis,* and *Ichthyornis,* would be placed as the *stammgruppe* within Aves. The problem, as Patterson and Rosen point out, is with those groups with a long history of diversity before the origin of clades which are now extant. Teleosts, to use Patterson and Rosen's (1977) example, have a long history with many species groups which became extinct before the extant clades arose. The family Equidae, within Mammalia, and the trilobites, within Arthropoda, are additional examples. In such cases, Hennig (1969) suggested that intermediate categories be inserted to express the relationships of the fossils to the Recent organisms but that the ranks of these fossil groups should not influence the ranks of the Recent taxa. The basic problem with the *stammgruppe* is that, unless it is monotypic (contains only one species) it may be (and perhaps usually is) paraphyletic, a grade taxon. Therefore, the concept of the *stammgruppe* is unsatisfactory for showing the exact relationships demanded of the phylogenetic system (Wiley, 1979c).

Nelson (1972a) suggested a convention which meets the needs of the problem as perceived by Hennig (1969). Nelson (1972a) suggested that fossil groups be tagged with the traditional dagger (†) which indicates that they are fossil groups. These groups would then be listed in the order of their branching pattern (i.e., the listing convention discussed previously). Finally, if fossil groups were ranked, this rank would be given only after their Recent relatives have been classified and the fossil group(s) in question would have the same rank as their nearest Recent relatives. For example, two classes, A and B, are sister groups and within B there are orders C and D and two fossil groups E and F. Their phylogenetic relationships are shown below in Fig. 6.7. The following is Nelson's classification of these groups:

Recent classification
    Class A
    Class B
        Order C
        Order D

Combined classification
    Class A
    Class B
        †Order E
        †Order F
        Order C
        Order D

This convention would work well with some types of groups but would not work so well with other groups because (1) within many fossil groups

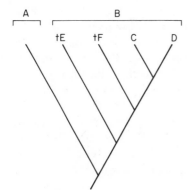

Figure 6.7 The phylogenetic relationships between classes A and B and among four orders of B. (From Wiley, 1979c.)

only one or two species are well enough known to be placed exactly, and (2) it requires fossils to be ranked, and this produces large numbers of monotypic taxa of high rank which are redundant (see Patterson and Rosen, 1977; and Farris's, 1976, requirement of minimal redundancy). Patterson and Rosen (1977) give an example of these problems in classifying teleost fossil fishes. Because of these difficulties, Patterson and Rosen (1977) suggested that fossil groups should be sequenced (listed) but not ranked. Their rank would be "plesion," a category which substitutes for Linnaean rank. One might think that a simple sequencing convention with the designation of the fossil groups via a dagger would suffice and that plesion was not necessary. This is true. However, the traditional use of the dagger has always been in concert with a specific categorical rank (Patterson and Rosen, 1977). Certain difficulties can be avoided by the use of plesion. For example, rather than say †order Ichthyodectiformes we might say plesion Ichthyodectiformes. Note that the ending "iformes" is conserved and therefore the form of the traditional name of the group is also conserved. But, if Nelson's convention were followed, the Ichthyodectiformes would have to be raised several categorical levels and changed to †supercohort Ichthyodectimorpha or to †Ichthyodectimorpha because the ending "iformes" is traditionally used for ordinal names in fish classifications. Even if we dispensed with the ranking, we would still have to change the traditional name ending of the group because the classification within the group would have to proceed from the subcohort level whether or not the group is ranked in the combined classification (Wiley, 1979c). But, by substituting "plesion" for "order" or "supercohort" we can avoid all such problems, the traditional spelling of the group name can be conserved and the classification within Ichthyodectiformes can begin with the traditional categories (suborders) associated with the group. Additionally, no higher category need be erected to place a fossil genus, for example, with its Recent sister group which happens to be an order (a problem in redundancy perceived by Farris, 1976). Thus, the use of "plesion" is a conservative means of classifying fossils with Recent taxa while at the same time conserving the standard classifications within the fossil groups no matter how many times the plesion's phylogenetic position may change in relation to its Recent relatives. As such, plesion is a powerful tool in erecting combined Recent and fossil classifications. I incorporate its use in Convention 7 (from Wiley, 1979c).

Convention 7.   Fossil groups will be classified in a different way than Recent groups. The status "plesion" will be accorded to all monophyletic fossil taxa. When a plesion is sequenced in a combined Recent-fossil classification, it is the sister group of all other terminal taxa within its clade and below it in that classification. Plesions may also stand *sedis mutabilis* or *incertae sedis* relative to other plesions and to Recent taxa.

For example, in an ascending dichotomy of fossil and Recent taxa, a plesion may be listed:

Class A
  Plesion X
  Order Y
  Order Z

In contrast, a polytomy between two plesions and two Recent taxa may be classified:

Class B
  Plesion C, *sedis mutabilis*
  Plesion D, *sedis mutabilis*
  Order E, *sedis mutabilis*
  Order F, *sedis mutabilis*

To demonstrate the robustness of the plesion convention in concert with sequencing, *sedis mutabilis*, and *incertae sedis*, I present a scenerio from my earlier discussion (Wiley, 1979c) of possible phylogenetic changes within the elopomorph teleosts (Figs. 6.8). Patterson and Rosen (1977) viewed Elopomorpha as consisting of three monophyletic extant orders and a single fossil species of uncertain relationships to these orders. Their phylogenetic hypothesis (Fig. 6.8a) was classified thus:

Cohort Elopomorpha
  Elopomorpha, *incertae sedis†* "*Anaethalion*" *vadali*
  Order Elopiformes
  Order Megalopiformes, new
  Order Anguilliformes

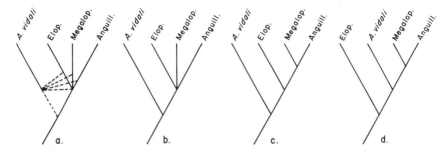

Figure 6.8 A hypothetical scenario of changing relationships among four taxa of elopomorph teleost fishes: (*a*) the current hypothesis of Patterson and Rosen (1977); (*b–d*) three possible changes in that hypothesis. (From Wiley, 1979c.)

Applying the *sedis mutabilis* convention to denote the trichotomy, Fig. 6.8*a* is translated into the following classification:

Cohort Elopomorpha
    Elopomorpha, *incertae sedis*: plesion *Anaethalion vadali*
    Order Elopiformes, *sedis mutabilis*
    Order Megalopiformes, *sedis mutabilis*
    Order Anguilliformes, *sedis mutabilis*

Let us imagine that new specimens of *Anaethalion vadali* are found and that they indicate that this species is the sister group of all other elopomorphs (Fig. 6.8*b*). The classification would then be:

Cohort Elopomorpha
    Plesion *Anaethalion vadali*
    Order Elopiformes, *sedis mutabilis*
    Order Megalopiformes, *sedis mutabilis*
    Order Anguilliformes, *sedis mutabilis*

Let us say further that subsequent analysis showed a dichotomous arrangement among the extant orders as shown in Fig. 6.8*c*. This hypothesis would translate:

Cohort Elopomorpha
    Plesion *Anaethalion vadali*
    Order Elopiformes
    Order Megalopiformes
    Order Anguilliformes

Finally, let us say that another worker found a completely different arrangement, as depicted in Fig. 6.8*d*. This hypothesis would translate:

Cohort Elopomorpha
    Order Elopiformes
    Plesion *Anaethalion vadali*
    Order Megalopiformes
    Order Anguilliformes

This theoretical exercise demonstrates that a number of alternate hypotheses can be presented using the same array of names. In this case, the array includes only the terminal taxa, and no intermediate categories of taxon names are needed. The robustness of these conventions warrants their use.

The Classification of Ancestors

One of the major problems Hennig (1966) saw in handling fossil organisms was the problem of classifying fossil species. Hennig (1966: 71–72) outlined the problem with great clarity:

> From the fact that . . . the boundaries of a "stem species" coincide with the boundaries of the taxon that includes all of its successor species, it follows that the "stem species" itself belongs in this taxon. But since, so to speak, it is identical with all of the species that have arisen from it, the "stem species" occupies a special position in this taxon. If, for example, we knew with certainty the stem species of the birds (and it is only from such a premise that we can start in theoretical considerations), then we would no doubt have to include it in the group "Aves." But it could not be placed in any of the subgroups of the Aves. Rather we would have to express unmistakably the fact that in the phylogenetic system it is equivalent to the totality of all species of the group.

For Hennig (1966), ancestors occupy a special position, but other phylogeneticists have a more pessimistic view. For example, Patterson and Rosen (1977: 154), citing Hennig (1966), Brundin (1966), Crowson (1970), and Griffiths (1974), have suggested that it is impossible to classify ancestral species in the same hierarchy as their descendants "since the ancestral species is equivalent to the taxon containing its descendants." Yet Nelson (1974) and Griffiths (1976) have shown that ancestors can be classified in hierarchies with descendants. Unfortunately, if this was done by Griffith's (1976) suggestion, we would have to abandon the Linnaean system. And if it was done by Nelson's (1974) suggestion (a sequencing convention), we would have to give up the sequencing convention for nonancestral taxa arranged in highly asymmetrical trees (Wiley, 1979c). Nelson (1974) was correct, in my opinion, in rejecting a sequencing convention, and because no other recommendation was made by him, he was left with simply rejecting that ancestors be classified. Because there is some reason to think that most ancestors can neither be identified nor scientifically corroborated or rejected by analyses now at the disposal of phylogeneticists (Hennig, 1966; Farris, 1976; Engelmann and Wiley, 1977; Platnick, 1977b), Nelson (1974) has a very good point. Hennig (1966: 72) takes the same position:

> Naturally, in practice this (ancestor recognition) meets basically insurmountable difficulties because it is scarcely ever possible to determine with certainty whether one (and in this case which) of the known species of *Archaeopteryx* (to continue our example) is the stem species of all other known species of Aves.

However, it is always possible that in the future some ancestor might be identified by some other criterion (for example, a biogeographic cri-

terion, Wiley, 1979a; Chapters 2, 4). And a general system of classification must be capable of handling *all* organisms and not just descendants (Wiley, 1979c). If it can not do this it will eventually be replaced
by a system that can handle all organisms. Can the Linnaean system,
with its conventions, be used to classify ancestors regardless of what we
may think about the practical problems of recognizing such ancestors?
I suggested that it could, by the use of Convention 8 (Wiley, 1979c).

Convention 8.   A stem species of a suprageneric taxon will be classified
in a monotypic genus and placed in the hierarchy in parentheses beside
the taxon which contains its descendants. A stem species of a genus will
be classified in that genus and placed in parentheses beside the generic
name.

For example, if one of the species of *Archaeopteryx* were, in the future,
by some criterion now unknown to us, actually identified as the stem
species of Aves, it would be classified thus (the remaining part of the
classification follows the implied relationships of Storer, 1971; note that
it fits with the fourth classification of ratites presented earlier:

Supercohort Aves (*Avus ancestorcus*)
   Plesion Archaeornithes
   Cohort Neornithes
      Plesion Hesperonithiformes
      Subcohort Palaeognathae
      Subcohort Neonathae

As another example, *if* the ancestor of mammals is identified the ancestor
would be classified as:

Subdivision Mammalia (*Mammalius primus*)
   Infradivision Prototheria
   Infradivision Theria

The consequences of adopting such a convention are two-fold. First,
where ancestors are hypothesized to be known, a sequencing convention
can not be employed at that level in the hierarchy because in a sequencing convention, there is no place on the hierarchy to put the ancestral
species. Thus, *if* all ancestral species were known, a completely ranked
classification would have to be employed. Of course, the advantages of
knowing ancestral species more than makes up for the increased complexity of the classification. In other words, the classification would be
more complex as a result of greatly increased knowledge (some might
even argue "final" knowledge). Second, in a mixed fossil-Recent classification, every plesion-Recent sister-group system where an actual an-

cestral species is hypothesized to be shared in common, a higher taxon must be erected to include the ancestral species and its descendants, or an additional convention of labeling ancestors and sequencing them (Bonde, 1977 would have to be employed.

Some interesting properties of phylogenetic classifications emerge from these considerations. First, the classification or nonclassification of ancestral species is no longer a controversy since the classification of such ancestors does not affect the classification of their descendants. Second, I submit that only a phylogenetic classification has the inherent quality of "making space" for ancestral species in such a way that they can be incorporated into classifications with minimum changes in existing classifications of hypothesized descendant species and monophyletic species groups. Phylogenetic classifications do this without distortion because every supraspecific taxon name stands as a substitute for an unknown ancestral species. In the simplest case, where ancestral species become extinct at speciation, there are n terminal species and $n - 1$ ancestral species. Only a completely ranked phylogenetic classification provides space in the classification for these n-1 ancestral species. In cases where the ancestral species survives a speciation event, then it is ancestral to two taxa, one of higher rank and the other of the next lower rank in the classification. Further, the convention adopted here places ancestral species where they belong, as coordinate names of the higher taxon that contains their descendants. Thus, the Linnaean system, with this convention, can place ancestral species into classifications of their descendants in a completely satisfactory manner.

To demonstrate the robustness of this convention, I present four hypothetical phylogenies (Fig. 6.9) and their classification (taken from Wiley, 1979c). The first case (Fig. 6.9a) involves extinction of ancestral species 1 and 2 at the speciation events.

Family Xidae (1)
   Subfamily Xinae
      Genus *Xus*
   Subfamily Yinae (2)
      Genus *Yus*
      Genus *Zus*

The second phylogeny involves a continuation of species 1 after the first speciation event within the family; i.e., species 1 is ancestral to both the Family and Genus *Xus* (Fig. 6.9b):

Family Xidae (1)
   Subfamily Xinae (1)
      Genus *Xus*
   Subfamily Yinae (2)
      Genus *Yus*
      Genus *Zus*

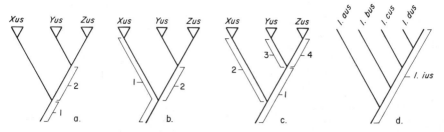

Figure 6.9 (*a–c*) Three possible phylogenetic trees of the hypothetical family Xidae, numbers represent ancestral species; (*d*) a known phylogenetic tree of the hypothetical genus *Ius*. Note that a cladogram of (*d*) would be a multifurcation of five branches. (From Wiley, 1979c.)

The third phylogeny is more complicated, species 1 survives the initial speciation event but not the second speciation event (Fig. 6.9*c*):

Family Xidae (1)
    Subfamily Xinae (2)
        Genus *Xus*
    Subfamily Yinae (1)
        Genus *Yus* (3)
        Genus *Zus* (4)

Figure 6.9*d* presents a different scenerio, a species is hypothesized to have given rise to four other species and yet remained unaffected and is still extant. Note that the ancestral status of *Ius ius* is indicated by a combination of sequence and annotation:

Genus *Ius* (*Ius ius*)
    I. aus
    I. bus
    I. cus
    I. dus
    I. ius

Classification of Reticulate Evolution and Taxa of Symbiotic Origin

Species and Higher Taxa of Hybrid Origin

Species of hybrid origin may be relatively rare in Metazoa(White,1978a) but they are considered rather common in Metaphyta (Grant, 1971). A general system of classification must be able to classify species and higher taxa of hybrid origin unambiguously if it is to fulfill its role as a general system. I have left the classification of reticulate evolution to this section because my proposed solution is similar to the classification of ancestral species. Figure 6.10*a* shows the phylogenetic relationships

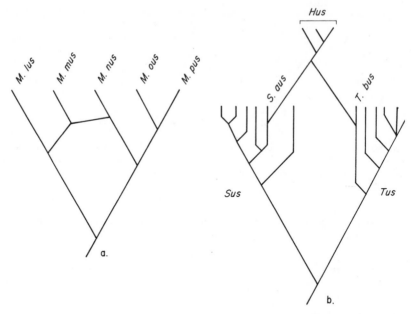

Figure 6.10 (*a*) The relationships among species of the hypothetical genus *Mismus*; (*b*) the relationships of the hypothetical genus *Hus* to its parental species *Sus aus* and *Tus bus*. (Adapted from Wiley, 1979c.)

of five species. One species, *M. mus*, is of hybrid origin. I assume that sufficient evidence of a reasonable nature supports this hypothesis. I propose to classify *Mismus mus* and all other hybrid taxa, with a convention that clearly specifies its hybrid origin (modified from Wiley, 1979c):

Convention 9. Taxon of hybrid origin will be classified with one or both parental taxa and its hybrid nature will be indicated by placing the names of its parental species in parentheses beside the hybrid's name. *The sequence of the hybrid taxon carries no connotation of branching relative to subsequently sequenced taxa of nonhybrid origin.*

Considering Fig. 6.10*a*, the phylogenetic hypothesis would be classified:

Genus *Mismus*
    M. lus
    M. nus
    M. mus (*M. lus* × *M. nus*)
    M. ous
    M. pus

An example of a supraspecific taxon of hybrid origin is shown in Fig. 6.10*b*. A population of species *Sus aus* hybridizes with a population of

Tus bus to produce a hybrid species. This species, in turn, speciates to produce several descendant species. This phylogeny is classified, at the generic level, thus:

Genus *Sus*
Genus *Tus*
Genus *Hus* (*Sus aus* × *Tus bus*)

The internal classification of these genera can then follow the other rules and recommendations. Note that the exact phylogenetic origin of the hybrid genus (family, order, etc.) is exactly specified by the binomials of the parental species. It would not be necessary to classify the parental species within the parentheses further since the various rules of nomenclature prohibit valid generic synonyms from being available names. Thus, an indication of the genus of each parental species is sufficient to indicate where they are found in the larger phylogeny.

The only other proposal I am aware of for phylogenetically classifying hybrid taxa was made by Nelson (1974). He suggested that hybrid species be sequenced once under one parental species and again under the other parental species. This has one advantage in that a monotypic taxon would not have to be erected to cover the origin of a single species from parental species of, say, two different genera. However, this is outweighed by the fact that this single lineage would have to be put in two genera. And it would not be apparent that the species was of hybrid origin unless both genera were always considered together (Wiley, 1979c).

The Classification of Hybrid Taxa: An Example

Humphries (1979) studied the phylogenetic relationships of the Compositae genus *Anacyclus* L. Of the twelve recognized species, three were of hybrid origin (Fig. 6.11). The classification presented here uses convention 9 to indicate the exact hypothesized origin and relationship of each of these hybrid species. Additional conventions include the listing convention and the use of informal taxon names. Note that the reticulate pattern of the phylogenetic tree is more complicated than that presented for *Mismus* in that two of the hybrid species (*A. officinarum* and *A.* × *valentinus*) have resulted from hybridization between species that are not adjacent to each other on the phylogenetic tree. These taxa have been arbitrarily sequenced without the loss of information concerning their ancestry.

Genus *Anacyclus* L.
  Sect. *Pyretharia* DC.
    Anacyclus pyrethrum (L.) Link
    Anacyclus officinarum Hayne (*A. pyrethrum* × *A. radiatus*)

Sect. *Anacyclus* L.
  Anacyclus monanthos (L.) Thell.
  Anacyclus marroccanus (Ball) Ball
  Anacyclus radiatus Loisel.
  Anacyclus × *valentinus* L. (*A. radiatus* × *A. homogamos*)
  Anacyclus clavicus species group
    Anacyclus linearilobus Boiss. & Reuter
    Anacyclus homogamos (Maire) Humphries
    Anacyclus clavicus (Desf.) Pers.
    Anacyclus × *inconstans* Pomel (*A. homogamos* × *A. clavicus*)
  Anacyclus nigellifolius species group
    Anacyclus latealatus Hub.-Mor.
    Anacyclus nigellifolius Boiss.

The Classification of Taxa of Symbiotic Origin

Taxa hypothesized to be of symbiotic origin may be classified in exactly
the same manner as taxa hypothesized to be of hybrid origin, with two
exceptions: (1) the actual species involved do not have to be specified
because reasonable evidence of symbiotic origin may not depend on

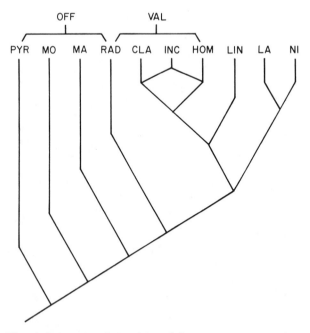

Figure 6.11 The phylogenetic relationships of the composite genus *Anacyclus* L. after
Humphries (1979). Abbreviations: CLA, *Anacyclus clavatus*; HOM, *A. homogamos*; INC,
A. × *inconstans*; LA, *A. latealatus*; LIN, *A. linearilobus*; MA, *A. marrocanus*; MO, *A.
monanthos*; NI, *A. nigellifolius*; OFF, *A. officinarum*; VAL, *A.* × *valentinus*.

identification of particular species but of particular groups, and (2) the multiplication sign (×) would be replaced by an addition sign (+) to denote the additive nature of the organisms. Thus, the larger aspects of organismic classification following Margulis's (1970) hypothesis of eucaryotic origins could be classified thus:

Superkingdom Procaryota
Superkingdom Eucaryota (Mycoplasmata + Spirochete + Eubacteria)
    Eucaryota, *incertae sedis*: "Protista"
    Kingdom Plantae ("Protista" + Cyanophyta)
    Kingdom Fungi
    Kingdom Animalia

Classifying Dendrograms and Trees

In Chapter 4 we saw that in certain cases phylogenetic trees are not strictly comparable in topology to phylogenetic dendrograms (the cladograms of Nelson, 1979; Platnick, 1977b; and others). Cladograms may depart from the topology of phylogenetic trees when (1) an incorrect assessment of the number of evolutionary species or monophyletic species groups is made, and (2) when there is a multifurcation between at least one species and two or more additional taxa. In terms of discussing differences in classification, condition 1 cannot be discussed profitably because both the dendrogram and the tree are incorrect. Thus, I will concentrate on point 2.

The difference between a phylogenetic dendrogram and a phylogenetic tree is illustrated in Fig. 6.12. The hypothetical order Omiformes contains four terminal taxa. Of these, three are in trichotomous relationship to each other. One of these trichotomously related taxa is the extinct species †*Zus lus*. The phylogenetic dendrogram is classified:

Order Omiformes
    Suborder Ominiiformes
        Family Omidae
    Suborder Xiniiformes (†*Zus lus*)
        Plesion *Zus lus, sedis mutabilis*
        Family Xidae, *sedis mutabilis*
        Family Yidae, *sedis mutabilis*

Because the trichotomy involves a single species and two supraspecific taxa, there are two possible phylogenetic trees (given no errors of the first type discussed previously). The first of these trees has the same topology as the phylogenetic dendrogram (Fig. 6.12*a*). The second states that †*Zus lus* is the common ancestor of families Xidae and Yidae (Fig. 6.12*b*). The classification of this tree involves moving †*Zus lus* from

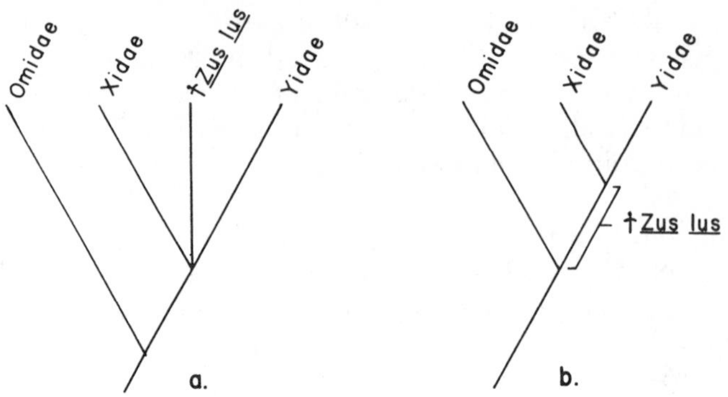

Figure 6.12 (*a*) A hypothesis of the relationships among *Zus lus* and its relatives. (*b*) a hypothesis of relationships showing *Zus lus* to be the ancestor of Xidae and Yidae. All taxa hypothetical.

plesion status to ancestral status and the removal of *sedis mutabilis* from the two families (because their phylogenetic relationships are now assumed to be known):

Order Omiformes
  Suborder Ominiiformes
    Family Omidae
  Suborder Xiniiformes (†*Zus lus*)
    Family Xidae
    Family Yidae

Other Information

Thus far I have discussed an array of conventions specifically designed to recover genealogical information. It is possible to attach other types

Table 6.1 Anagenetic distances (δ) within the family Aidae

	Taxon						
Taxon	1	A	2	B	3	C	D
1	0						
Aus	2	0					
2	2	4	0				
Bus	6	8	4	0			
3	6	10	6	10	0		
Cus	7	11	7	11	1	0	
Dus	7	11	7	11	1	2	0

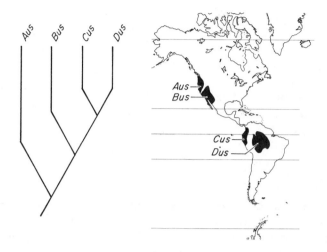

Figure 6.13   The phylogenetic relationships (left) and geographic ranges (right) of four hypothetical plant genera.

of information to a strictly phylogenetic classification without loss of information. I have already attached extra information in the rattite classifications in the form of the common names of the major groups. I did so because more readers of this book will be familiar with the common names of these birds than with their scientific names. It is also possible to attach biogeographic information or even character data. In Fig. 6.13 an array of information is presented for the hypothetical family Aidae. In Table 6.1 these taxa are shown with their morphological gaps.[2] The biogeography of the group is incorporated into the classification:

> Family Aidae (NA, SA)
>   Subfamily Ainae (NA)
>   Genus *Aus*
>   Subfamily Binae (NA, SA)
>   Tribe Bini (NA)
>   Genus *Bus*
>   Tribe Cini (SA)
>   Genus *Cus*
>   Genus *Dus*

The anagenesis of the group is incorporated into the following

---

[2] There are a variety of ways of computing such gaps, including purely phenetic and anagenetic (Brothers, 1975) equations. For purposes of this discussion, we shall not be concerned with how the values were arrived at, only that they are estimates of distance between taxa.

classification:

Family Aidae
    Subfamily Ainae (δ = 2)
        Genus *Aus*
    Subfamily Binae (δ = 2)
        Tribe Bini (δ = 4)
            Genus *Bus*
        Tribe Cini (δ = 6)
            Genus *Cus* (δ = 1)
            Genus *Dus* (δ = 1)

The problem with incorporating such information as anagenetic estimates of divergence arises when the investigator wishes to sequence. To accomplish both sequencing and presentation of anagenetic information we need another convention.

Convention 10. When a sequenced classification is annotated to reflect such information as taxon distinctiveness the information between terminal taxa in the sequence may be indicated by placing this information in parentheses between the taxa and at the same indentation level.

Using this convention we could combine the information in Table 6.1 and Fig. 6.13*a* to produce the sequenced classification:

Family Aidae
    Genus *Aus* (δ = 2)
    (δ = 2)
    Genus *Bus* (δ = 4)
    (δ = 6)
    Genus *Cus* (δ = 1)
    Genus *Dus* (δ = 1)

I would not think that such information would normally be associated with a formal classification. Farris (1980) has suggested that such information be associated with the diagnosis or description of the taxa classified. Two points are in order. First, such information can be incorporated into an annotated classification. Second, Farris (1980) has pointed out that only a phylogenetic classification can contain such information in a fully recoverable form. This has relevance to the claims of evolutionary taxonomists concerning information content, and classification, as discussed in Chapter 7.

Applications of the Annotated System

In Wiley (1979c) I discussed several examples of the annotated system, including a classification of Vertebrata from Wiley (1979d) and a classification of Mammalia modified from McKenna (1975). The purpose of these exercises was to demonstrate that strictly phylogenetic classifications could be produced without destroying the familiar names and even ranks associated with many taxa. I should like to enlarge that exercise by classifying Primates down to selected genera. Although Primates is a rather easy test of the annotated system (rodents or beetles would be harder), I selected it because Schwartz et al. (1978) have recently provided a strictly phylogenetic classification of primates with which to work. Using this group I can demonstrate the full application of the system of conventions from subphylum to genus. No doubt groups with more complicated phylogenetic histories will require more rank and taxon changes. The more complicated the history, the more complicated the classification which reflects that history.

Presented here are a series of classifications. First is a classification of Vertebrata reflecting the phylogenetic hypothesis shown in Fig. 6.14.

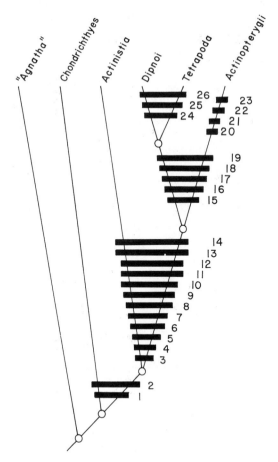

Figure 6.14 A hypothesis of phylogenetic relationships among Recent vertebrates. (For characters see Wiley, 1979d.)

As noted by Wiley (1979c,d), the hypothesis and classification are based on previous workers such as Nelson (1969), Patterson (1973), Rosen (1973), Wiley (1976), Miles (1977), and Patterson and Rosen (1977).

    Subphylum Vertebrata
        Infraphylum Myxiniodea, *sedis mutabilis*
        Infraphylum Petromyzontia, *sedis mutabilis*
        Infraphylum Gnathostomata, *sedis mutabilis*
            Superclass Chondrichthyes
            Superclass Teleostomi
                Class Actinistia
                Class Euosteichthyes
                    Subclass Actinopterygii
                    Subclass Sarcopterygii
                        Infraclass Dipnoi
                        Infraclass Choanata
                            Superdivision Amphibia
                            Superdivision Amniota
                                Division Anapsida, *sedis mutabilis*
                                Division Sauropsida, *sedis mutabilis*
                                Division Theropsida, *sedis mutabilis*
                                    Subdivision Mammalia

Given that Mammalia is a subdivision of Theropsida, the following classification expresses the placement of Primates within Mammalia as presented by Wiley's (1979c) modification of McKenna's (1975) classification. The entire classification and a comparison of it to McKenna's is shown in Wiley (1979c).

    Subdivision Mammalia
        Infradivision Prototheria
        Infradivision Theria
            Plesion *Kuehneotherium*
            Plesion Symmetrodonta
            Plesion Dryolestoidea
            Plesion *Peramus*
            Supercohort Marsupialia
            Supercohort Eutheria
                Cohort Edentata
                Cohort Epitheria
                    Epitheria *incertae sedis*: order Rodentia
                    Subcohort Ernotheria
                    Subcohort Preptotheria
                        Preptotheria *incertae sedis*: Order Pholidota
                    Plesion Deltatheria

     Supersection Archonta, *sedis mutabilis*
       Order Primates, *sedis mutabilis*
      (Other orders of Archonta, all *sedis mutabilis*)
    (Other supersections of Preptotheria, all *sedis mutabilis*)

We may now consider Primates. Schwartz et al. (1978) proposed a new classification of Primates based on their knowledge of the phylogenetic relationships within the order. Their classification was relatively efficient in terms of the number of categorical ranks employed because they used both sequencing and the plesion convention (although not exactly as used here). From their classification I reconstructed a phylogenetic tree (Fig. 6.15). The following is the classification of this tree, using the full array of conventions:

Order Primates
  Suborder Plesitarsiiformes
    Plesion Microsyopida
    Plesion Plesiadapiformes
     Superfamily Anaptomorphoidea
      Family Teilhardinidae
      Family Microchoeridae
       Tribe Loveinini
       Tribe Shoshonini
       Tribe Washakiini
       Tribe Microchoerini
      Family Anaptomorphidae
     Superfamily Plesiadapidoidea
      Family Paromomyidae
      Family Plesiodapidae
       Subfamily Pronothodectinae
        Genus *Pronothodectes*
       Subfamily Carpolestinae
       Subfamily Plesiadapinae
  Infraorder Tarsiiformes
    Plesion Unitasoricidae
    Plesion Macrotarsiinae
    Family Omomyidae
     Plesion *Omomys*
     Genus *Tarsius*
     Plesion *"Uintanius ameghini"*
     Plesion *Chumashius*
     Plesion *Pseudoloris*
  Suborder Simiolemiformes
    Infraorder Stepsirhini
     Superfamily Lorisioidea
      Family Lorisidae

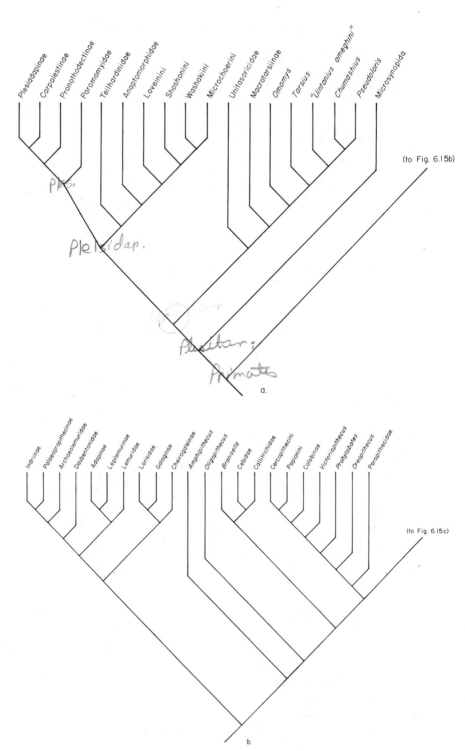

Figure 6.15 A hypothesis of the relationships among primates based on the classification of Schwartz et al. (1978).

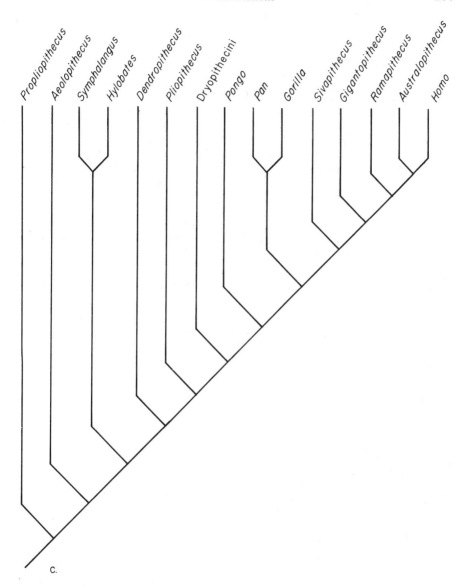

Figure 6.15   (*Continued*)

Family Galagidae
   Subfamily Galaginae
   Subfamily Cheirogaleinae
Superfamily Lemurioidea
   Plesion Adapidae
     Subfamily Adapinae
     Subfamily Lepilemurinae
   Family Lemuridae

Superfamily Indrioidea
  Family Daubentonidae
    Genus *Daubentonia*
  Plesion Archaeolemuridae
  Family Indriidae
    Plesion Palaeopropithecinae
    Subfamily Indriinae
Infraorder Anthropoidea
  Plesion *Amphipithecus*
  Plesion *Oligopithecus*
  Superfamily Platyrrhinoidea
    Plesion *Branisella*
    Family Cebidae (all genera *sedis mutabilis*)
    Family Callitrichidae (all genera *sedis mutabilis*)
  Superfamily Catarrhinoidea
    Plesion Parapithecidae
    Plesion *Oreopithecus*
    Family Cercopithecidae
      Plesion *Prohylobates*
      Plesion *Victoriapithecus*
      Subfamily Colobinae
      Subfamily Cercopithecinae
        Tribe Papionini
        Tribe Cercopithecini
  Superfamily Hominoidea
    Plesion *Propliopithecus*
    Plesion *Aeolopithecus*
    Family Hylobatidae
      Genus *Hylobates*
      Genus *Symphalangus*
    Plesion *Dendropithecus*
    Plesion *Pliopithecus*
    Family Hominidae
      Plesion Dryopithecini
      Tribe Pongini
        Genus *Pongo*
      Tribe Panini
        Genus *Pan*
        Genus *Gorilla*
      Tribe Hominini
        Plesion *Sivapithecus*
        Plesion *Gigantopithecus*
        Plesion *Ramapithecus*
        Plesion *Australopithecus*
        Genus *Homo*

VERACITY AND PHYLOGENETIC CLASSIFICATION

Veracity refers to the "explanatory power" of a hypothesis. As we have seen in previous sections, the phylogenetic system can express a variety of phenomena. Purely hierarchical phylogenetic classifications (including those using some of the conventions listed in this chapter) permit the recovery of all types of genealogical information. With appropriate conventions, this includes the relationship between descendants, between ancestral species and their descendants, and the general mode of origin of these taxa (i.e., whether speciation was divergent or reticulate). Further, in using the annotated system, we should expect that nonphylogenetic classifications can be converted with a minimum of taxonomic change which is directly related to increased knowledge about the groups classified. This point was illustrated using the Mammalia (Wiley, 1979c) and the Primates examples. Finally, following Farris (1980), we have seen that suitable annotations can be made to express other information directly in the classification in a form that can be exactly recovered. Information such as anagenetic distance has been claimed to be the special information incorporated into evolutionary classifications (Brothers, 1975; Ashlock, 1980). However, Farris (1980) has made the point that *only* a phylogenetic classification can both incorporate and retrieve this anagenetic information in its exact form whereas evolutionary classifications can not accomplish this task. This leads to the conclusion (reached by Hennig, 1966, and discussed in Chapter 7) that only phylogenetic classifications can serve as a general reference system for the diverse knowledge we now have and are gaining about the evolution of organisms.

Chapter 7

Taxonomic Alternatives to the Phylogenetic System

The phylogenetic system is one of four basic approaches currently practiced. The other three are (1) "traditional" taxonomy, (2) evolutionary taxonomy, and (3) phenetic taxonomy. Traditional taxonomists provide no particular theoretical framework and suggest that taxonomy is what one does, not what one thinks about (Blackwelder, 1967). Although this makes traditional taxonomy theoretically uninteresting, it cannot be denied that the majority of taxonomic works are atheoretical and yet have made immeasurable contributions to our knowledge of organismic diversity. A scholarly taxonomy is a real and lasting contribution. A badly thought out theoretical framework is doomed to replacement and may actually cause more harm than good. The remaining alternatives, evolutionary taxonomy and phenetics, provide theoretical frameworks that purport to be superior to, and result in better classifications than, the phylogenetic system. Phenetics is often termed "numerical taxonomy." I prefer "phenetics" because both phylogeneticists and evolutionary taxonomists use numerical methods (i.e., varous multivariate algorithms, computerized Wagner tree analysis, etc.). In the following sections I shall explore both approaches to systematics and discuss what I see as the difficulties of both as contrasted with phylogenetics.

EVOLUTIONARY TAXONOMY

I think it would be fair to state that most theoretically minded systematists who reject phylogenetics and phenetics count themselves among the evolutionary taxonomists. This approach to systematics is often termed the "Mayr-Simpson School" (Nelson, 1972b; Ashlock, 1973; Ashlock and Brothers, 1979). This label is an injustice to both Mayr and Simpson, because they differed widely on such concepts as species (biological vs. evolutionary), speciation (allopatric vs. allochronic), the relationships

240

implied by taxonomic groupings (genetic vs. genealogical), and, lately, monophyly (strict vs. minimal).[1] These differences and the additional differences of such workers as Michener (1977) and Ashlock and Brothers (1979) point to the fact that evolutionary taxonomy is a heterogeneous discipline or an array of different points of view more than it is a method or system united by a single body of theory. Evolutionary taxonomists do share at least two common beliefs: (1) they reject both phenetics and phylogenetics and (2) they accept paraphyletic groups as valid and meaningful taxonomic entities (i.e., as natural taxa or some equivalent of natural taxa). The basis for accepting paraphyletic groups, however, differs widely as we shall see.

The very diffuseness of evolutionary taxonomy makes this approach hard to criticize in a short or straightforward fashion. However, criticism is warranted if for no other reason than that evolutionary taxonomists have been among the most vocal critics of phylogenetics (the "cladistics" of Mayr, 1969, 1974). Thus I should explain why their criticisms are misplaced. Before I address these problems I would like to discuss some concepts that are held in common between the two approaches. This will clear the boards for the differences.

Given the articles one reads in the pages of *Systematic Zoology*, one might get the impression that evolutionary taxonomists and phylogeneticists have little or nothing in common. This might have been so in the past, but lately at least some evolutionary taxonomists have adopted some strictly phylogenetic ideas, including: (1) the principle that only synapomorphies should be used to reconstruct phylogenetic trees (Ashlock, 1974; Mayr, 1974); (2) the idea that critical biogeographic analysis demands such a phylogenetic tree (Ashlock, 1974); (3) the principle that monophyletic groups must be derived from a single ancestral species (Mayr, 1942; 1974) *and* that this species must be a member of the higher taxon which includes its descendants (Tuomikoski, 1967; Ashlock and Brothers, 1979); and (4) the principle that natural organismic classifications must agree with critically formulated evolutionary theory and that the ultimate justification for any particular taxon's existence is based on its conformance to that theory (Bock, 1977; Ashlock and Brothers, 1979). Given these areas of agreement, one might wonder why evolutionary taxonomists are not phylogeneticists. The answer lies with the goals evolutionary taxonomists set for their classifications.

The Goals of Evolutionary Taxonomic Classifications

Evolutionary taxonomists such as Mayr (1969, 1974), Ashlock (1973, 1974), Simpson (1975), Bock (1974), and Ashlock and Brothers (1979) are

[1] Mayr (1969) follows Simpson's (1961) minimal monophyly but Mayr (1974) follows Ashlock's (1971) monophyly, discussed later.

not satisfied with genealogical (phylogenetic) classifications because they believe phylogenetic classifications cannot fully reflect all the various aspects of evolution. They are absolutely correct in two respects. Phylogenetic classifications can reflect only a hypothesized pattern of speciation and group membership. It gives no clues as to the causal agents responsible for this speciation nor can it imply the actual characters used for grouping species in higher taxa. But then, neither can an evolutionary classification (Farris, 1977). We know that we may recover an inferred history of speciation from a phylogenetic classification and that a particular grouping of species shares unique evolutionary innovations. What may we recover from an evolutionary classification? Cracraft (1974a) suggests that nothing can be recovered from those parts of the classification that are not identical with a phylogenetic classification. This is true, but perhaps evolutionary taxonomists are not really interested in recovering anything in a particularly precise way (Hull, 1964). Rather, they prefer to put various "evolutionary phenomena" into their classifications without worrying about recovering them. The recovery of the history of speciation is made directly from the phylogenetic hypothesis whereas the classification reflects other things. For example, Michener (1977) wishes to "recover" phenetic similarity from the classifications whereas Mayr (1974) prefers to recover genetic similarity.

The principal differences between evolutionary classifications and phylogenetic classifications revolve around this extra information. Evolutionary classifications differ from phylogenetic classifications in one primary aspect. Evolutionary taxonomists feel justified in naming and ranking paraphyletic groups (grades). They recognize such grades as significant evolutionary entities. Evolutionary taxonomists use one or more of the following criteria in justifying grade taxa:

1. *Morphological gaps.*   If a sufficient morphological gap exists between two sister taxa, they may be separated and placed in two coordinate groups (Mayr, 1969). One of these groups is paraphyletic, the other may be paraphyletic or monophyletic. Ashlock and Brothers (1979) tighten this criterion by requiring that the discontinuities between such coordinate taxa be greater than the discontinuities within each taxon. An example of the separation of genealogically close relatives (if not the demonstration of a large morphological gap) is the separation of therapsids (advanced mammallike reptiles) from mammals and the relegation of therapsids to Reptilia.

2. *Species richness.*   A very speciose group may be given a higher rank than its species-poor genealogical relative. The sister taxon is relegated to a paraphyletic group including more distant relatives. This should be done in such a way that the two resulting groups either contain about the same number of species (Mayr, 1969), or are "reasonably uniform phenetically" (Ashlock and Brothers, 1979).

3. *Adaptive zones.*   A group judged to occupy a unique adaptive

zone may be given high rank while its sister taxon may be relegated to a paraphyletic group composed of more distant relatives. Conversely, the paraphyletic group may itself be justified because it occupies a unique adaptive zone. Raising Aves to a class and placing crocodiles in Reptilia is an example of the first type of justification.

4. *Monophyly.* Taxa must be monophyletic (Simpson, 1961; Mayr, 1969; 1974). However, the concept of monophyly differs depending on the evolutionary taxonomist, as I shall discuss in a later section. Basically, evolutionary taxonomists view grade or paraphyletic groups as "monophyletic."

In the sections following I shall discuss each of these criteria.

Morphological Gaps

A **morphological gap** may be characterized as a difference in one or several characters between two or more taxa. The gap may be absolute, as in the observed differences between two sister species or as in the hypothetical differences between the hypothetical common ancestral species of two monophyletic sister groups. Further, as Mickevich (1978) has pointed out, such an absolute gap must be based on a sample of characters which typifies the true differences between these sister groups. Any gap that is not absolute is due to sampling error. That is, a gap that is not absolute is a gap measured between two species that are not each other's closest relatives. Or, the gap is based on a single array of characters and another array of characters does not show the gap (Mickevich, 1978). In this case one or more taxa or one or more characters intermediate between the two taxa measured were not included in the gap analysis. Such sampling error may be unavoidable as in the case of unknown fossil species or species groups. From what I can tell, evolutionary taxonomists use gaps in two ways. First, they use them to justify certain taxonomic rankings. This produces problems with evolutionary theory as well as with classification stability. Second, they use gaps to delimit certain strictly monophyletic groups. This causes no particular problems to a strict phylogenetic system. I shall discuss both uses.

Gaps, Ranks, and Evolution

Evolutionary classifications frequently differ from phylogenetic classifications because two morphologically dissimilar sister groups are placed in different, and rank-coordinated, higher taxa. The "primitive" taxon is placed in a paraphyletic group whereas the "very different" advanced taxon is raised in rank to reflect its distinctiveness. The classic example in Vertebrata is the Recent Archosauria. Aves is accorded the taxonomic rank of class. Crocodiles are relegated to Reptilia as an order.

That crocodiles are different from birds is true. It is also true that the morphological gap between crocodiles and birds can be expressed by any number of quantified coefficients (phenetic, anagenetic, or a simple listing of the synapomorphies of birds on the one hand and crocodiles on the other hand). But, the significance of the gap does not lie in its quantification. Rather, it depends on how we view this gap from an evolutionary perspective. To those evolutionists who hold to the theory that evolution is gradual (i.e., Darwinian or neo-Darwinian), morphological gaps are an anathema and are always artificial. That is, any observed gap is due to sampling error. This extends even to the gaps between two sister species which should disappear if it were possible to sample the entire spectrum of the two lineages to their separation from their common ancestor. In the case of birds and other archosaurs, the observed gap may be due to extinction of intermediate species (Simpson, 1961; Michener, 1970) and their subsequent loss to science because they were not preserved. Nevertheless, the gap would be an artificial gap and not a by-product of the evolutionary process of diversification. Gradualist evolutionary theory predicts that the gap will eventually be filled (albeit always incompletely). As the gaps disappear, so does the basic rationale of ranking on the basis of gaps. For evolutionary taxonomists who hold to gradualism the problem with gaps is that the theory is not amenable to the taxonomic application.

The alternate to gradualism is saltation, the rapid morphological divergence of species from their ancestral morphology. To the saltationist any morphological gap between two sister species is a potential absolute gap. That is, saltationists allow that some morphological gaps are real and not the by-product of sampling error. But the significance of such absolute gaps depends very much on what kind of saltation the investigator espouses. Saltation according to Eldredge and Gould (1972; Gould and Eldredge, 1977; punctuated equilibrium) or Mayr (1963; genetic revolutions) or Simpson (1944; quantum evolution) produces *small* gaps. For lack of a term, we may name this phenomenon **microsaltation**. It is the type of saltation that produces new species not new classes. Further, it is firmly coupled with the history of speciation so that such gaps can only be appreciated if the classification documents speciation (i.e., if it is phylogenetic). Put simply, this type of saltation is complemented only by a strictly phylogenetic classification (1) because the stability of such a classification is relatively unaffected by the discovery of new group members, and (2) because the presence of a large gap that is due to sampling error is immediately apparent—relatively dissimilar organisms are placed in the same taxon whereas relatively similar organisms may be placed in different, coordinately ranked, taxa. The potential error lies with missing species and thus missing speciation events which couple the microsaltations with lineage divergence. None of the types of microsaltation coupled with speciation allows for sequential anagenetic saltation. Indeed, Eldredge and Gould disallow anagenesis as a major evolutionary force at all. The result is predictable. Gaps larger than those observed between relatively close branch points are probably due to sampling error and thus the larger gaps are artificial. The result for evolutionary classifications is the same—the theory does not complement the classification.

The second type of saltation involves major morphological changes of a type observed between major clades. These may be termed **macrosaltations**. Such saltations have been advocated by such workers as Goldschmidt (1952), Waddington (1957), Ohno (1970), and Løvtrup (1974). If at least some saltations are macrosaltations, then at least some large morphological gaps are absolute gaps. I see no reason to deny the existence of macrosaltations. They have an increasingly firm genetic basis (Britten and Davidson, 1971; Davidson and Britten, 1979) and a firm ontogenetic basis (Goldschmidt, 1940; Waddington, 1957; Løvtrup, 1974). There probably were never a great number of adult morphological intermediates between parazoans (sponges) and eumetazoans or between oligomeres and protostomes. In cases where large morphological gaps may reasonably be ascribed to macrosaltation, the resulting groups are usually accorded equal taxonomic rank compatible with a phylogenetic classification. One would think that it would be easy to relegate Parazoa to the Protista, or protochordates to the Invertebrata based on the large gaps exhibited by Eumetazoa and Chordata respectively. That such things are not advocated speaks for the basic *phylogenetic* integrity of higher classification rather than the need to create grade taxa because of large morphological gaps. It would seem that although at least some major morphological gaps are real, they have not been used to create the type of classifications advocated by evolutionary taxonomists. Further, at the levels of classification that evolutionary taxonomists have used large gaps, they have never (at least to my knowledge) invoked macrosaltation as the causal factor in producing the observed gaps.

Artificial Gaps and the Stability of Classifications

If the gaps used by evolutionary taxonomists are mostly artificial, then what does this mean for evolutionary classification? It makes evolutionary classifications inherently unstable (Mickevich, 1978). Theory predicts that the gap between one paraphyletic group and another paraphyletic group or a monophyletic group will be filled with either newly discovered fossils or Recent taxa. As the gap becomes increasingly smaller, the paraphyletic group becomes less and less justified and it can only be saved by adopting a typological attitude. This destroys the relationship between the classification and its underlying theory of evolution (if it had one in the first place). I shall demonstrate this with a hypothetical example and then cite some actual examples.

In Fig. 7.1a we have four sampled genera, A, B, C and D. Each genus is strictly monophyletic. Thus no species in B actually fills the genealogical gap between B and C + D. This condition prevents an artificial morphological gap due to a paraphyletic B. A phylogeneticist might classify this hypothesis in one of two ways:

Ranked Phylogenetic Classification	Sequenced Phylogenetic Classification
Family Aidae	Family Aidae
Genus A	Genus A
Family Biidae	Genus B
Subfamily Biinae	Genus C
Genus B	Genus D
Subfamily Ciinae	
Genus C	
Genus D	

In contrast, an evolutionary taxonomist might recognize the gap and incorporate it into the classification, producing the following classification:

Family Aidae
  Genus A
  Genus B
Family Ciidae
  Genus C
  Genus D

Now let us imagine that two fossil genera ($\dagger X$ and $\dagger Y$) were found which filled the genealogical gap such that the morphological differences between A, B, $\dagger X$, $\dagger Y$, C, and D were of approximately the same magnitude (see Fig. 7.1b). A phylogeneticist, utilizing the array of conventions discussed in Chapter 6, might revise the two alternate phylogenetic classifications thus:

Ranked Phylogenetic Classification	Sequenced Phylogenetic Classification
Family Aidae*	Family Aidae
Genus A*	Genus A*
Family Biidae	Genus B*
Subfamily Biinae*	Plesion Y
Genus B*	Plesion X
Plesion Y	Genus C*
Plesion X	Genus D*
Subfamily Ciinae*	
Genus C*	
Genus D*	

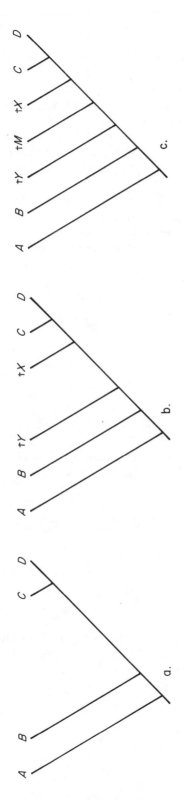

Figure 7.1 A series of phylogenetic trees showing the discoveries of fossil genera in the hypothetical family Aidae.

Note that in both cases these classifications are minimally different from
the original classifications. *The starred taxa (*) have exactly the same
rank and contain the same subtaxa.* And, change in composition of Biidae (left) and Aidae (right) involves adding taxa without removing taxa.
The classifications have changed minimally to incorporate new information on diversity. How might an evolutionary taxonomist revise his
classification? One possibility would be to recognize that the gap was
artificial and adopt a phylogenetic classification. *If so, then it would
have been more efficient and more conservative to have begun with a
phylogenetic classification.* There are, of course, other possibilities. Let
us say that some rule of thumb was adopted such that if the gap is of
quantity X, then a family is expected, whereas $X - 1$ denotes a subfamily,
$X - 2$ a genus, and so on. Perhaps we could group †Y in Aidae and †X
in Biidae:

Family Aidae
    Genus A
    Genus B
    †Genus Y
Family Biiade
    †Genus X
    Genus C
    Genus D

The problem here is consistency. If the morphological gap between †X
and †Y is less than X, then they must be placed in different subfamilies
or perhaps supergenera, *not* separate families. Thus, if the morphological
gap $= X - 1$, the resulting classification would be:

Subfamily Ainae
    Genus A
    Genus B
    †Genus Y
Subfamily Biinae
    †Genus X
    Genus C
    Genus D

Applying the principle *consistently*, if there were discovered a fossil
genus (†M) between †Y and †X (see Fig. 7.1c), such that the morphological gap separating †M from †X was $X - 2$, then the classification would
be:

Supergenus A
    Genus A
    Genus B
    †Genus Y
    †Genus M

Supergenus *C*
  †Genus *X*
  Genus *C*
  Genus *D*

Put simply, the more information one gains, the *less the resulting classification looks like the original classification*. Change is not inherently bad and evolutionary taxonomists could couple specific ranks with specific magnitudes of morphological dissimilarity. But even doing this would not solve their problems *because the resulting classification cannot indicate between which species the gap occurs*. In our example, it is not obvious nor even discernible that the judgement of morphological dissimilarity between supergenus *A* and supergenus *B* was made on the basis of the differences between †*M* and †*X*.

The last resort is to adopt an essentialistic attitude and group intermediate taxa based on a typological notion of what should be placed where. If †*M* is judged to have the essential character(s) of Biidae, then †*M* belongs in Biidae. If †*M* has neither the essential characters of Biidae nor of Aiidae, then it is placed in its own family (perhaps Miidae).

An essentialist attitude is what makes therapsids "reptiles" rather than theropsids and crocodiles "reptiles" rather than archosaurs. We might ask (as did Hull, 1964) what such classifications have to do with evolutionary theory? Nothing, unless evolution is essentialistic.

Gaps as Useful Taxonomic Tools

In view of the previous discussion one might have the idea that phylogeneticists object to gaps being used in any manner. Actually, phylogeneticists only object to gaps being used to erect paraphyletic taxa. Michener (1970: 23) finds another use for gaps: "in making formal classifications . . . the best (from all standpoints, probably) place to separate two taxa will be at the largest discontinuity between them." This attitude is not necessarily contrary to the aims of phylogenetic classification. Figure 7.2 shows two groups, A–E and X–Z. Let us say that the species of A–E are very different morphologically from the species in X–Z. In other words, there is a large morphological gap between species A and species Z. We may wish to place these two species groups in different genera simply because they look so different from each other. So long as we do this without creating a paraphyletic taxon at a higher level of universality (i.e., out of the relatives of A–E and X–Z), our decision is neutral with regard to the phylogenetic system because (1) each group is monophyletic in the strict sense and (2) their sister group relationships are clearly apparent. We may expect the gap to be filled or left unfilled depending on our evolutionary perspective. Either way, the generic classification need not be modified. This is consistent with the idea that monophyletic groups sensu Hennig are natural because we would expect

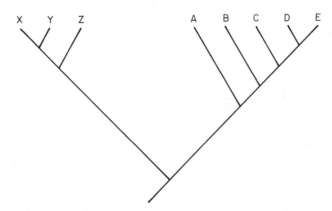

Figure 7.2   Two monophyletic groups separated by a real or artificial morphological gap.

that such groups would not change with respect to each other, but only with respect to their separate contents (i.e., their included species).

I leave the topic of morphological gaps not doubting that they may be quantified in some manner, either phenetically (Michener, 1978) or anagenetically (Brothers, 1978). Further, I do not doubt that gaps may be produced by extinction (Michener, 1970). However, I fail to see the relevance of gaps to classification when the use of such gaps obscures the phylogenetic relationships of the historical course of speciation.

Speciose Groups

The idea that genera should be composed of the same number of species and families about the same number of genera is a very old idea and not particularly characteristic of evolutionary taxonomy. Many convenience classifications strive for such balance and there is certainly something to say in favor of such symmetrical classes in terms of information theory. Indeed, I would advocate such balance in information retrieval systems so long as the information stored is retrievable (see discussion by Cracraft, 1974a). I see real problems, however, in trying to justify paraphyletic groups on the basis of the number of included species. Before discussing this point, I would like to give an example in which the number of included species might be taken into account in making a classification.

Let us say that we have a very large genus of organisms, say 1,000 species (Fig. 7.3a). Because all species are usually discussed in terms of their binomens, we would have to do quite a bit of memorizing to remember that *Mitus mitus* was more closely related to *Mitus minimus* than to *Mitus maximus*. After all, we must keep in mind some 997 other species named *Mitus*. Let us say that we studied the phylogenetic relationship of *Mitus* and found that it comprised two monophyletic

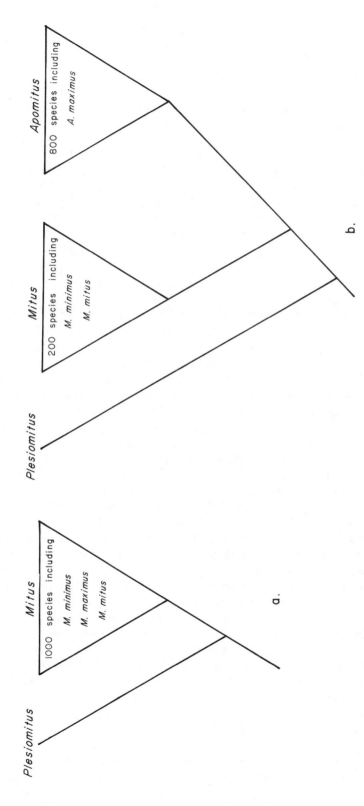

Figure 7.3 (*a*) The phylogenetic relationships of *Mitus* and *Plesiomitus*. (*b*) splitting *Mitus* into two genera to show more clearly that *M. mitus* and *M. minimus* are more closely related to each other than either is to *Apomitus maximus*.

251

groups; one group composed of 200 species, including *M. mitus* and *M. minimus*, and one comprising 800 species, including *M. maximus* (Fig. 7.3*b*). If we gave these two monophyletic groups different generic names we could tell at a glance the comparative relationships of the three species because two are in one genus (*Mitus*) and one is in the new genus (*Apomitus*). Further, such division would be useful in making the group as a whole somewhat more comprehensible.

I have no objection to using the number of species in this way. In fact, I advocate it. Any chance to make evolutionary statements about organisms in a classification should be taken so long as it leads to clarity. However, would we be justified in placing *Mitus* in family Mitidae along with the next genus (*Plesiomitus*) down the line (see Fig. 7.3) and elevating *Apomitus* to familial status because *Apomitus* has 800 species whereas *Mitus* has only 200 species? No, and the reason is rather simple.

Ashlock and Brothers (1979) state unequivocally that higher taxa must be the descendants of a single species and that this species must be classified in that higher taxon. This means that the higher taxon must have originated concurrently with the stem species. Given that a higher taxon originates with the stem species, then *subsequent speciation cannot affect that higher taxons origin relative to that taxon's nearest genealogical relative* (also see Chapter 4). If this is so, then how can the number of species affect the rank of a higher taxon? Both Simpson (1961) and Mayr (1969) agree that subsequent speciation may change the rank of a higher taxon, but it is not clear to me whether they are talking about the rank of an already established taxon or the origin of the taxon itself. For example, Simpson (1953) states, "a higher category (= taxon in this case) is higher because it [later] became distinctive, varied or both to a higher degree and not directly because of the characteristics it had when it was arising." Ashlock and Brothers disagree: "Valid taxa are furthermore delimited by characters derived from a single stem ancestor, thus ensuring that the taxon is a valid evolutionary unit." This concept is clear—higher taxa must be justified by the characters of the stem species. If evolutionary taxonomists are to follow Ashlock and Brothers (1979) on this point, then later speciation and thus the ultimate number of species in a natural taxon cannot effect the rank of that taxon. Indeed, if evolutionary taxonomists change the ranks of certain taxa because of the number of species they contain and erect paraphyletic taxa for those species left over, this will affect their estimates of morphological gap and thus the evolutionary deductions they may make about the origin and significance of the taxa involved. The resulting gaps will be artificial gaps rather than estimates of evolutionary divergence.

Adaptive Zones

The term "adaptive zone" is frequently used by evolutionary taxonomists in reference to supraspecific taxa. An adaptive zone may be thought of as a collective niche (Simpson, 1961: 222–223; Mayr, 1969: 234). As used by Van Valen (1971), an adaptive zone is a multidimensional abiotic place in nature which may or may not be inhabited by organisms. Elements of this concept are also found in Mayr (1969).

Mayr (1969: 234) suggests that the distinctness of an adaptive zone should be taken into account when determining the rank of a taxon. A taxon whose ancestral species has "shifted into a promising new adaptive zone ... usually receive higher rank than a taxon which lacks such ecological significance." It is this concept rather than the reality or unreality of an adaptive zone as an ecological phenomena which distinguishes the evolutionary taxonomist from the phylogeneticist.

Ecology and evolution comprise an array of interlocking processes. There can be no comparative species ecologies nor interspecific competition without different species. And, no species is without an ecology. Speciation may result in daughter species with similar or different ecologies. So far as I am aware, potential empty niches (if such things exist) or the subdividing of an existing niche by a number of species must await speciation. Thus, species-level adaptation in relation to ecology is usually produced by speciation, adaptation rarely causes speciation (certain sympatric speciation phenomena are exceptions, see Chapter 2). Certain morphological innovations, such as wings, may make certain parts of an environment available for use by a daughter species which was not available (or fully available) to ancestral species. And, this ecological shift may be passed on to that daughter species' descendants. In this context, the term "adaptive zone" has some meaning and is fully coupled with evolution.

The number of species inhabiting a particular ecosystem depends on niche availability, niche subdivision, and the number of available species. The number of species in a monophyletic group depends on the number of successful speciation events which have occurred during the history of that clade and the number of species that have become extinct. Since speciation is largely allopatric or parapatric, the ecological interactions of these daughter species are usually confined to other species and not to each other. In other words, ecological interactions are phenomena of biotas and ecosystems and not phenomena exhibited by clades as taxonomic entities. Rather, the species of a clade participate in the ecological interactions present in the biotas they inhabit. These may include members of their own clades or members of other clades.

Adaptation and Adaptive Zones

Evolutionary taxonomists are divided in their opinions concerning the relationship between adaptations and adaptive zones. On the one hand there are such workers as Van Valen (1971) who conceive of adaptive zones as being filled with clades which come across key characters, that is, characters that permit use of a previously unavailable niche such as bat wings or tetrapod limbs. Put simply, the origin of an evolutionary innovation or a series of evolutionary innovations makes a part of the environment newly available. On the other hand workers such as Mayr (1969: 593–595) conceive of the niche as the causal agent which stimulates the adaptation. He, and Bock (1965), seem to imply that species seek out and take advantage of unfilled niches. If we forgive the teleology, which I assume was only in the interest of clarity and does not imply true teleology, we can see that these are quite distinct viewpoints. Which is relevant?

Van Valen's Adaptive Zones. As an *a posteriori* statement concerning the "ecology" of a higher taxon I have no objections to Van Valen's concept. In other words, the fact that the key character of wings in bats reflects certain chiropterid habits is not a matter of dispute. My criticism with the adaptive zone as used by Van Valen stems from the fact that I do not think supraspecific taxa have a niche *per se*. The word "niche," like the word "character," defies adaquate definition, probably because it is a primitive term. Yet, when applied to species or populations, ecologists have little trouble in communicating the concept to each other. The problem is that higher taxa are not ecological units any more than they are active evolutionary units (Hull, pers. comm.). A grouping of organisms that does not have an ecology can not have a niche. If niche is equated with adaptive zone (Van Valen, 1971: 421), then higher taxa cannot occupy adaptive zones except as these zones are the byproduct of evolution on the species level. Perhaps this is what Van Valen means. But it has the same consequences as a strict definition of monophyly— *only monophyletic groups sensu Hennig can occupy an adaptive zone of consequence to both evolution and comparative ecology.*

Mayr's Concept of Causal Adaptive Zones. Organisms and species must be adapted if they are to survive. An adapted organism may or may not leave offspring. An adapted species has the reproductive capacity to maintain lineage continuum or it becomes extinct. Recent species are the result of an unbroken chain of (at least minimally) adapted species extending back to the origin of life. Extinct species share this unbroken chain up to the point where they were either outcompeted or otherwise lost their capacity to maintain a reproductive continuum.

If species had some mechanism to help them identify how they should become adapted, then I could believe that "empty niches" were causal agents in the evolutionary process and I could believe that the invasion of empty niches was an evolutionary phenomena. But if daughter species are randomly adapted with respect to their ancestral species (Wright, 1955), then daughter species find their ways into new niches by chance, or they are pushed into new niches by necessity; they do not actively seek them out. In other words, if Wright's rule applies as a general phenomena, then the invasion of a new adaptive zone is a random phenomenon constrained only by a clade's history of descent.

By saying that adaptations, and thus niche shifts, are random with respect to species divergence, I do not mean to imply that such shifts are irrelevant to ecological theory. Such phenomena as niche breath, competing and predictor species, species packing, and resource partitioning are interesting ecological phenomena. But niches cannot be packed without the species to pack them, and I doubt that species somehow know (or can sense) that a niche can be subdivided yet again and then accomplish this by speciating.

The second objection I have to Mayr's (1963, 1969) concept of adaptive zone concerns its influence on determining the ranks of taxa. Both Simpson (1961) and Mayr (1969) seem to suggest that an investigator determines the presence of an invasion of an adaptive zone *not by ecological principles, but by the presence of morphological gaps.* We may determine that an adaptive threshold has been surmounted by finding a large morphological gap. And, we can tell the success of the adaptation from the number of species of the taxon.

Using this concept, birds occupy an adaptive zone because of their distinctive morphology, not because Aves occupies a particular, unique niche in relation to other clades. Indeed, birds do not occupy a collective niche nor do they occupy a coherent hierarchy of related niches. Their biology, as a function of their history, does limit them to certain ways of life. But these limitations certainly cannot be described as constituting a collective niche. I conclude that the concept of adaptive zone, when used by some evolutionary taxonomists, is not a criterion independent from morphological gap and relative species richness. In fact, it seems to be completely subsumed by these other criteria and thus should not be used at all. In saying this I do not imply that many clades do not have well defined collective niches in the ecological sense. Cestodes and fleas are examples that do. But I see no relationship between these groups' ecologies and some ecological criterion which would rank them differently from their respective sister groups.

Would it be possible to formulate some ecological criterion? Certainly, but if it was applied consistently I doubt that evolutionary systematists would like the results any more than phylogeneticists. If the ability to fly and the ecological consequences of flight are of such importance that Aves is elevated to the rank of class, then it should be just as important to bats (Chiroptera) and hatchet fishes (Gasteropelecidae).

Monophyly

Evolutionary taxonomists, like phylogeneticists, feel that taxa in a classification should be monophyletic. However, evolutionary taxonomists agree neither among themselves nor with phylogeneticists on the concept of monophyly. Phylogeneticists hold that monophyletic taxa above the species level must include an ancestral species and all descendants of that ancestor. Evolutionary taxonomists hold one of two concepts. The first, Simpson's (1961) concept of minimum monophyly, requires only that an ancestral species be postulated to exist in some other higher taxon of equal or lesser rank. The second, Ashlock's (1971), supposedly allows only for species-level ancestors. I shall discuss both in turn.

Simpson's Minimum Monophyly

Simpson (1961: 124) defined a monophyletic group as one derived "through one or more lineages . . . from one immediately ancestral taxon of the same or lower rank." Simpson (1961) viewed monophyly as a relative concept with degrees of monophyly ranging from barely minimal (as in the derivation of one supraspecific taxon from another of equal rank) to strict (as in the derivation of a higher taxon from a single evolutionary species). This principle was widely accepted. For example, Mayr (1969) uses this definition of monophyly in his text. However, several problems with the definition have lead to its abandonment by many evolutionary taxonomists:

1. The relationship between a given phylogeny and one or more classifications based upon the concept of minimal monophyly is obscure. What groups are recognized and how they are ranked follow no particular or well stated biological principles but rather suit the taste or tastes of the classifier. Hull (1964) suggests that the vague correspondence between classifications based on minimum monophyly and the phylogenetic trees they purport to classify greatly decreases the information content of the classifications.

2. Simpson's definition is unrestrictive; literally any taxon can be justified as monophyletic (Hennig, 1966; Tuomikoski, 1967; Ashlock, 1971) because the concept is tied to categorical rank. Reptilia is monophyletic because it is the descendant of Amphibia, a taxon of equal rank. Yet Homotherma (birds and mammals) can also be justified as monophyletic by appropriate juggling of the taxonomic rank of Reptilia, thus:

Superclass Reptilia
Superclass Homotherma
    Class Aves
    Class Mammalia

We may conclude one of two things. Either any proposed grouping is monophyletic, or minimum monophyly must be rejected as a taxonomic concept.

3.   The definition results in taxa whose monophyletic or nonmono-phyletic status changes from one level of universality to another in the same classification (Wiley, 1979c). Simpson (1961: 124) states "*Mery-chippus*[2] is considered monophyletic at the generic [level] but poly-phyletic at the specific level." Taxa monophyletic at one level but polyphyletic at another level can hardly be considered natural and if we wish to preserve the naturalness of the term "monophyletic," minimum monophyly must be rejected.

4.   Supraspecific taxa can not be ancestral to other supraspecific taxa (Wiley, 1978, 1979b,c). The concept of "common ancestral taxon of equal or lower rank" is tenable as a critical evolutionary statement only if supraspecific taxa can be biologically meaningful ancestors (Simpson does not suppose that they are, as discussed later). But supraspecific taxa are composed of individual lineages that evolve independently of each other (Lewontin, 1974) within the constraints provided by historical and environmental factors. In contrast, supraspecific taxa do not respond to the forces of evolution as units. Individual organisms may be ancestors because they have the potential capabilities of reproduction and differ-entiation. And individual species may be ancestors because they have the potential to speciate. Thus, they have the potential to give rise to another species. But supraspecific taxa have no such capabilities because there is no known process by which they can leave descendants unless, of course, they are composed of a single species (and that type of descent is strictly a phylogenetic construct). Since taxa that cannot evolve cannot reasonably be expected to give rise to anything, I conclude that supras-pecific "ancestors" are not biologically meaningful entities. In other words, they may be considered evolutionary artifacts (Hennig, 1966; Engelmann and Wiley, 1977).

Although these criticisms are correct in a real biological sense one should not suppose that Simpson (1961) actually thinks that supraspecific taxa evolve and are actual biological ancestors. Rather, it is an imprecise statement relating to species-level ancestry. If it is stated "genus X is the ancestor of genus Y," and if no particular species of X said to be the actual ancestor, then this can only mean that the ancestor of Y is similar to, but not identical to, one or more species in genus X. In other words, if the ancestor of Y were found, it would presumably be more similar to the members of X than to its descendants in Y. Such statements are useful only in systems of classifications which require ancestor-des-cendant statements at *every* level of classification. But they do not con-tribute anything critical to evolutionary theory and, in fact, tend to ob-scure common ancestry relationships on the species level.

---

[2] A grade genus of fossil horses.

Ashlock's Monophyly

Ashlock (1971) suggested that there are two types of monophyletic groups: paraphyletic groups and monophyletic groups sensu Hennig (Ashlock's "holophyletic" groups) are both monophyletic (Ashlock's definitions are given in Chapter 3.). Objections to Ashlock's definitions have come from two basic sources.

1. As pointed out by Nelson (1973b), the definition is not based on pattern of descent but on the characteristics of the unobservable ancestral species (also see Platnick, 1977b). This criticism is valid, but it allows the possibility that if some criterion could be established for identifying ancestral species, Ashlock's definition could be used.

2. Ashlock's definitions render "polyphyletic" any holophyletic group whose sister group is paraphyletic (Wiley, 1979c). In my earlier discussion of Ashlock's (1971, 1973, 1974) monophyly, I suggested that we set aside the various objections to seeking ancestral species (see Hennig, 1966; Nelson, 1973b; Engelmann and Wiley, 1977; Wiley, 1978) and see if Ashlock's definitions could be applied given known ancestors. I found they could not and I illustrate this point with my 1979c discussion:

Consider two phylogenetic trees, as shown in Fig. 7.4. These trees consist of known ancestral and descendant evolutionary species. The branch points are speciation events. As earlier, I assume that ancestral species become extinct at each speciation event. This is characteristic of these particular trees and is not meant as a general rule because extinction at the branch point is not a biological necessity (see Chapter 2; and Wiley, 1978; Bell, 1979). In making this assumption, I am simply giving Ashlock the benefit of the doubt because if an ancestral species survives a spe-

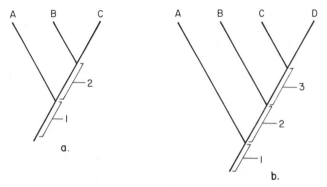

Figure 7.4 (a) The phylogenetic relationships of three descendant and two ancestral species; (b) the phylogenetic relationships of four descendant and three ancestral species. (From Wiley, 1979c.)

ciation event it is possible, by his definition, to place this single evolutionary entity into two taxa of equal rank. Such a classification is illogical (it violates the principles of group membership; see Buck and
Hull, 1966).

I will use Fig. 7.4 to examine Ashlock's (1971) definitions on the
species level. The discussion follows Wiley (1979c). Fig. 7.4a contains
two ancestral species (1, 2) and three descendant species (A, B, C). Using
Ashlock's definitions we might designate a paraphyletic taxon AB and
justify its existence as monophyletic because ancestral species 1 is included within AB. Taxon AB consists of four species (A, B, 1, and 2).
Given this, what is the status of species C? *Species C is polyphyletic by
Ashlock's* (1971) *definition because the ancestor of species C* (ancestral
species 2) *is cladistically a member of a taxon* (AB) *which does not
include C*. I conclude that Ashlock's definitions are unacceptable on the
species level because they permit a single unitary evolutionary species
to be polyphyletic. But, might the definitions work on the supraspecific
level?

Figure 7.4b contains seven species, three ancestors (1, 2, and 3) and
four descendants (A, B, C, and D). I will apply Ashlock's (1971) definitions by recognizing two taxa, AB and CD. Taxon AB is a paraphyletic
group containing species A, B, 1, and 2. It is monophyletic by Ashlock's
definitions because species 1 is a cladistic member of the group. Taxon
CD is a holophyletic group containing species C, D, and 3. It is supposedly monophyletic by Ashlock's definitions because species 3 is a
cladistic member of the group. At first glance all seems well. But when
the status of CD is examined by strict application of Ashlock's definitions,
certain problems arise which Ashlock (1971, 1973, 1974) did not address
(Wiley, 1979c).

Both Ashlock (1971) and Hennig (1966) would agree that the ancestor
of all descendant taxa in CD is classified in CD. But, where in this
arrangement is the ancestor of species 3? The ancestor of species 3 is
species 2. Species 2 is classified in taxon AB. Thus, the ancestor of 3 is
classified in a taxon which does not include 3. Although a taxon including
only C and D might be monophyletic by Ashlock's definitions, a taxon
including 3, C, and D would be polyphyletic by these same definitions.
This may seem a semantic trick to some, but I do not think it is. By any
definition of "cladistic membership," ancestral species 2 is equally related to both B and 3 and, therefore, must be a member of a group
including both species. By what criterion do we place 2 in AB rather
than in taxon BCD? By no criterion that I know other than the ad hoc
desire to call paraphyletic taxa "monophyletic," and thus to connote that
they are natural. The inclusion of 2 in AB must be an arbitrary decision
(or based on an arbitrary rule of thumb) and arbitrary decisions do not
result in taxa that can be defended as natural by their proponents.

Some might see a paradox in this line of reasoning. It may be stated thus: If species 2 is the immediate common ancestor of 3, then why is 2 not placed in CD? And if 2 can be placed in CD, then why not 1 as well and so on down the line until the system collapses under the weight of an infinite regress? To this apparent paradox there is a simple solution: Species 2 is not only the common ancestor of 3, it is also the common ancestor of B. Thus, 2 cannot be placed in CD without B also being placed in CD. This would eventually result in CD being the taxon "Living Organisms." Rather than do this, 2 and B are placed with 3, C and D in the taxon BCD and no paradox exists.

I conclude that if we have ancestral species with which to make decisions, Ashlock's definitions would fail. Therefore, any claim (e.g., Michener, 1978) that such definitions result in natural taxa is unjustified (Wiley, 1979c).

Concepts of Grades and Horizontal Relationships

A **grade** is a taxon characterized by a general level of organization (Huxley, 1958). As developed by Simpson (1953, 1961), grades are composed of independent lineages progressing by parallel (nonhomologous) development through sequences of adaptive zones toward increasingly effective organization levels. One problem of grade is that we must accept the concept of minimum monophyly to justify grades in formal classifications (Simpson, 1961: 128). Another problem is that grades are based on nonhomologous (parallel) characters or, in the case of such grades as Reptilia, on plesiomorphic characters. In practice, grades are not created by evolutionary processes but by taxonomic decisions. Reptilia is a grade because evolutionary taxonomists prefer to exclude Mammalia and Aves. Because nonmammalian and nonavian amniotes must be placed in *some* taxon (they have to be classified), they are relegated to Reptilia.

Simpson (1961) attempts to develop a theoretical basis for grades in terms of the concepts of horizontal and vertical relationships. Vertical relationships, on the most precise level, are phylogenetic relationships. Horizontal relationships are the relationships among "contemporaneous taxa of more or less distant common origin" (Simpson, 1961: 129). Simpson claims that both types of relationship are equally phylogenetic and thus both can be incorporated into phylogenetic classifications. However, horizontal relationships would seem to be either phenetic (similarity between species or species groups) or ecological. If we consider "horizontal relationship" to mean phenetic similarity, then the incorporation of both vertical and horizontal relationships is simply an attempt to combine phenetics and genealogy (the usual evolutionary taxonomic goal). If we consider "horizontal relationship" to mean "ecological relationship," then we would have to attempt to combine ecology and genealogy. Since these two phenomena are based on quite different natural processes, we could not expect a mix of the two to produce a single coherent system. It would be more productive to produce separate ecological and genealogical classifications that would complement and perhaps illuminate each other.

Finally, as Ghiselin (1980) has pointed out, grades are really classes of things, not individuals or (as Wiley, 1980, and Chapter 3 would term them) historical groups. As classes, grades have no place in taxonomies that purport to be phylogenetically natural.

Genetic and Genealogical Relationships

Although Simpson (1961) holds (at least in part) to the concept of genealogical relationship, Mayr (1969, 1974) and other evolutionary taxonomists such as Bock (1974) and Ashlock (1971) hold that the term "relationship" does not necessarily connote genealogical relationship. Rather, it connotes genetic relationship (Nelson's, 1972b, analysis of "Simpsonian" and "Mayrian" relationships is useful here). Mayr has criticized the phylogenetic concept of genealogical relationship as defective (or at least novel) because it does not contain the concept of genetic relationship.

Genetic relationship is an estimate of genetic similarity as evidenced by either phenetic similarity or an actual sample of the genome of two or more organisms. The reason Mayr (1974) thinks the genealogical concept of relationship is incomplete is that he believes that genealogical relations and genetic relations are not necessarily concordant. That is, of two sister species, one may be genetically more similar to a third species than it is to its genealogically closest relative.

For example, Mayr (1974: 102) states:

> But, cladistic kinship alone, for an evolutionist, is a completely one-sided way of documenting a relationship, because it completely ignores the fate of phyletic lines subsequent to splitting.
>
> Let me explain. Since a person receives half of his chromosomes from his father, and his child again receives half of his chromosomes from him, it is correct to say that a person is genetically as closely related to his father as to his child. The percentage of shared genes (genetic relationship), however, becomes quite unpredictable, owing to the vagaries of crossing over and of the random distribution of homologous chromosomes during meiosis, when it comes to collateral relatives (siblings, cousins) and to more distant descendants (grandparents and grandchildren, etc.). Two first cousins (even two brothers for that matter) could have 100 times more genes in common with each other than they share with a third cousin (or brother) (among the loci variable in that population). The more generations are involved, the greater becomes the discrepancy between genealogical kinship and similarity of genotype. . . .

If both the analogy and the assertion were correct I think phylogeneti-cists would have some problems. However, I find neither correct.

The analogy between the genetic relationship between parent, child, and cousin *within* a species and the genetic relationship *between* des-cendant species is not convincing. Mayr equates the vagaries of genetic discorrespondence between genealogically related individual organisms within Mendelian populations with supposed discorrespondence be-tween populations and species. The two are quite different. Within pop-ulations genealogical relationships between individuals may not corre-late with genetic similarity owing to the Mendelian processes of independent assortment, recombination, random mating, and so on. This process is manifested by what may be termed "genealogical panmixia," the inability of an investigator to unravel the genealogical relationships among individuals of the same Mendelian population. But between-pop-ulation genetic relationships are determined by common ancestry, the Hardy-Weinberg equilibrium, and various evolutionary forces that work on that equilibrium. This fact makes population genetics different from Mendelian genetics and cytogenetics. Mayr cannot cite one set of genetic factors to make a point when the factors are not directly applicable to the point at hand. The analogy fails.

In spite of the failure of the analogy, is Mayr (1974) correct in his conclusion? In some cases he might be correct. Consider, for example, a genealogical relationship between species A, B, and C as shown in Fig. 7.5. Let us assume that A and B can and do hybridize but neither A nor B can hybridize with C. It may be that A and B are genetically more similar to each other because of the retention of plesiomorphic genomes. But this does not provide a justification for classifying A and B together any more than some criterion of phenetic similarity.

At the level that Mayr (1974) wishes to justify certain groups I find genealogy and genetics concordant and not discordant as he asserts. Because there is a Pongidae recognized by Mayr (1969, 1974) which includes *Hylobates* (gibbons), *Pongo* (the orang), *Pan*, and *Gorilla* but excludes *Homo*, we should expect that pongids are genetically more

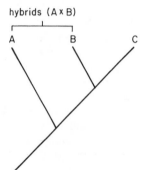

hybrids (A × B)

A   B   C

Figure 7.5 The relationships of three species and a popu-lation of hybrids.

similar to each other than any one of them is to *Homo*. This is a natural prediction and the only one, in Mayr's context, that escapes an overtly typological Pongidae. But the prediction is falsified. Those estimates of genetic similarity made from the genome itself indicate that *Pan* and *Gorilla* are genetically more closely related to *Homo* than to either *Pongo* or *Hylobates* (a sampling of references includes Sarich and Wilson, 1966; Goodman and Moore, 1971; Kohne et al., 1972; Read, 1975; and King and Wilson, 1975). In other words, the genetic evidence is concordant with the postulated genealogical relationships not with the families recognized by Mayr. Further, these works demonstrate that overall phenetic similarity can be a poor estimator of genetic similarity. What, then, are we to think of Pongidae as Mayr (1969) conceives this taxon. It fits neither the criterion of phylogeny balanced against overall genetic similarity (Bock, 1974; Mayr, 1974) nor the criterion of relationship as outlined by Simpson (1961).

As another example, we examine Reptilia. Mayr, a classifier of genotypes, must base at least some of his justification of a natural Reptilia on the supposed (or hypothesized) genetic similarity of lepidosaurs (lizards and snakes), turtles, and crocodiles, and the supposed genetic dissimilarity of crocodiles and birds. Yet, the only estimates of genetic similarity of these taxa show crocodiles and birds to be more genetically similar in terms of actual loci than crocodiles are to other reptiles. Gorman et al. (1971: 177) state: "Every antiserum (of *Alligator*) gave weak reactions in immunodiffusion and microcomplement fixation with bird albumins. . . . Lizard and turtle albumins failed to react with anti-*Alligator* albumin." Similar results were obtained at the LDH locus. Needless to say, albumins and LDH do not a genome make. But the supposed drastic differences between genealogical and genetic data alluded to by Mayr (1974) are not supported by the evidence. Given that there are genetic differences between groups of organisms, the evidence indicates that genetic and phenotypic data are both concordant with hypothesized genealogical descent (another demonstration of this is given by Mickevich and Johnson, 1976, for fishes of the genus *Menedia*). This is no great surprise to me since genealogy is, after all, the special relationship that passes similarities and differences between generations and between ancestral species and their descendants.

Even if overall genetic similarity were discordant with cladistic relationships, it is still not a valid criterion for grouping taxa. The claim that overall genetic similarity results in natural groups is closely akin to the assertion by pheneticists that overall phenetic similarity produces natural groupings. As I have discussed in Chapter 3, set theory and concepts of overall similarity as used to define sets are not compatible. Thus, overall similarity, whether it is phenetic similarity or an estimate of genetic similarity, is invalid as a grouping criterion for natural, non-overlapping sets of organisms.

Classifications and Information Content

We may now examine Mayr's criterion of classification and the information content of evolutionary taxonomic classifications. Mayr (1969) suggests that the major criterion of classification is to combine maximal information content with maximal ease of information retrieval. He states (1969: 78):

> The evolutionist believes that a classification consistent with our reconstruction of phylogeny has a better chance of meeting these objections than any other method of classification. Taxa delimited in such a way as to coincide with phylogenetic groups (lineages) are apt not only to share the greatest number of joint attributes but at the same time to have an explanatory base for their existence.

There is much to agree with here. I agree that phylogenetic groups (lineages, clades) have an explanatory bases for their existence. But, I do not agree that such clades should be expected to share the greatest number of joint attributes. *Archaeopteryx* shares more features in common with certain lower archosaurs than with other Aves. Crocodiles probably share the greatest number of joint attributes with other reptiles than with their Recent sister group, Aves. *Pan* and *Gorilla* may share a larger number of joint attributes with *Pongo* than with *Homo* (although this has not been demonstrated, to my knowledge). The statement that classification should be consistent with phylogeny and that such a classification should coincide with lineages (clades) is simply inconsistent with his assertions concerning the number of joint attributes supposedly shared among members of such natural groups. However, members of clades do share the characters of *evolutionary interest*, the novelties we term apomorphies; that is what makes them informative to the study of modification during descent. On the one hand Mayr recommends what are essentially cladistic classifications, on the other hand he rejects the basis for such cladistic classifications: the concept that monophyletic groups are natural groups defined by synapomorphies and which include all of the descendants of the ancestor of the group. Mayr (1974) criticizes Hennig for opting for a strict definition of monophyly. It has, says Mayr, caused Hennig to transfer the qualification "monophyletic" from a taxon to that taxon's mode of descent. Setting aside the fact that Mayr's connotation transfers the qualification "monophyletic" from taxon to some vague and unsupported assertion that genetic similarity is inconsistent with genealogical descent, we will examine a test case to see if the Archosauria (including Aves) is a useless assemblage (Mayr, 1974: 104) or whether it might contain some useful information, and the alternate Reptilia contains misinformation. The analysis presented is similar to Farris's (1977) analysis of Pongidae (including *Pan*) and Hominidae.

In Fig. 7.6 we have four classifications of the major terminal taxa of amniote vertebrates. Two are expressed by Linnaean hierarchies, two are expressed as dendrograms. I stress that neither dendrogram is necessarily a phylogenetic tree. Rather, both are classifications of the terminal taxa. We now examine the implied information content of the two evolutionary taxonomic classifications.

1.   If the criterion for grouping is morphological distinctiveness, then both evolutionary classifications imply that (a) turtles, lepidosaurs, and crocodiles are equally distinct from each other, (b) that crocodiles are as distinct from birds as they are from mammals, and (c) that crocodiles are as distinct from birds as are turtles and lepidosaurs. Implication (a) is true, the three groups of reptiles are distinctive. That is, each may be adequately diagnosed. Implications (b) and (c) are false as evidenced empirically by the suite of morphological characters shared by birds and crocodiles. Thus, the evolutionary classification is misleading in terms of two of the three implications expressed by both classifications.

2.   If the criterion for grouping is genetic similarity, then the evolutionary classification imply that (a) crocodiles, lepidosaurs, and turtles are equally distinct genetically, (b) that crocodiles are genetically more similar to lepidosaurs and turtles than to birds, and (c) that birds and mammals are equally distinct, in terms of genome, from all reptiles. Implication (a) has not been empirically examined. Implications (b) and (c) are empirically false given the little information we have (see previous discussion). Thus, the evolutionary classification is misleading in at least two of the three implications.

3.   If the evolutionary classifications are meant to portray distinctiveness of adaptive zones then they imply that (a) reptiles occupy a unique adaptive zone equally different from both birds and mammals, and (b) crocodiles are more similar in their ecologies to turtles and equally different from birds and mammals. These implications may or may not be borne out by empirical evidence; there are simply no good data one way or the other.   .

In summary, the evolutionary classification is misleading with respect to both phenotype and genotype and untested regarding ecology. In contrast, the two phylogenetic classifications are informative with respect to both genotype and phenotype. It would also seem to be informative even in regard to behavior because crocodiles and birds, among all Recent sauropsids, build nests, provide parental care for their young, and establish and defend territories by singing songs (although the songs of alligators are certainly less pleasing to us than the songs of meadowlarks). Finally, if the phylogenetic classification is considered to reflect a phylogenetic tree, we see that it indexes and documents the history of speciation for the terminal taxa shown. This information cannot be

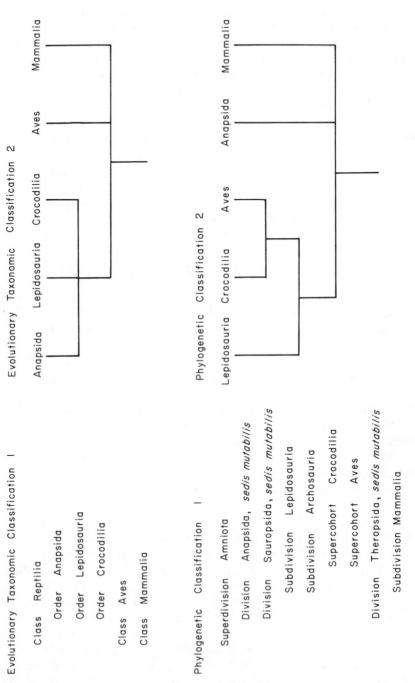

Figure 7.6 Four classifications of amniote vertebrates. See text for discussion.

retrieved from the evolutionary classification (Hull, 1964; Hennig, 1966; Cracraft, 1974a). Further, evolutionary taxonomists have claimed that information about the degree of divergence can be incorporated into their classifications, presumably in such a way that it can be retrieved. (Brothers, 1975; Michener, 1977). This claim is misplaced because Farris (1980) has shown that a strictly phylogenetic classificatha is capable of retrieving such information whereas an evolutionary classification is not. It would seem that the phylogenetic classification is far from being a useless assemblage of names. It would also seem that Mayr's criterion that classifications should be consistent with phylogeny is satisfied with a phylogenetic rather than an evolutionary classification.

But can we nevertheless justify retaining Reptilia as a grade group in spite of the inconsistency of the classification with the (hypothesized) known morphological, genetic and phylogenetic relationships? Mayr (1969, 1974) suggests that we can because reptiles form a "well integrated adaptive complex." In asserting this, Mayr must fall back on a concept of monophyly sensu Simpson (1961). Reptilia is monophyletic because it is descended from Amphibia. But, Reptilia is not retained because of any intrinsic merits of the group. Rather, it is retained as a contrast with its two descendant groups, Aves and Mammalia. To quote Mayr (1974: 112–113):

> A rigid application of their dogma forces cladists [phylogeneticists] to break up the reptilian grade into many separate "classes" and to designate particular reptilian lineages as the "sister groups" of birds and mammals. The fact that no one would place the crocodiles outside the reptiles if birds did not exist reveals how artificial and arbitrary this procedure is. The essential unity of the reptiles is best illustrated by the continuing argument among paleontologists as to which orders of reptiles are most closely related to which others.

I respond with the following points:

1. The biological reality of grade groups has just been refuted herein, they are classes, not natural entities.

2. Of course, phylogeneticists wish to break up grades; that is the business of cladists, but it is also the business of classifiers in general. It was not the cladists who broke up the Vermes (a perfectly good "monophyletic" group sensu Ashlock), but taxonomists (beginning with Lamarck) who realized that this group was a grade. It was Simpson (1960) who rigorously defended keeping *Hyracotherium* ("*Eohippus*") in the Equidae rather than return it to the grade Condalartha (see Nelson, 1972b). I submit that the general history of taxonomic revision has been directed toward breaking grades into clades (the pheneticists excluded) from Vermes to Reptilia.

3.   The fact is, if Aves and Mammalia did not exist, the Reptilia would be a clade and a natural group (albeit not named because there would be no taxonomists around to name it). If mammals existed and Aves did not exist, we would still have three Recent lineages, Anapsida, Theropsida, and Sauropsida, and crocodiles would still be sauropsid archosaurs. Even if mammals did not exist, we would have a clade composed of the theropsid pelecosaurs and therapsids (mammallike reptiles). The point of view expressed by Mayr goes back, once again, to the general contention of evolutionary taxonomists that higher taxa are inventions of the taxonomist's mind and have no reality in nature. I reject this line of reasoning because it denies there is a real history to recover in a logical form.

4.   Mayr seems to be saying that the reality of a grade should depend on ignorance about the interrelationships among the members of the grade. It almost sounds as if Mayr contends that the *less* we understand about a taxon the more we can trust its "essential unity."

Evolutionary Taxonomy—Summary Remarks

The evolutionary taxonomic approach is not a viable alternate system of classification. Evolutionary taxonomic classifications seem at variance with what we know about evolutionary process. In those instances in which such evolutionary classifications differ from phylogenetic classifications, the resulting groups do not reflect the hypothesized process and direction of descent with modification (such as we can understand the processes involved and the pattern of descent). Thus, evolutionary taxonomic classifications cannot be used as a basis for criticizing hypothesized mechanisms that produced the pattern of descent observed. The underlying reason for this noncorrespondence, I suspect, is that evolutionary taxonomists either (1) postulate that higher taxa originate from supraspecific taxa of equal or lower rank (Simpson, 1961), (2) view higher taxa as somewhat artificial entities (convenience classes) rather than real historical entities (Mayr, 1963: 600), or (3) view lower taxa as developing into higher taxa sometime after the origin of the stem species of those higher taxa (Simpson, 1953; Mayr, loc. cit.).[3] Also, in those evolutionary taxonomic systems that allow only species-level ancestors, every attempt to recognize a paraphyletic group as having biological reality results in the removal of an ancestral species from the group containing at least one of its descendants. This, by formal definition of the terms employed by Ashlock (1971), renders holophyletic groups polyphyletic. No evolutionary taxonomist has demonstrated that ge-

---

[3] In all fairness I should point out that some who align themselves with Hennig (1966) have asserted in the past that higher taxa originate *after* the ancestral species (see Platnick, 1976b).

nealogy and genetic similarity are discordant. Therefore, grade taxa can not be justified on the basis of supposed and undemonstrated genetic similarity. Lastly, and a point that I discussed in Chapter 3, neither estimates of overall genetic similarity nor estimates of overall phenetic similarity can be used to justify a taxon as natural because overall similarity is not a valid set characteristic of natural organismic groups.

PHENETICS

Phenetics has been defined as any manipulation of phenotypic data that results in a classification (Sneath and Sokal, 1973). However, most workers consider phenetics to comprise an array of techniques that group individuals into taxa on the basis of an estimate of overall similarity. This is done without initial weighting of characters. The relationships expressed are similarity relationships with no necessary connotation of phylogenetic relationship (although they may be so interpreted). The result is a phenogram, usually associated with a scale of phenetic similarity and/or phenetic distance. The results are rarely translated into a hierarchy of names. The major recent synthesis of phenetics is Sneath and Sokal (1973).

The major conflict between phenetics and phylogenetic systematics does not stem from the fact that pheneticists use mathematical algorithms. As we saw in Chapter 5, there are a number of such algorithms especially designed for phylogenetic analysis. And we shall see in Chapter 10 that various multivariate analyses can be used profitably by all systematists. The conflict between the two approaches stems from the assertion by pheneticists that grouping by overall similarity results in stable and natural classifications whereas phylogenetic classifications are inherently unstable and unnatural. Naturalness in this sense is **Gilmour naturalness**, a concept based on character predictions that can be made from a classification (Gilmour, 1961). The purpose of the following sections is to examine these claims.

Stability, Phenetics, and Phylogenetics

Early pheneticists (Sneath, 1961; Sokal and Sneath, 1963; Michener, 1963) hoped that phenetic methods would result in stable classifications containing or implying a maximum amount of information about the classified groups. It soon became apparent that phenetic dendrograms were anything but stable except in the trivial case where the same data matrix was analyzed by the same algorithm. When a single data set was analyzed by two different algorithms or when two data sets taken from the same specimens were analyzed by the same algorithm, the results obtained were invariably different (see Minkoff, 1965; also see Farris,

1971; Sneath and Sokal, 1973). This should have been disconcerting for pheneticists because if the methods employed were all estimates of a natural system, then the results obtained should be concordant within the limits imposed by the characters themselves. Alternately, we should expect that some reason would be advanced to explain why one data matrix or algorithm was not as good an estimator of the natural system as another matrix or algorithm. Sneath and Sokal (1973: 25) addressed this problem by asserting that one important characteristic of a natural classification is its stability in the face of new information. I fully agree, but we shall see that such stability is a characteristic of phylogenetic classification, not of phenetic classification.

Mickevich (1978) investigated the comparative stabilities of several phenetic and phylogenetic algorithms. In an earlier paper Michevich and Johnson (1976) had found that the results of separate phenetic clustering of allozyme and morphological data for several populations of the silverside genus *Menedia* produced discordant phenograms but that a Wagner parsimony algorithm produced largely the same results from both data sets. Mickevich and Johnson suggested that phylogenetic methods produced stable classifications whereas phenetic methods produced unstable classifications. Mickevich (1978) extended this comparison to several real data sets and eight different clustering algorithms. The data sets included morphological and alloenzyme data for different biochemical data sets for various groups. The clustering routines included three purely phenetic algorithms, two phenetic algorithms using overall similarity to infer phylogenetic relationships, and three algorithms specifically designed for inferring phylogenetic relationships. For each set of taxa, classifications were produced separately from each data set. The results were then compared by generating a "consensus classification" derived from that information common to both original classifications (Adams, 1972). Because widely different classifications will have little in common, the amount of common information provided a measure of the similarity between the two originals. The amount of distortion was then measured based on Farris's (1973) model as modified to give a measure of "information content." "Information content" was related to the complexity of the hypothesis with a phylogenetic "bush" (multifurcation) having no information content and with a completely resolved dichotomous classification having maximum information content. Mickevich's conclusions are summarized as follows:

1. In terms of stability, there is little difference between pure phenetic methods and phenetic methods which draw phylogenetic conclusions based on measures of overall similarity. However, all three phylogenetic methods were significantly more stable than the phenetic methods. Wagner tree classifications were the most stable of the three phylogenetic methods.

2.   Adding information in the form of extra characters did not increase the stability of phenetic or phylogenetic classifications. The supposition on the part of pheneticists that larger matrices of characters would lead to more stable classifications (Sneath, 1961; review by Sneath and Sokal, 1973) was falsified.

3.   The affects of nonhomologous characters (homoplasies) were more pronounced on phenetic classifications than on any of the phylogenetic classifications. This, in turn, affected the stability of the phenetic classifications. The result is intuitively obvious. Coding two nonhomologous features as the same character state results in two relatively unrelated taxa appearing more similar to the algorithm. Increasing homoplasy results in increasing overall similarity. Because it is unlikely that two totally different data sets (such as cytochrome c and hemoglobin sequences) will show the same amount of homoplasy distributed in the same way among the same taxa, the result is a lack of congruence between classifications based respectively on the two data sets. The fact that increasing the number of characters does not seem to help the situation tends to leave phenetic algorithms at the mercy of these similar but nonhomologous characteristics.

4.   Mickevich concluded that the reason the phylogenetic algorithms, and especially the Wagner algorithm, showed congruence of classification between data sets was because they separated the observed characters into plesiomorphic, apomorphic, and nonhomologous states and employed them at their appropriate levels of universality. Put simply, all of the phylogenetic methods approximated phylogenetic analysis sensu Hennig (1966) and the Wagner method was the closest approximation of the three.[4]

Information Content, Phenetics, and Phylogenetics

It is not clear what is meant by "information content" in a classification (Farris, 1977). I have previously discussed this problem in relation to evolutionary classifications and concluded that classifications are an index which may lead or mislead the user in terms of group characteristics of evolutionary interest (i.e., evolutionary novelties). Workers such as Jardine and Sibson (1971) and Sneath and Sokal (1973) imply that phenetic classifications contain information about overall similarity. Farris (1977) and others before him have made the point that *no* classification can express anything about those entities classified except information about group membership. Only when the groups are associated with a diagnosis and/or description can we gain information about the actual features of the organisms classified. Thus, *neither phenetic nor phylo-*

---

[4] This was impressive in that the Wagner method employed by Michevich (1978) was a rootless Wagner method without benefit of out-group comparisons.

genetic classifications convey character information in and of themselves. But given that either can be used as a reference system, Farris (1977) asked which would meet the goal set out by the pheneticists themselves, the goal of naturalness sensu Gilmour (1961).

Gilmour Naturalness and Phenetics

Gilmour (1961) suggested that a natural classification would contain taxa that would explain the distributions of as many characters as possible. Such a taxon would be distinguished by tightly correlated characters largely or exclusively confined to it and no other taxon (Sneath, 1961; Farris, 1977). Sneath and Sokal (1973) suggested that phenetic classifications best approximated Gilmour-natural groupings. They further suggested that various phenetic algorithms could be compared for their relative worth in estimating Gilmour-natural groupings through the use of the **cophenetic correlation coefficient**. The cophenetic correlation coefficient measures the correspondence of a classification to the actual overall similarities among taxa as calculated from the original character data matrix. Implied similarities for groups of terminal taxa are measured by the level of similarity at which they are linked. Put simply, the coefficient checks the implied similarities evidenced by the classification against the actual overall similarities exhibited by the organisms. The higher the cophenetic correlation coefficient, the closer the classification supposedly reflects the original data.

Farris (1977) suggested that if it were true that the cophenetic correlation yielded higher values as a classification was more natural sensu Gilmour, then this would provide an evaluation between phylogenetic and phenetic classifications as well as simply between various phenetic classifications.[5] Farris used a number of hypothetical and real data sets to explore this suggestion. His results are summarized here (they are also summarized in Farris, 1978b):

1. It is not generally possible for every feature of every organism to contribute to distinguishing a natural taxon sensu Gilmour. Thus there must be features which are either uninformative or perhaps misleading in making decisions about uniting organisms into natural taxa. Conversely, there must also be features which are more informative.

2. Since there are features which may be either uninformative or misleading, it is possible that measures of overall similarity may not lead to Gilmour-natural taxa. Farris (1977) designed a data matrix which would yield a cophenetic correlation coefficient = 1 when analyzed by

---

[5] Farris (1977) did not suggest that the cophenetic correlation coefficient was an actual measure of Gilmour-naturalness. He simply took the "charitable view" that such was the case (Farris, 1978b: 235).

the UMPGA phenetic clustering algorithm. This demonstrated that a phenetic algorithm could produce a Gilmour-natural classification for that data set. Thus, measures of overall similarity may, in certain circumstances, yield natural classifications. This matrix is shown in Table 7.1. Note that if each character state coded as 1 is taken as a synapomorphy, then a phylogenetic solution is immediately apparent. In this case, that phylogenetic solution is identical with the phenetic solution shown in Fig. 7.7a. A second data matrix was generated from the first by duplicating or repeating some characters three or five times (Table 7.2). This has no effect on the phylogenetic analysis except to make some groups better justified. For example, the taxon A + B is now justified by three synapomorphies rather than by one synapomorphy (Fig. 7.7b, characters 9.1–9.3). And taxon A is more distinctive in having five autopomorphies (1.1–1.5) rather than one. Yet, when this data matrix is analyzed by the phenetic method it results in an unresolved polytomy and has a cophenetic correlation coefficient = 0 (Fig. 7.7c). Thus, Farris (1977) concluded that clustering by overall similarity *might* or *might not* result in a Gilmour-natural classification as evidenced by the cophenetic correlation coefficient.

   3.   If phenetic methods may or may not produce natural classifications, will phylogenetic methods do any better? Farris investigated this question using a *coefficient of special similarity* developed by Farris,

Table 7.1   A hypothetical data matrix adapted from Farris (1977)

	Taxa							
Variables	A	B	C	D	E	F	G	H
1	1	0	0	0	0	0	0	0
2	0	1	0	0	0	0	0	0
3	0	0	1	0	0	0	0	0
4	0	0	0	1	0	0	0	0
5	0	0	0	0	1	0	0	0
6	0	0	0	0	0	1	0	0
7	0	0	0	0	0	0	1	0
8	0	0	0	0	0	0	0	1
9	1	1	0	0	0	0	0	0
10	0	0	1	1	0	0	0	0
11	0	0	0	0	1	1	0	0
12	0	0	0	0	0	0	1	1
13	1	1	1	1	0	0	0	0
14	0	0	0	0	1	1	1	1

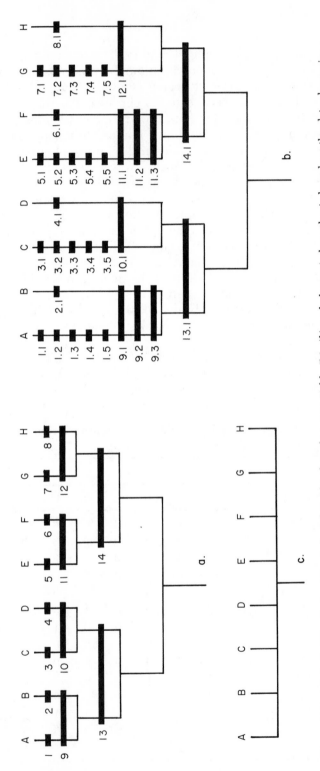

Figure 7.7 (a) A phylogentic hypothesis based on the data shown in Table 7.1; (b) a phylogentic hypothesis based on the data shown in Table 7.2; (c) a phenetic hypothesis based on the data shown in Table 7.2. See text for discussion. (From Farris, 1977.)

Table 7.2 The same data matrix as Table 7.1 with
certain replicated characters. Data from Farris (1977)

				Taxa				
Variables	A	B	C	D	E	F	G	H
1.1	1	0	0	0	0	0	0	0
1.2	1	0	0	0	0	0	0	0
1.3	1	0	0	0	0	0	0	0
1.4	1	0	0	0	0	0	0	0
1.5	1	0	0	0	0	0	0	0
2.1	0	1	0	0	0	0	0	0
3.1	0	0	1	0	0	0	0	0
3.2	0	0	1	0	0	0	0	0
3.3	0	0	1	0	0	0	0	0
3.4	0	0	1	0	0	0	0	0
3.5	0	0	1	0	0	0	0	0
4.1	0	0	0	1	0	0	0	0
5.1	0	0	0	0	1	0	0	0
5.2	0	0	0	0	1	0	0	0
5.3	0	0	0	0	1	0	0	0
5.4	0	0	0	0	1	0	0	0
5.5	0	0	0	0	1	0	0	0
6.1	0	0	0	0	0	1	0	0
7.1	0	0	0	0	0	0	1	0
7.2	0	0	0	0	0	0	1	0
7.3	0	0	0	0	0	0	1	0
7.4	0	0	0	0	0	0	1	0
7.5	0	0	0	0	0	0	1	0
8.1	0	0	0	0	0	0	0	1
9.1	1	1	0	0	0	0	0	0
9.2	1	1	0	0	0	0	0	0
9.3	1	1	0	0	0	0	0	0
10.1	0	0	1	1	0	0	0	0
11.1	0	0	0	0	1	1	0	0
11.2	0	0	0	0	1	1	0	0
11.3	0	0	0	0	1	1	0	0
12.1	0	0	0	0	0	0	1	1
13.1	1	1	1	1	0	0	0	0
14.1	0	0	0	0	1	1	1	1

Kluge, and Eckardt (1970). This coefficient effectively eliminates those shared character states which are uninformative (i.e., it considers only those character states which may effectively be considered apomorphic). Using this coefficient of special similarity and the same basic clustering routine (UPGMA), Farris compared the performance of it compared to a coefficient of overall similarity against 50 real data sets. The average cophenetic correlation coefficient using special similarity was .89, whereas that using overall similarity was .63. Thus, clustering by phylogenetic methods yielded a higher cophenetic correlation coefficient and if this is equated with higher Gilmour-naturalness, it yielded more natural classifications.

   4.   Farris (1977) concluded that phenetic classifications were not optimal under the criteria provided by either the cophenetic correlation coefficient or Gilmour-naturalness. Further, *phylogenetic classifications better describe the overall similarity between organisms than do phenetic classifications* (Farris, 1980).

Phenetics—Summary Remarks

It would seem that the results obtained by Mickevich (1978) and Farris (1977) refute the claims of stability and naturalness made for phenetic methods and their resulting classifications. In particular, Farris's (1977) deductions concerning overall vs. special similarity closely parallel the objection I expressed in Chapter 3 concerning naturalness and overall similarity. This is no doubt due to our similar outlook and probably implies that Gilmour-naturalness is the equivalent of what I term naturalness at the level of character analysis. This is intuitively reasonable because if phylogenetic methods are good estimates of descent with modification we would expect that they would be most informative concerning the morphological course of evolution as well as the course of speciation.

Chapter 8

Biogeography

Biogeography is the study of the distribution of organisms in space through time. The biogeography of plants is usually termed **phytogeography** or plant geography. The biogeography of animals is usually termed **zoogeography**. Unfortunately there has been little interchange between botanists and zoologists and the primary literature is thus largely divided into these two fields. I view biogeography as a holistic science and feel that biogeographic studies should combine both plant and animal information. Thus biogeography as a whole will be stressed.

Biogeography is currently undergoing scrutiny regarding its aims and methods. Among the approaches which may be termed "interpretive biogeography" (Cain, 1944), we shall see that ecological and historical biogeography constitute complementary and largely nonoverlapping fields of inquiry. A lack of appreciation for the differences and limits of these subdisciplines has led to some confusion and I shall attempt to discuss their differences. Historical biogeography is of direct interest to systematists and I shall concentrate on this aspect of biogeography throughout the chapter. There are various approaches to the study of historical biogeography. This is obvious when one samples such divergent discussions as Simpson (1947, 1965), Cain (1944), Camp (1947), Darlington (1957, 1970), Good (1964), Croizat (1952, 1964), Brundin (1966, 1972), Raven and Axelrod (1974), Vuilleumeir (1975), Keast (1977), Rosen (1978, 1979), and Platnick and Nelson (1978). There can be no doubt that there is a lack of consensus on what constitutes historical biogeography much less a consensus of the methods employed in studying the subject. The major sections of this chapter explore these divergent views and address the question of a general method and strategy for analyzing distribution patterns using a vicariance model of analysis based on Croizat's (1964) track analysis and Hennig's (1966) phylogenetic analysis. Before addressing this subject, I will first discuss the major divisions of biogeography and some basic terminology.

DESCRIPTIVE AND INTERPRETIVE BIOGEOGRAPHY

There is a great difference between illustrating or describing the geographic distribution of a taxon and attempting to explain how that distribution has come about. It is the difference between collecting and analyzing data. To draw an analogy to phylogenetic systematics, it is the difference between describing a species and showing the relationship of that species to other species. Following Cain (1944), I shall term these two activities "descriptive" and "interpretive" biogeography, respectively. Interpretive biogeography includes what some have termed "analytic," "historical," "ecological," and "predictive" biogeography (see Vuilleumeir, 1975; Keast, 1977).

Descriptive Biogeography

Descriptive biogeography includes the documentation of the ranges of taxa, the summarization of taxic compositions for various geographic regions and the formulation of phytogeographic and zoogeographic regions. The documentation of ranges provides baseline data for interpretive biogeography.

The practice of delimiting "realms" for plants and animals also constitutes a part of descriptive biogeography. This practice began with A. P. de Candolle's (1820) division of the world into 20 geographic regions. The regions used by phytogeographers today number some 35 (Good, 1964). Zoogeographers have recognized fewer, having settled essentially on Sclater's (1858) bird regions (Wallace, 1876; Darlington, 1957).

Interpretive Biogeography

Interpretive biogeography involves attempts to synthesize the base line data of descriptive biogeography into narrative explanations, hypotheses, predictions based on hypotheses, or a combination of the three. Within the broader area of interpretive biogeography I prefer to distinguish between historical and ecological biogeography.

Historical biogeography. Historical biogeography is the study of the spatial and temporal distributions of organisms (usually on the taxic level) which attempts to provide explanations for these distributions based on past historical events.

Ecological biogeography. Ecological biogeography is the study of the dispersion of organisms (usually on the individual and deme levels) and the mechanisms which maintain or change this dispersion. By dispersion I mean the pattern of demes and individuals within demes that exists at any particular time span for a species. In terms of phytogeography, ecological biogeography is concerned with the mechanisms of dispersal of **diaspores** (Cain, 1944: 255), which are the units of individual dispersal.

The separation of these two fields was recognized by de Candolle (1820; see Nelson, 1978b). Unfortunately, biogeographers have often mixed these two fields with resulting confusion. This is not to say that these two areas of inquiry are mutually exclusive. Many historical patterns have, ultimately, ecological bases, and dispersion has a historical as well as an ecological component. However, the goals and interests of the ecological biogeographer lie more with ecology than systematics whereas the opposite is true of the historical biogeographer. The former is interested in the dynamics of biotas as ecological units, the latter is interested in the origins and relationships between biotas.

I shall not dwell on the goals and methods of ecological biogeography because I do not feel that systematics can contribute directly to the field. For those interested in this field I recommend Good (1964), MacArthur and Wilson (1967), MacArthur (1972), Udvardy (1969), and reviews by Vuilleumeir (1975) and Keast (1977).

SOME BIOGEOGRAPHIC TERMS AND CONCEPTS

Any particular species is characterized by the general area of the earth where it occurs and the particular places within that area where its individuals actually live. These two parameters are usually termed "range" and "locality." These terms are defined in reference to Fig. 8.1.

1. **Range.** The range of a species is that general biogeographic area where its demes naturally occur.
2. **Locality.** A locality is a particular geographic place where a deme of a species actually lives.

The range of a species is the minimum area which encompasses all of the localities of its deme. No species is found in every part of its range.

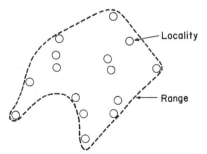

Figure 8.1 The range of a species and the localities from which it has been collected.

It is found only where certain suitable habitat exists that will support one of its demes. Further, habitat and locality are not synonomous because not every species has a deme in every suitable habitat throughout its range. The absence of a species at a suitable habitat may be due to historical factors, geographic barriers, or the fact that chance dispersal has missed the habitat (Udvardy, 1969). Further, the concept of range is complicated by such phenomena as regular migration. For purposes of discussion, I include the entire migratory route of such species within its range. A species may come to occupy its range in one of several ways:

1.   It may have expanded its range from a smaller area to a larger area.
2.   It may have contracted its range from a larger area to one or more smaller areas.
3.   It may occupy an original subportion of the range of its ancestor without major expansion or contraction.

For purposes of analytic biogeography, two additional terms are introduced:

1.   **Track.**   A track is a line drawn on a map that circumscribes the total range of a monophyletic taxon. In the case of a species, the track is a graphic representation of its range. For a supraspecific taxon, the track circumscribes all the ranges of its component species. A track may circumscribe taxa that are either contiguous or disjunct (Fig. 8.2). The

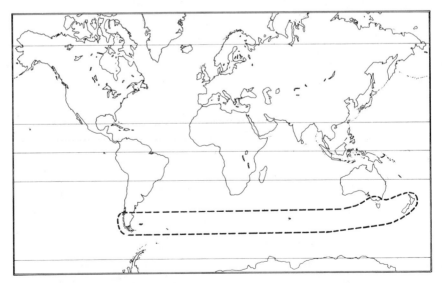

Figure 8.2   The distribution of chironomid midges of the subfamily Podonominae. (Data from Brundin, 1966.)

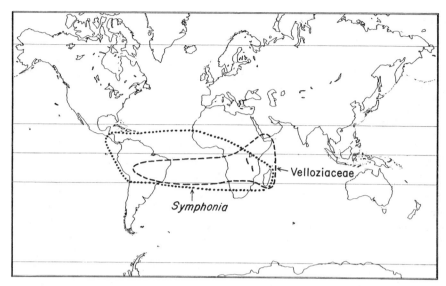

Figure 8.3 Transatlantic tracks of the plant family Velloziaceae and the plant genus *Symphonia*.

term, so far as I am aware, was coined by Camp (1947). We shall see that a track may estimate the distribution of an ancestral species.

2. Generalized track. A generalized track is the combination of two or more *congruent* tracks that respectively circumscribe two or more different monophyletic taxa (Rosen 1975). The concept is illustrated, using two plant taxa, in Fig. 8.3.

We shall see that track analysis and generalized tracks are the heart of the panbiogeographic method.

Biotas and Areas of Endemism

Biogeographers use the terms fauna (in zoogeography) and flora (in phytogeography) to denote the array of organismic diversity found in a particular geographic region. A combination of the fauna and flora may be termed the **biota** of a region. Although we may speak of the biota of some arbitrary geographic area (such as a county or state) the term has more meaning if we restrict it (for biogeographic purposes) to what may be termed a "natural geographic area," defined in terms of its unique biotic components (endemic species) or by its unique physical attributes. For example, in speaking of the biota of Kansas we must realize that Kansas is a political subdivision that does not constitute a single naturally defined assemblage of organisms; whereas in speaking of the Great Plains we can see that it is a biogeographical unit which may be defined by a natural assemblage of plants and animals (not necessarily endemic) that operates as an ecological whole. In turn, the Great Plains may be sub-divided into several smaller units.

Biotas, like organisms, contain a blend of characteristics. Some taxa are shared with other biotas, others are endemic only to the biota in question. Further, these endemics rarely (if ever) occupy the entire area of the biota, being found only in parts of it. Such areas are termed **areas of endemism** and are the basic working unit of track analysis. An array of areas in which one or more species of a monophyletic group is found may be termed a **set of endemic areas**. When more than one monophyletic group is considered, such areas of endemism rarely correspond exactly owing to chance and differing ecological requirements. In some cases, such as isolated valleys or lakes, endemic areas between two monophyletic groups may be exactly congruent. If they are not exactly congruent but are largely congruent, then they may be considered equivalent for purposes of biogeographic analysis. Figure 8.4 illustrates this point with four areas of endemism and two monophyletic groups found in two biotas.

Dispersal

The word "dispersal" has two different biogeographic connotations (Udvardy, 1969). First, it may be applied to describe the movement of individual organisms within the normal range of their species, that is, from deme to deme. This includes migration if the species is normally migratory. Second, it may be used to describe the movement of individ-

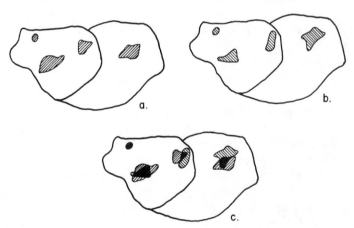

Figure 8.4 The phenomena of complete and incomplete range congruence among two groups inhabiting the same range: (a) the distribution of genus A in two adjacent geographic regions; (b) the distribution of genus B in the same regions; (c) the two distributions superimposed on the same map, with distributional congruence shown by the black areas. Complete congruence is achieved only in one area, the other three show incomplete congruence.

uals or populations beyond the species' range into areas previously unoccupied by that species. Platnick (1976c) suggested that the term "dispersion" be used for movements within the species' border. I prefer to
distinguish these phenomena by the use of adjectives applied to dispersal because dispersion has other connotations.

1. **Organismic dispersal.** The movement of individuals from deme
to deme or to previously unoccupied habitat *within* the range of the
species. Organismic dispersal is a major field of inquiry for ecological
biogeographers. As pointed out in Chapter 3, this type of dispersal is
one major factor in promoting species cohesion in sexually reproducing
organisms. Thus, it affects mode of speciation and, indirectly, historical
biogeography.
2. **Species dispersal.** The movement of individual organisms from
a habitat within a species' present range to one or more habitats outside
that species' present range. There are two types of species dispersal:
(a) Dispersal into geographic areas never previously occupied by the
species. Examples of this phenomenon are the species dispersal of the
cattle egret and house sparrow over North America since their intro-
ductions in the 19th Century and the dispersal of the coyote more recently. (b) Species may also disperse into geographic areas occupied at
some earlier time by the species or its ancestor but unavailable for oc-
cupation in the immediate past because of a change in environment.
Many examples of this phenomena may be drawn from those species
which have reoccupied areas previously covered by glaciers at various
times in the Pleistocene of Europe and North America.
3. **Biotic dispersal.** The spreading of an entire biota to occupy geographic areas previously occupied by other biotas. In many cases species
dispersal is only a single component of the larger phenomena of biota
dispersal (as in the Pleistocene example). It may involve both 2a and 2b
types of species dispersal. Species dispersal is always part-and-parcel
of biotic dispersal, but may also occur independently of biotic dispersal.

Centers of Origin

The classic view of centers of origin is that new species originate only
in certain areas of the earth and that all other areas of the earth are the
recipients of species originating within these centers. In other words,
centers of origin are "species factories" which export their species to
other parts of the world. Wallace's (1876: 159) statement below was used
by Croizat et al. (1974) to illustrate this point:

> The north and south division [of the earth] truly represents the fact, that
> the great northern continents are the seat and birthplace of all the higher
> forms of life, while the southern continents have derived the greater part,
> if not the whole, of their vertebrate fauna from the north.

As this idea caught on, the claims of such workers as Hooker (1853, 1860) that some of the southern floras, and of Huxley (1868) and Hutton (1873) that some of the southern faunas were the remnants of an ancestral panaustral flora were set aside in favor of the idea that although these floras and faunas were certainly similar, they actually originated in the northern hemisphere and dispersed southward.

Another, more recent, view does not restrict centers of origin to any particular land mass such as one or more northern continents. Rather, several areas of the earth have acted as centers of origin depending on the group in question (Darlington, 1957; Briggs, 1974). This point of view is associated with what some have termed "classical" or "evolutionary" biogeography under certain assumptions and "phylogenetic" biogeography under others. Each is best understood within the context of these approaches to historical biogeography and will be discussed in the following section.

APPROACHES TO HISTORICAL BIOGEOGRAPHY

As I alluded to in the introduction to this chapter, there are several different approaches to the study of historical biogeography. These may be defined by their basic tenets. At the risk of oversimplifying, I shall outline and discuss three of these approaches. I do not suppose, nor should the reader, that every historical biogeographer can be unambiguously identified with one of these approaches. My purpose is not to classify biogeographers. Many biogeographic papers that deal with historical issues contain a blend of attitudes concerning the means by which taxa have come to occupy the areas they now occupy. Thus one should not suppose that vicariance biogeographers believe that all present disjunctions between members of a taxon must be the result of vicariance or that all evolutionary biogeographers believe that all present disjunctions are the result of long distance dispersal from restricted centers of origin. Some workers, such as Wulff (1950) and Cain (1944), should not be identified with any approach, although Wulff (1950) was a forerunner of what we now know as vicariance biogeography and Cain (1944) stands at the intellectual junction between vicariance and phylogenetic biogeography. Some, such as Simpson (1947) and Darlington (1957), are firmly associated with evolutionary biogeography principally because of their rejection of the idea that continental drift could have influenced the distribution of the organisms they studied. (Simpson, as judged from his autobiography, no longer holds this view). The acceptance of continental drift has made most recent workers postulate that many disjunctions are the result of vicariance. However, vicariance biogeography is not simply a method to discover continental drift. Rather it purports to furnish a general method of biogeographic analysis applicable at any level. Thus the acceptance of continental drift as a causal mechanism for producing disjunctions does not imply that the investigator follows the methods of vicariance biogeography. With these points in mind, I will briefly outline the major approaches to historical biogeography.

Evolutionary Biogeography

Evolutionary biogeography is so named because its recent adherents tend to be evolutionary taxonomists. However, adherence to evolutionary taxonomy is not a prerequisite to adherence to evolutionary biogeography. Darwin, for example, was an evolutionary biogeographer but advocated phylogenetic classifications. This approach is based on the following tenets:

1. Higher taxa arise in certain limited areas of the earth. Subsequent speciation continues to produce species within these centers (Darwin, 1859; Wallace, 1876; Matthew, 1915).

2. As new species are produced and disperse they displace older, more primitive species toward the peripheral areas and away from the center of origin (Wallace, 1876; Matthew, 1915; more recently, Darlington, 1957, 1970; Briggs, 1974). *Thus, older extant members of a taxon will be found at the edges of that taxon's range while younger extant members will be found nearer the center of origin.*

3. The center of origin of a taxon may be estimated by certain principles derived from tenet 2. Specific rules for finding centers of origin may be formulated and applied (Darlington, 1957; but see Cain, 1944 for comments). The most important of these rules for Recent groups are (a) the most derived Recent members of a taxon will be found at the center of origin, and (b) that area with the most species is probably the center of origin.

4. The distribution of fossils is essential to understanding fully the historical biogeography of a group because the oldest fossils are probably located near the center of origin (Darlington, 1957; Simpson, 1965). Darlington (1959: 314) states:

> In a really good fossil record the earliest, most primitive fossils of a group will be at the place of origin, and later and more derivative fossils will clearly show directions of movement.

5. Animals and plants will disperse as widely as their abilities and the physical conditions of the environment permit (Darwin, 1859; and others). This tenet coupled with the idea of progressive evolution explains the tendency for derived taxa to "push" primitive taxa toward the edges of the groups' range.

Comments on Evolutionary Biogeography

The notion that taxa arose at a certain place, a restricted center of origin, and dispersed out of that center was based on the idea that the continents were either stable or had not moved so recently as to effect the distributions of extant groups (Darlington, 1957, 1970; Mayr, 1952; Hubbs, 1958; Simpson, 1965; also see Croizat et al., 1974). The notion is understandable given continental stability, for one could hardly hold to the idea that widely disjunct groups had a single origin unless one was willing to invoke some sort of dispersal. The alternates to Darwin and Wallace were land bridges, lost continents, or creationism. Land bridges and lost continents were rejected because too many were required to explain known disjunctions and Darwin could see little evidence for their existence except as ad hoc explanations of these disjunctions.

The notion that "advanced" or derived members of a taxon would be found in or near the center of origin of that taxon was originally based on the idea that evolution was progressive. That is, the idea that there was increasing perfection within a lineage as evolution proceeded (Matthew, 1915). Since newly evolved species were better adapted than older species, these better adapted species would tend to displace the older species from their original range toward the peripheral areas of the earth.

Considering the time and place and the culture of its originators (which affects all of us to some degree), one can understand the basis for evolutionary biogeography as practiced before the idea of continental drift. It is harder to understand the adherence of some recent proponents (e.g., Briggs, 1974) to the idea that primitive members of a group will be found near the periphery of the groups' range because so far as I know progressive evolution has not been taken seriously for years. Cain (1944) rejected the notion because his knowledge of angiosperms pointed to the opposite conclusion.

The phenomenon of long-range dispersal as a common dispersal mechanism is also suspect. Wulff (1950: his Chapter 8) devoted considerable attention to the problems of long-range dispersal of plants. There is a variety of older literature concerning experiments and observation involving plant dispersal. The result in most cases was that short-range dispersal, as exemplified by biotic dispersal, was within the means of many plants, but that long-range dispersal was not. Wulff extended his arguments even to cocoa palms and considered such dispersal mechanisms as wind-borne seed adaptations and dispersal by birds. Cain (1944) summarized the arguments against long-range dispersal in two ways (Cain, 1944: 154, 242):

> That long-distance dispersal has resulted in migration and accounted for discontinuous areas seems rarely to have been the case. This conclusion is based primarily upon existing distributions which largely show sym-

metrical replicate patterns mostly unrelated to chance and other elements of dissemination, and upon evolutionary phenomena, such as endemism in general and the occurrence of local races in particular. . . .

Minor discontinuities of areas probably frequently result from recent migrations, but major disjunctions seem almost exclusively to have resulted from historical causes which have produced the disjunctions, in a once more nearly continuous area, through destruction or divergent migrations caused by climate or some other changes.

Wulff (1950: 134–135) concludes:

Hence, we must seek an explanation of those movements in the geography of plants that are incomprehensible in light of present-day factors *not* in the action of chance factors but in the connection existing between plant distribution and the past history of our globe.

The major problems of organismic or species dispersal as a primary causal agent involved in present disjunct distribution is not merely the lack of empirical evidence for this dispersal. There is also a methodological problem. If we suppose that long-range organismic or species dispersal is the primary causal factor of plant and animal distributions, then we must employ it as the primary explanation. Because every organism has some capability for dispersal, we can explain every disjunction in terms of dispersal. The philosophical problem with this line of reasoning is that dispersal, in explaining everything, really explains nothing. If we allow that even 5 percent of the present biogeographic patterns exhibited by organisms are the result of vicariance rather than dispersal, then we will never discover which 5 percent it is if we used a dispersalist model of analysis. The larger the percentage of vicariance patterns, the larger the amount of lost information. Thus, from a heuristic standpoint alone the evolutionary biogeographic mode of inquiry is not satisfactory.

Phylogenetic Biogeography

Phylogenetic biogeography has been largely a development associated with phylogenetic systematics, principally as developed by Hennig (1960) and Brundin (1966, 1972). The method is based on considering a detailed phylogenetic hypothesis for a group of organisms and inferring the biogeographic history from that phylogeny. In many respects, phylogenetic biogeography foreshadows vicariance biogeography. The basic tenets of this approach are:

1. Closely related species tend to replace each other in geographic space (from Wallace, 1855; see Hennig, 1966; Brundin, 1966). Higher

taxa may also vicariate but frequently show degrees of geographic overlap (Hennig, 1966: 169–170).

2. If different monophyletic groups show the same biogeographic patterns, they probably share the same biogeographic history (Hennig, 1966: 169–170). This tenet definitely foreshadows vicariance biogeography. It is not used extensively except on the level of explanation of continental drift. Both Hennig (1966) and Brundin (1966) as well as their followers (Cracraft, 1973, 1975; Ball, 1975) were aware of the phenomenon of widespread ancestral biotas that were divided into smaller descendant biotas. However, phylogenetic biogeographers have utilized this tenet less frequently than have vicariance biogeographers because they concentrate on single groups rather than on congruent distributions shared by many groups inhabiting the areas studied.

The Progression Rule

The progression rule is central to the phylogenetic biogeographer's concept that phylogenetically primitive members of a taxon (Cracraft, 1975) will be found near the center of origin of a group. To understand this rule, we must understand (1) the chorological method and (2) the predominant mode of speciation as viewed by phylogenetic biogeographers.

Hennig (1966) analyzed the geographic races of the fly *Mimegralla albimana* whose ranges are shown in Fig. 8.5*a*. Hennig (1966: 135–136) considered this species to be composed of natural groups of subspecies, as shown in the phylogeny, Fig. 8.5*b*. The most morphologically primitive subspecies inhabits Area 1 and progressively more apomorphic groups are found as one proceeds east from Area 1. Hennig conceived of speciation as generally an allopatric phenomenon involving small peripheral isolates. Further, he believed that dispersal and speciation were causally connected. Taking these two operating principles as true, one can see that Area 1 is the center of origin and that dispersal/speciation has proceeded from Area 1 eastward.

Hennig's (1966) example is rather simple. A more complicated example is Brundin's (1966) analysis of chironomid midges. Two monophyletic groups, *Podonomus* and tribe Heptagyiae will serve as examples. Figure 8.6*a*, *b* shows that both groups inhabit Australia, New Zealand, and southern South America. Within Heptagyiae, all of the earliest taxa inhabit southern South America. In contrast, the earliest Recent species of *Podonomus* are found in New Zealand. Applying the progression rule, dispersal routes were worked out as shown by the arrows in Fig. 8.6. Brundin (1966) concluded that both groups inhabited Gondwana (i.e., long-distance dispersal was not a factor in establishing the present distributions, a view distinctly different from Darlington, 1970). Nevertheless, he identified New Zealand as the center of origin

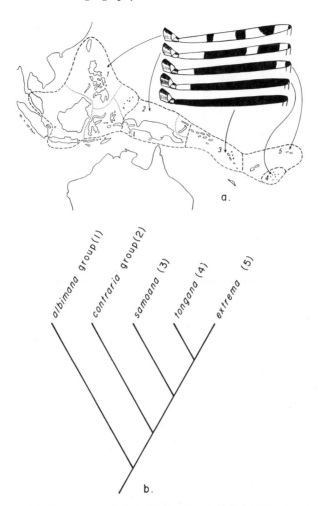

Figure 8.5 Hennig's Progression Rule I: (*a*) the distributions and characters of four sub-species groups of the tylid fly *Mimegralla albimana*; (*b*) the phylogenetic relationships among these groups. (*a* from *Phylogenetic Systematics* by Willi Hennig. Copyright 1966 by the University of Illinois Press. Used with permission.)

for *Podonomus* and South America as the center of origin for Heptagyiae, based on the location of phylogenetically primitive taxa for each group.

Comments on Phylogenetic Biogeography

There is some tendency for vicariance biogeographers (e.g., Croizat et al., 1974; Rosen, 1975) to lump phylogenetic and evolutionary biogeography into the same approach because both are based on a center of origin concept. Although advocating the vicariance approach myself, I

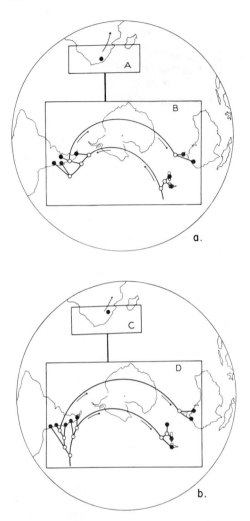

Figure 8.6 Hennig's Progression Rule II: phylogenetic relationships and distributions of chironomid midges of the subfamilies Diamesianae (*a*) and Podonominae (*b*). (From Brundin, 1972. Used with the author's permission.)

think phylogenetic biogeography represented a real advance over evolutionary biogeography. Ball (1975) has pointed out that phylogenetic biogeography represents an advance over the largely *a priori* tenets of the evolutionary approach. The evolutionary approach consists largely of descriptive enumeration, scenario building around suspect principles, and rather gross analysis of area relationships in what can only be termed a "phenetic mode." If Hennig, Brundin, and their adherents, such as Ball and Cracraft, have done nothing else, they have demonstrated that biogeographic analysis is best accomplished with reference to explicit phylogenetic hypotheses. The problems with phylogenetic biogeography are two. First, the progression rule was formulated following an explicit speciation model, peripheral isolation (Allopatric Model II, Chapter 2). If other models of speciation are allowed, the progression rule cannot be taken as a basic methodological tool because we would then tend to overestimate the number of times peripheral isolation was the actual speciation mode. In other words, presupposing peripheral isolation followed by or coordinated with speciation holds the same methodological perils as supposing long-distance dispersal. Second, the method is usually applied only to the groups of interest to the investigator, thus the second tenet of distributional congruence is often not used. This is probably because the supposition of process is assumed *a priori*, thus supposedly freeing the investigator from critically examining other groups.

Vicariance Biogeography

The problem with biogeography is largely the problem of specialization. Each systematist studies his or her group. At the end of the study some story must be presented giving reasons why the group is distributed as it is. The worker usually searches for some explanation based on the intrinsic qualities of the organisms studied. However, we might ask a question. What are the chances that two groups with different intrinsic qualities have come to occupy largely the same regions of the earth as a direct result of their intrinsically different qualities of, say, dispersal capability? If we are willing to answer that the chances are good for two groups, what about five or perhaps 50 groups?

Vicariance biogeography, as practiced today, is a method which begins by asking questions about common patterns of distribution between groups before asking questions about causal factors affecting the distribution of any one group. In other words, it is a method which attempts to sort out series of distributional patterns that require general explanations (such as the vicariance of a single ancestral biota into two or more fragments) from distributional patterns that require particular, unique, explanations (such as intrinsic mechanisms for long-distance dispersal). Vicariance biogeographers argue, cogently, that if we take as an operating principle the proposition that most disjunctions are the result of dispersal from a center of origin, then we will never discover those disjunctions that resulted from vicariance. Likewise, assuming that all disjunctions have resulted from vicariance will result in an underestimation of long-distance dispersal (Platnick and Nelson, 1978). To combat these problems, vicariance biogeographers have adopted the following tenets:

1. Historical biogeographers should not begin an analysis by assuming anything concerning dispersal, centers of origin, the comparability of various ranked taxa, or even vicariance (Rosen, 1978, 1979; Platnick and Nelson, 1978).
2. The first phase of analysis should be a data collecting phase. The

distribution of various monophyletic groups should be plotted on maps (i.e., tracks drawn). A search should be made for replicated patterns of distribution. This phase may be termed *track synthesis* and it is a basic data reduction phase (from Croizat, 1952, 1958, 1964).

3. From the data collected in the track-synthesis phase, areas of endemism are identified. Natural sets of endemic areas may then be analyzed with respect to the interrelationships of their respective faunas and floras. The questions asked are two:

a. What are the interrelationships among the organisms inhabiting these areas of endemism?

b. How does this relate to the geographic and geologic histories of the areas themselves?

4. To answer these questions we need phylogenetic hypotheses of the organisms inhabiting the areas in question.

5. From an array of phylogenetic hypotheses we search for congruence in relation to area inhabited. To accomplish this, the area in which a species lives is substituted for that species name in the phylogenetic hypothesis. The resulting dendrogram has been termed an **area cladogram** (Rosen, 1978, 1979). For example, in Fig. 8.7 we see two groups, fish and moss, confined to three areas of endemism, 1, 2, and 3. The phylogenetic relationships of fish and moss are analyzed. The area of each species is then substituted for the species name (i.e., substituted for A and X, etc.). In this example, there is complete congruence between the hypothesized phylogenetic relationship of both groups and the areas in which they occur.

6. The frequency with which we find complete congruence is related to the frequency with which common or general factors affected the

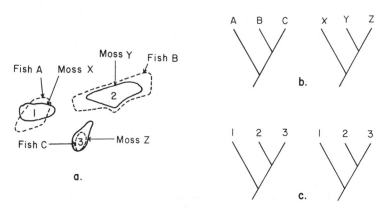

Figure 8.7 Area Cladograms I: (*a*) the distributions of three species of mosses and three species of fishes; (*b*) the phylogenetic relationships among the two groups; (*c*) area cladograms produced by substituting a taxon's geographic area (numbers) for its name on the phylogenetic tree.

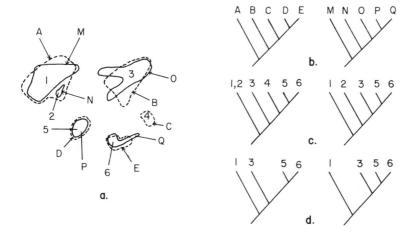

Figure 8.8 Area cladograms II: (*a*) the distributions of groups A–E and M–Q in six geographic areas (1–6); (*b*) the phylogenetic relationships among members of the two groups; (*c*) the area cladogram for taxa; (*d*) the reduced area cladogram produced by deleting incongruent parts of each area cladogram.

evolution and distribution of two or more groups of organisms. To search for such congruence we must separate unique factors particular to one group from common factors related to the evolution of all groups considered. Rosen (1978, 1979) accomplishes this by deleting unique factors from the area cladogram. For example Fig. 8.8*a* shows the distributions of two monophyletic groups (A–E and M–Q) among six areas (1–6). The phylogenetic relationships of these groups are shown in Fig. 8.8*b*. The area cladograms are shown in 8.8*c*. Note that the area cladograms are not fully congruent.

For example, Areas 1 and 2 are both inhabited by species A, but inhabited by two different species of group M–Q (M in 1 and N in 2). Also, Area 4 is inhabited by C but not inhabited by any member of M–Q. We attempt to find the congruence between these two groups and the areas they inhabit by deleting the unique factors of each group (Fig. 8.8*d*). This results in what Rosen (1978) termed a **reduced area cladogram**. These particular reduced area cladograms show congruence in that the areas of one group are arranged exactly like those of the other group.

7. Reduced area cladograms imply that congruences observed are due to factors that affected both groups. The next step would be to search for a **geographic cladogram** that specifies the relationships of the various endemic areas without reference to the organisms that inhabit these areas. Such geographic cladograms are easier to construct at the continental level (Fig. 8.9) but are more difficult for smaller geographic areas because of the greater precision of geologic or geographic data needed.

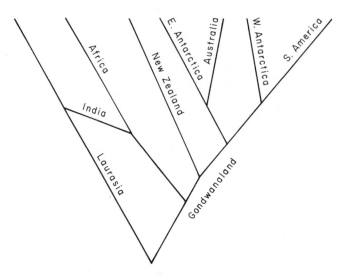

Figure 8.9 An area cladogram of continents. (From Rosen, 1976. Copyright American Museum of Natural History. Used with permission.)

If there is a correspondence between a geographic cladogram and one or more biological area cladograms, then we may infer that the causes associated with the geographic cladogram also caused the sequence of vicariance seen in the phylogenetic hypothesis. Lacking a geographic cladogram, we may provide a hypothesis of area relationships for the geologist to test using the evidence provided by geology. We may also hypothesize about the geographic history of the groups studied if we accept that congruence of complicated phylogenies and areas are probably associated with vicariance rather than dispersal.

Vicariance Methodology—An Example

The application of phylogenetic hypotheses to geographic areas is a demanding method. Most workers have enough to do simply working on the organisms they are familiar with much less on those they are unfamiliar with. Yet the vicariance method demands many phylogenetic hypotheses about widely different groups. As a result, while the theory and methodology is available, it is not surprising that they are little applied. There are two reasons for this: (1) the method as advocated now (Rosen, 1978, 1979; Platnick and Nelson, 1978) is rather new, and (2) there are very few phylogenetic hypotheses available. However, Rosen (1979) has done an analysis of two genera of Middle American poeciliid fishes which can serve as a practical example of applying the method.

Poeciliids are killifishes distinguished by the fact that they have a unique method of internal fertilization and give birth to young fish in an advanced state of growth. Common poeciliids include the guppies, swordtails, and platyfishes found in home aquaria. The genera *Heterandria* and *Xiphophorus* (which includes platyfishes and swordtails) were the subject of Rosen's analysis. Although both genera have representatives elsewhere (Fig. 8.10), each has a monophyletic subgroup inhabiting the same general areas in southern Mexico south to eastern Honduras (and in the case of *Heterandria* further south to eastern Nicaragua).

Rosen's hypothesis of relationships among members of the two genera is shown in Fig. 8.11. We shall be concerned only with the swordtail group of *Xiphophorus* and the Middle American species of *Heterandria*. Figure 8.12 shows all of the areas in which species of the two groups occur together, with the exception of areas 1 and 3, which are north of the area shown. Area 11 is an area of intergradation between two species in each group and will not concern us in this review. The phylogenetic hypothesis of relationship for Middle American *Heterandria* is converted into an area cladogram by substituting the area inhabited by each species for that species' name on the phylogenetic hypothesis (Fig. 8.13). The same is done for *Xiphophorus* and the additional step of producing a simplified area cladogram is taken for this genus (Fig. 8.14). Reduced area cladograms are then produced for each genus by deleting the unique factors from each area cladogram (Fig. 8.15). The reduced area cladograms (Fig. 8.15*b,d*) indicate four major congruences shared by members of both genera. If we examine the history of the two groups together, we may obtain a clearer picture of why congruence denotes common ancestry and why the deleted unique distributions are not informative about the common history shared by these species groups and the region they inhabit (although the unique aspects may certainly be of interest to a specialist in either group). I shall do this by going back and plotting the ranges of various ancestral species within each of the species groups of each genus.

If we take the view that the ancestral species of each genus was rather widely spread across southern North America and eastern Middle America (Rosen, 1979) we may infer that the ancestral species of Middle American *Heterandria* and *Xiphophorus* were distributed approximately as shown in Fig. 8.16*a*. This assumption is reasonable in view of the fact that relatives of both genera are found in and north of the area Rosen analyzed and the entire fauna and flora has a North American generalized track connection (Rosen, 1978).

The first speciation events inferred from the original phylogenies were unique for both *Xiphophorus* and *Heterandria*. There was speciation within Area 1 for *Xiphophorus* (the origin of the ancestral species that gave rise to *X. pyqmaeus* and *X. nigrensis* from the ancestor of all other swordtails) and a unique event which marked the origin of *H. attenuata* in Area 6. The event for *Xiphophorus* is not of direct interest to track

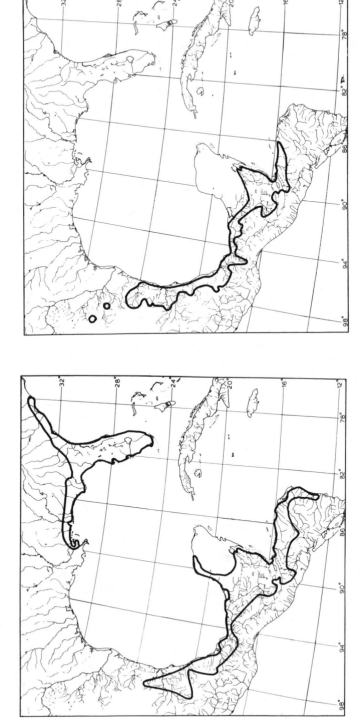

Figure 8.10 The distributions of the fish genera *Heterandria* (left) and *Xiphophorus* (right). (From Rosen, 1979. Copyright American Museum of Natural History. Used with permission.)

296

congruence between the two genera and thus is combined on the simplified area cladogram. This produces the areas and cladograms shown in Fig. 8.16*b*. The second speciation event was common to both genera and involved the vicariance of Area 1 from the more southern portion of the remaining ancestral range (Fig. 8.16*c*). This resulted in the origin of *H. jonesi* and the ancestor of the two remaining Area 1 *Xiphophorus* (*X. montezumae* and *X. cortezi*). The subsequent speciation within Area 1 for *Xiphophorus* was unique to that genus and will not be considered further (all will be shown as Area 1 on subsequent figures). It should be noted that one of the two speciation events in Area 1 did not affect *Heterandria* but that both affected *Xiphophorus* as evidenced by that fact that Area 1 *Xiphophorus* form a paraphyletic assemblage rather than a monophyletic group.

Two separate and unique events then occurred in each genus. *Xiphophorus clemenciae* originated in Area 3 and *Heterandria litoperas* originated in Area 9 (Fig. 8.16*d*). The next inferred event was common to both genera and involved the separation of a western central part of the remaining ancestral range (Fig. 8.16*e*). This vicariance represents the second common speciation event of the two genera, as depicted on the cladograms of area below the map in Fig. 8.16*e*.

The fifth event involved the separation of the southern portion of the remaining ancestral range and the origin of the undescribed *Xiphophorus* designated "PMH" and *Heterandria anzuetoi* (Fig. 8.16*f*). The sixth event was unique for *Heterandria*—the origin of *H. cataractae* in Area 7 (Fig. 8.16*g*). The last event was the peripheral isolation and origin of *Xiphophorus signum* and *Heterandria dirempta* in Area 8 and the origin (or continuation) of *X. helleri* and *H. bimaculata* in Area 2 (Fig. 8.16*h*).

Comments on Vicariance Biogeography

The heuristic value of the vicariance approach is evident from Rosen's (1979) analysis. The sorting out of the biogeographic history of a number of groups inhabiting a series of endemic areas into patterns which can be reasonably considered due to common events and patterns that may be unique for a particular group would seem a reasonable first step for further study. I do not mean simply further biogeographic study, although this should be done with additional groups from the same endemic areas. This type of analysis also yields valuable baseline information for further evolutionary studies. For example, two dichotomous cladograms showing both common and unique events could result from a combination of Model I and Model II allopatric (or parapatric) speciation. Discounting extinction, and noting that the incongruent components in both *Heterandria* and *Xiphophorus* swordtails occupy rather small ranges, we might conclude that both models were at work during the evolution of both groups (see Chapter 2). As another example, the

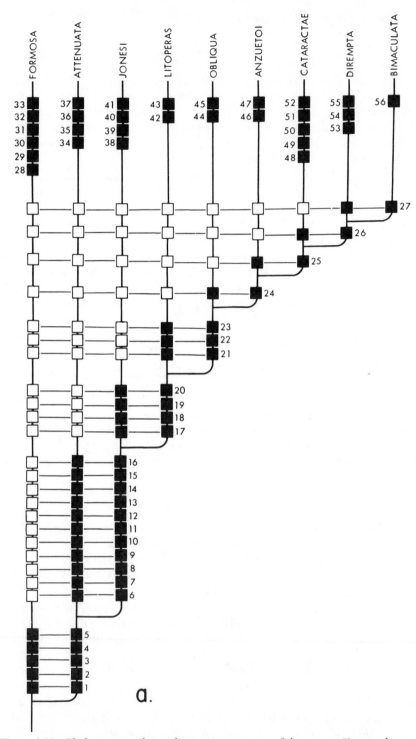

Figure 8.11 Phylogenetic relationships among species of the genera *Heterandria* (left) and *Xiphophorus* (right). Black squares are synapomorphies (for characters used see Rosen, 1979). (From Rosen, 1979. Copyright American Museum of Natural History. Used with permission.)

298

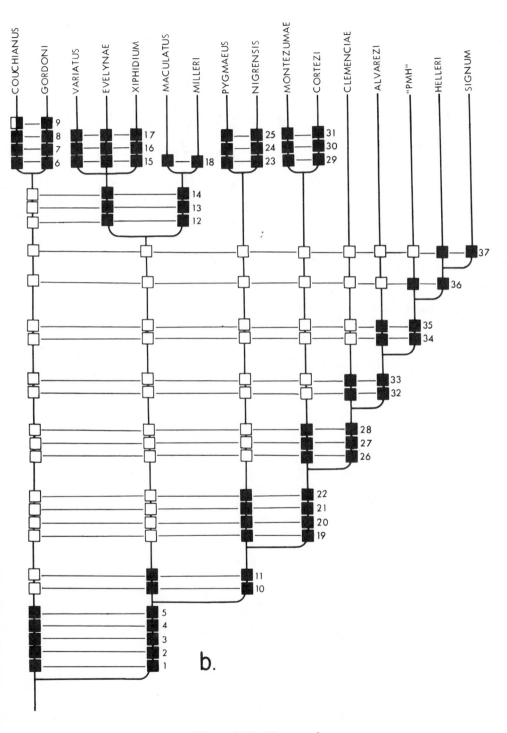

b.

Figure 8.11 (Continued)

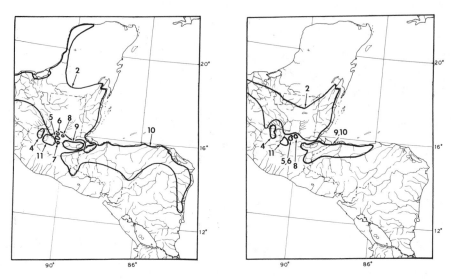

Figure 8.12 Detailed distributions of certain species of *Heterandria* (left) and *Xiphoporus* (right) from the southern portions of these genera's ranges. Areas 1 and 3 and the remaining part of Area 2 are north of the geographic region shown. (From Rosen, 1979. Copyright American Museum of Natural History. Used with permission.)

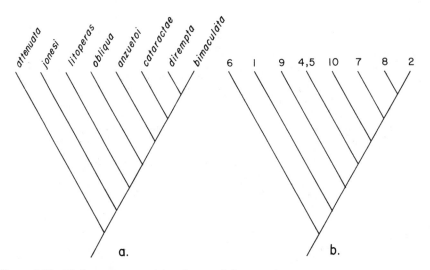

Figure 8.13 Phylogenetic tree (*a*) and area cladogram (*b*) for Middle American species of *Heterandria*. (Redrawn from Rosen, 1979.)

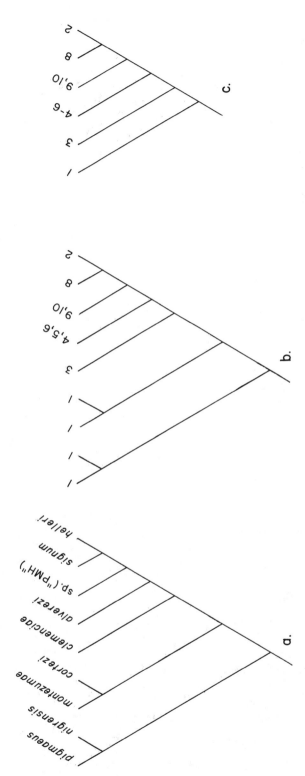

Figure 8.14 Phylogenetic tree (*a*), area cladogram (*b*), and simplified area cladogram (*c*) for swordtail *Xiphophorus* (Redrawn from Rosen, 1979.)

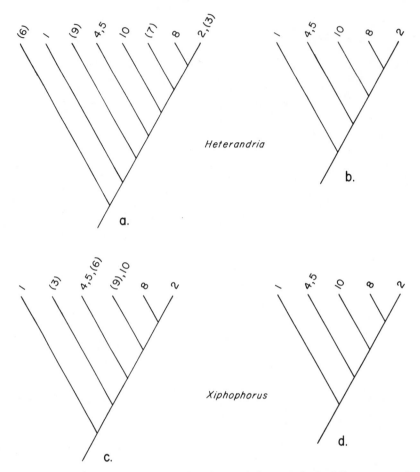

Figure 8.15   Area cladograms and reduced area cladograms for Middle American *Heterandria* (*a, b*) and swordtail *Xiphophorus* (*c, d*). (Redrawn from Rosen, 1979.)

vicariance hypothesis postulates that *X. signum* and *H. dirempta* have the same time of origin. Does *X. signum* have about the same amount of genetic divergence from *X. helleri* as *H. dirempta* does from *H. bimaculata*? If so, did this occur concurrently? If not, might we suspect molecular clock estimates which might be postulated for the group? Such questions have interest far beyond biogeography *per se*. Yet, the biogeography provides the basic data from which we may pose these questions.

Dispersal and Vicariance Biogeography

Thus far we have concentrated on the phenomenon of allopatric speciation via the fragmentation of ancestral species. However, vicariance biogeographers recognize the reality of biotic dispersal, species dispersal and even long-distance dispersal. Croizat et al. (1974) listed several situations where dispersal can be reasonably postulated:

1. If an allopatric speciation mode is postulated, then sympatry within a monophyletic group is evidence for dispersal.

2. The intersection of generalized tracks implies biotic dispersal.

3. A species or a monophyletic group that shows no congruence with any generalized track may have achieved their distribution by organismic or unique species dispersal.

4. For the modern world to show evidences of a fragmented Mesozoic

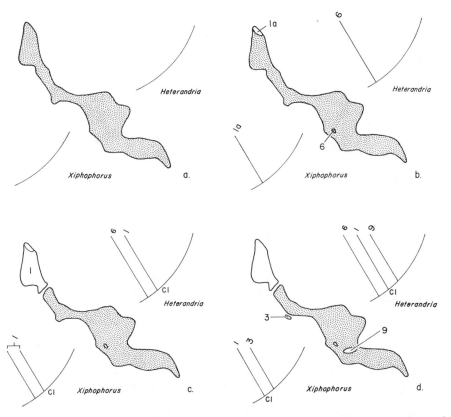

Figure 8.16 Inferred sequence of speciation and area cladograms for *Heterandria* and swordtail *Xiphophorus*. In each diagram the stippled area represents the inferred range of an ancestral species. The phylogenetic position of this ancestral species is represented by the unnamed branch in the area cladograms above and below the geographic figure. Named branches correspond to species represented by the number of the area they inhabit. Common speciation events in the history of the two groups are labeled on the area cladograms as C_1, C_2, and so on. Note that h corresponds to the original area cladograms of each species group.

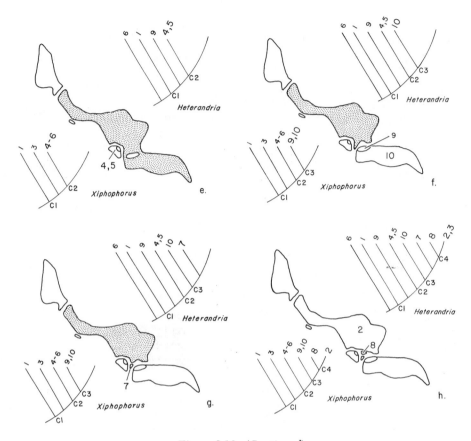

Figure 8.16 (*Continued*)

biota, it is a necessary assumption that a large-scale dispersal of an ancestral biota had first to have occurred on a predrift landscape (Rosen, 1978: 161).

In terms of our actual example, the more southern distribution of *Heterandria* may have been achieved by species dispersal. Thus the estimated southern range of both common ancestors (Fig. 8.16*a*) may represent an overstimate. The more northern distribution of *Heterandria* into the Yucatan (Fig. 8.12*a*) is certainly a case of dispersal because much of the peninsula was recently submerged. And considering Rosen's (1978) larger study of the North and Central American deciduous forests and their associated generalized track, we must postulate a Tertiary biotic dispersal to explain the present distributions of the vicariant groups inhabiting disjunct ranges in North and Central America. It should be noted, however, that Rosen's (1979) analysis included only two groups. It is possible that some of the apparently unique aspects of one or both genera will be found congruent with other groups analyzed in the future.

Vicariance Biogeography—Historical Precedents

The idea that present disjunctions should be studied by looking for replicated patterns of distribution is an old idea. Hooker's (1860) speculation that the southern continents were connected was derived from his observations of replicate plant distributions. More recently, Wulff (1950), Cain (1944), and Camp (1947) stressed the need for studying replicated patterns and expressed the opinion that long-distance dispersal could not account for present plant disjunctions. Camp (1947) even used the term "track" in its original sense and detailed major generalized tracks for 113 families of angiosperms. These authors stressed biotic dispersal and fragmentation, with an emphasis on continental drift. It remained for Croizat (1952, 1958, 1964) to firm up the study of replicated patterns and stress the idea that historical biogeography should begin with the search for generalized tracks. Croizat's aversion to centers of origin no doubt stemmed from his belief in Rosa's theory of hologenesis. He took the view (also espoused by Wulff, 1950) that the earth and its biota evolved together and therefore both could be studied in concert. The heuristic value of setting aside *a priori* rules for finding centers of origin has freed the historical biogeographers to seek basic data (Ball, 1975). Whatever one thinks about Leon Croizat (he is a controversial person), his works stimulated a new and interesting synthesis of historical biogeographic methods. Croizat's original method was purely track analysis without the particulars of explicit phylogenetic hypotheses. Croizat's writing is difficult, a factor which no doubt interfered with the understanding and acceptance of his methods. His discovery by the zoological community began with Brundin (1966) and Nelson (1973c), and a summary of the basic tenets of the original method is contained in Croizat et al., 1974. Applications of these ideas are included in Rosen (1974, 1975, 1978, 1979), Wiley (1976), Platnick (1976a), McCoy and Heck (1976). The current method of applying explicit phylogenetic hypotheses, thus yielding more detailed information, is detailed in Platnick and Nelson (1978b: theoretical considerations) and Rosen (1978, 1979: theory and practical application). It represents a melding of Hennig's (1966) concepts of relationship with Croizat's track analysis (Nelson, 1978).

Chapter 9

Specimens and Curation

In earlier chapters I have concentrated on the theory and practice of phylogenetic systematics and why I believe that the approach is superior to alternate systematic approaches. This and the following two chapters will concentrate on various matters of general systematic interest. In this chapter I will discuss specimens, how they are gathered for study, and how they are housed and cared for.

SPECIMENS AND SAMPLES

A **specimen**, as discussed in Chapter 5, is an individual organism. Two or more specimens hypothesized to belong to the same deme, taxon, or ecosystem comprise a sample of that deme, taxon, or ecosystem. A **systematic series** consists of one or more specimens which represent a population of organisms thought to exist in nature as either a deme or a taxon. For purposes of discussion, "series" is used here for "systematic series." Most series utilized by systematists are preserved in some manner such that some characteristics of the specimens can be observed. Occasionally a specimen may be preserved in several ways to increase the number of observable characters. For example, a fish specimen may be karyotyped, blood and tissue samples preserved for biochemical analysis, and the remainder of the individual fixed in formalin and preserved in alcohol. Occasionally, specimens are not preserved but observed while living.

Specimens may be selected from preserved series deposited in a systematic collection, or they may be collected or observed in the field. The aims of the investigator (i.e., the questions to be asked) and the availability of series will largely determine the adequacy of existing samples and the need for field collecting. Conversely, a lack of extensive series and the inability of the investigator to obtain additional samples in the field may limit the questions that may be asked. The list below summarizes some of these areas of research and the sample sizes needed to answer them.

1. *Geographic variation.* The investigator should examine series from all parts of the range of the species. The number of specimens per series should be statistically adequate because most studies of geographic variation are statistically oriented.

2. *Species-level studies.* An adequate sample of the variation of each species should be assembled. When working on the assessment of the number of species present in a species group, this should include type specimens of all nominal species as well as samples drawn from throughout the range of the nominal species whenever this is practical. Especially critical are series from areas at or near the periphery of a species' range and areas of sympatry between sister species. When studying the phylogenetic relationships of species whose species status is not questioned (because of previous work), an adequate number of specimens should be examined to assess the variation in characters hypothesized as apomorphic or plesiomorphic.

3. *Higher-level studies.* The nature of the samples and number of specimens per sample will vary depending on the problem. Generic studies often use species as samples. If several large genera are studied, sample species may be drawn from each to represent the limits and variation within any one genus. The number of specimens examined is frequently low because of the nature of character variation.

The lack of samples or large numbers of specimens might restrict the questions which may be asked by a systematist. Conversely, in systematic studies involving well known floras and faunas, there may be more specimens and series than the systematist should analyze. There is a diminishing return of useful information as more specimens are studied and one should always try to set a limit (based on character variation) to the number of specimens examined. In other words, one should not count and measure 1000 specimens in a sample simply because there are 1000 specimens available when 30 specimens would give an adequate representation of the series. And, one should not take 100 measurements without a clear idea as to what one wishes to accomplish with these measurements (25 or even 10 measurements might suffice).

Subsampling is easily accomplished in a number of ways. In fishes, which are catalogued in lots, I lay the entire sample out and take a random number of individuals of appropriate age classes or sex until my subsample is complete. In individually catalogued groups, one could put specimen numbers into a box and pull out the specimen numbers to be examined.

Access to Specimens

Before going to the field, one should survey the available holding of preserved material. This is important even if the objective of the study requires living specimens or specimens preserved in specialized ways because preserved series will give valuable locality information that will reduce the amount of time spent in collecting new material. There are two basic sources from which the availability of preserved specimens can be determined.

1. *Previously published literature.* Most recent taxonomic papers and most older papers list the specimens examined in the study and where these specimens are deposited. Those that do not should reference a source from which specimens examined may be obtained. In earlier publications where this information is not published, one can frequently determine where the specimens are deposited by later papers on the same group or by being familiar with the history of the particular author and his or her work.

2. *Systematic collections.* Specimens may be deposited in a variety of universities and museums, ranging from small local museums or university collections to large national museums. One learns through experience which museums are likely to have pertinent material. When starting a revision, it is worth sending letters of inquiry to major museums and to local museums within the distributional boundaries of the group studied. Further, a list of synonyms should be included in the request, especially when sending an inquiry to an older museum where specimens of interest may be catalogued under more than one name. An uninformed blanket loan request is likely to be disregarded (and with excellent reason). Only when one knows where specimens are located and what samples are needed for study should loans be requested. Where collections are extensive, the best way of determining what should be borrowed may be a personal visit to the institution.

Field Collecting

There are two basic reasons to go to the field (apart from the pure pleasure it affords). The first is to collect specimens with the object of increasing the resource base of the institution in particular or the region in general. This is the general objective of field work in poorly known regions of the world. The second reason is to collect specimens for a specific systematic problem. There are two basic rationales for doing this.

1. The available preserved specimens (from all sources) do not adequately cover the suspected geographic range of the group or the avail-

able samples contain too few specimens for the contemplated study. For example, the investigator might be interested in a group of nominal species. Samples from critical parts of the ranges of several species may not be available (perhaps in peripheral areas or zones of possible sympatry). Additional sampling would be required to reach decisions about the validity of these nominal species.

2. The characters of interest cannot be studied on the preserved specimens. This is frequently the case for biochemical characters and always the case for karyological characters. It may also be true for many histological characters. Thus, even in areas such as North America where extensive preserved collections exist for many groups, additional and extensive field collecting may be necessary. Further, the need for such collecting will broaden as systematists increase the base from which they draw characters.

I do not think there is any substitute for getting out in the field and seeing what the organisms are doing and where they are living. Field collecting has always helped me to understand better the organisms I work with and has frequently led to characters that were not available to me in preserved specimens. The difference between a successful and an unsuccessful field experience is frequently the amount of time spent in planning the trip. Poorly planned field trips are likely to result in misplaced effort. The following points should be addressed before proceeding to the field:

1. *The biology of the organisms.* Some time should be invested in learning about the ecological requirements and probable habitats in which the organisms to be collected are likely to be encountered. This is especially important if there are relatively few existing collections or if one is going into an area where the organisms are thought to occur but have never been collected. If there is little or no useful information available in the literature, then localities where the organisms have been collected previously should be visited and collected. Observations made at these localities will guide the worker in uncollected areas.

2. *Previous collections.* The available literature should be consulted to see where previous collections have been made. Given that previous field parties were conservation-minded and that the habitat has not been destroyed (neither of which, unfortunately, can be presumed), particular species are usually found at the same place at appropriate seasons year after year. If the object of collecting is to gather specimens for examination of new taxonomic characters (such as karyotypes or biochemical studies), knowledge of previous collecting sites and ecological habitat will almost ensure a successful field trip. If a previously unsampled region is the objective, previous collecting sites will define the uncollected area and field notes on these sites should give a clue as to exactly where the specimens are likely to be found.

3. *Maps.* Maps of the region should be studied before proceeding to the field. Good maps will frequently give clues to likely collecting sites.

4. *Regulations.* State, federal, and international laws now govern the collecting activities of many systematists. Unlawful collecting of specimens reflects badly on the individual, the institution, and the systematic community in general. And you might end up in jail. Before proceeding to the field, the systematist should (a) check to see if the groups to be collected are governed by law, (b) obtain any necessary permits to collect lawfully the species needed, and (c) be familiar with any particular species which may be protected when these occur along with species that may be collected. For those unfamiliar with various federal and state regulations, the Association of Systematic Collections (c/o ASC, Museum of Natural History, University of Kansas, Lawrence, Kansas, 66045, USA) has various publications summarizing these regulations.

5. *Collecting methods and preservation techniques.* The systematist should be familiar with the most efficient means of collecting and the best field preservation techniques available. If poisons are used, the systematist must ensure that they are used properly. A checklist of equipment should be made before proceeding to the field so that nothing vital is left behind.

6. *Sociological and political considerations.* It is always helpful to inquire about the attitudes and beliefs of the peoples one is likely to contact in the field. And there are certain regions of the world where political constraints may prevent collecting even if that region can be entered. As a general rule, always find out whether you will be welcomed or at least tolerated before you show up. It may save the cost of transportation.

Field Data

A certain amount of basic data must be associated with specimens if they are to be of scientific interest. If possible, these data should be directly associated with the sample and duplicated in a field notebook. The basic data includes:

1. *Locality.* The locality should be as specific as possible. For example: "U.S.: Kansas: Douglas Co.: 12 km. W JCT. U.S. Hwy. 59 and Kan. Hwy. 10 on K 10," rather than "12 km W Lawrence, Kansas." The first locality gives an accurate distance from a known point, the second depends on where one believes the investigator means when he or she says "Lawrence" (i.e., from the center of town or the western-most city limits?). In areas where the names of fixed geographic points are apt to change (as, for example, in developing countries) one should also record latitude and longitude as accurately as available maps permit. For collections made at sea or in terrestrial areas where no appropriate geographic land marks are available, latitude and longitude are the only possible form of geographic record. Another method for recording localities in certain parts of North America is to record the locality by the section/township method. I have a preference for assigning locality numbers. These are in numerical sequence by year (EOW 1979-1, EOW 1979-2, etc.). The locality number is placed on (1) the field label, (2) the field notebook, and (3) a map of the area. The field label and notebook also contain the locality information.

   2. *Date and time period.*   To avoid confusing day and month, I prefer to abbreviate the month—12 Mar 1980 rather than 12/3/1980 (which might be confused for December 3, 1980). Another method is to use a Roman numeral for the month (12 III 1980). The time period during which the collection is made should also be noted (1300–1500 hrs. or 1 p.m. to 3 p.m.).

   3. *Collectors.*   The names of the collectors should also be noted.

This list represents the *minimum* data necessary for specimens and is usually what is found on field tags. Additional data that should be included in the field notebook are:

   1. *Site description.*   A brief and concise description of the collecting site. This can include both physical and ecological characterizations. Many investigators photograph their collecting localities.

   2. *Specimen notes.*   Notes should be made of any characteristics of interest to the investigator which fade or change with preservation.

   3. *Collecting techniques and preservation.*

   4. *Disposition of living or specially preserved specimens.*   Appropriate entries might include specimen field numbers that match tissue or blood samples. The specimens from which such samples are taken should be preserved as voucher specimens.

   5. *Ecological notes.*   These might include habitat notes for different species, notes concerning relative abundance, and so on.

THE SYSTEMATICS COLLECTION

A systematics collection consists of many separate series of specimens which are properly documented to preserve their systematic value. Most systematics collections are housed in museums or universities and are generally separated into different collections which are curated in a similar manner. A systematic collection should be thought of in the same way as a research library. It provides an accessible record of the flora and/or fauna of the geographic region of coverage in the same way that a library provides a record of literature. Like a library, a systematic collection must:

  1. Be organized in such a way that its holdings are accessible to users.
  2. Be willing to make its holdings available to those properly qualified to study those holdings.
  3. Make a commitment to provide for the long-range safety of the specimens housed in the collection.

Many larger systematic museums strive for worldwide taxic coverage. Others concentrate on fuller coverage on a more limited geographic scale. No museum can provide collections suitable for all groups at all levels of analysis. The vast majority of specimens in such collections are preserved by means traditional for the group in question. This is changing, however, and it is not unusual for systematic collections to house voice recordings, tissue and blood samples, karyological slides, pollen, spores, and a variety of other nontraditional systematic material.

Loans and Exchanges

I prefer to think of systematic collections as belonging to the systematic community rather than to the particular institution where they are housed. One of the major jobs of the curator/systematist is to insure free access to the specimens under his or her care to those colleagues who are qualified to study them. Where possible, this can be accomplished by loans of material. Loans carry a dual responsibility for both the borrower and the lender. The person requesting the loan should (1) be careful to request only that material which is needed to accomplish the project, (2) gear the request to those specimens which can be analyzed in a reasonable amount of time, and (3) maintain all borrowed specimens as if they were still at the lending institution. In addition, any manipulation of the specimens (sectioning, dissection, etc.) should be done only with permission from the lending institution. The borrower is also responsible for returning loan material on time or requesting a loan extension. The lending institution is responsible for making the material available and for recovering it in a reasonable amount of time. "A reasonable amount of time" is subject to broad interpretation. I do not object to a borrower keeping specimens under my care for several years so long as (1) he or she is actively working on the specimens, and (2) no other investigator needs the specimens.

Exchanges involve the permanent exchange of specimens between institutions. That is, the recipients catalogue the specimens into their collections and take over permanent care of the specimens. Exchanges permit curators to broaden their research base without the expense of doing field work. Most active systematic collections have active exchange programs with other institutions and curators should stand ready to help other curators broaden their base by exchanging duplicate series or parts of large series. In our Division of Fishes we generally follow a few rules: (1) we exchange statistically significant samples when duplicate collections exist; (2) when requesting an exchange, we negotiate the general contents of our part of the exchange so that the recipient will know what to expect; (3) we attempt to maintain a small uncatalogued subcollection of available exchange material so that decataloguing is kept to a minimum; and (4) when a catalogued series is split for exchange, we also note the number of specimens removed and where they were sent in all catalogue entries pertaining to the specimens involved.

CURATION

Curation involves a series of activities from the initial receipt of specimens to the continuing processes of insuring that specimens in the collection do not deteriorate. Most systematists learn curatorial practices from their older colleagues. Thus a certain diversity of approaches exists from collection to collection as well as from taxonomic group to group. In general, the following activities are involved in curating a systematic collection.

   1. *Receipt of specimens and initial sorting.*   Curation begins with the receipt of specimens, either newly collected or as exchange material from another institution. Newly collected specimens should be accompanied by field data. The curator must ensure that the data do not become dissociated from the specimens. To ensure this, an **accession number** is frequently assigned to the entire collection. The specimens are then prepared for permanent storage if field preservation differs from storage preservation (as is frequently the case). Specimens may then be initially sorted into appropriate taxonomic groups.
   2. *Identification.*   Sorted specimens with associated field data are then identified. The level of identification varies from group to group depending on the ability of the curator to make accurate identifications. As a general rule, specimens should be identified to the taxon that the curator can be confident is correct. In insects this may be to family or subfamily. In fishes this may be to species. Identification should be done with keys (see Chapter 12) unless the curator feels fully qualified to make identification without keys. This process of identification is called **determination**. A label with the determination and the name of the person who made the determination is usually associated in some way with the specimens.
   3. *Cataloguing.*   In many collections (especially insect collections) catalogue numbers are not assigned to individual specimens or lots. In

collections where such catalogue numbers are assigned, each specimen
or each lot of specimens thought to belong to the same species (regardless
of whether determination is made at the specific level) is then assigned
a catalogue number. Essential data are recorded in the main catalogue
and the catalogue number is associated in some manner with the specimen. Cross-referenced catalogue entries may then be made out. The
most important consideration in this process is accuracy in making the
entries. A miscatalogued specimen or lot may be permanently "lost" in
a collection in which it is housed just as may a miscatalogued library
book.

Curators pay particular attention to the materials used to make catalogue entries and specimen labels. It is worth the extra expense of using
high quality paper and ink. The curator realizes that the recorded specimen data will be of interest far into the future. The systematist who
consults a catalogue entry made 100 or 200 years ago will appreciate the
difference quality materials can make.

4. *Storage.* Specimens should be stored in such a way that they are
accessible and well protected. Particular attention should be paid to
insuring that a specimen or lot is placed in the collection in its correct
place.

Arrangement of Collections

Collections should be arranged in such a way as to facilitate retrieval of
specimens. Most are arranged in approximate phylogenetic order. As
information on phylogenetic order changes one has the choice of moving
specimens around or leaving them as they are. Museums with immense
holdings have no choice—it is simply not worth trying to maintain a
large collection in the latest published phylogenetic order. For small
collections, or those used extensively for teaching purposes, rearrangement may be warranted for the educational benefits it confers on the
users. It is usually more important to provide easy access to a collection
than to attempt to keep step with changing ideas.

Type Specimens

Primary type specimens (holotypes, syntypes, lectotypes, neotypes) are
nomenclaturally important and thus require special attention. Holotypes
are usually set apart from the rest of the collection and receive special
curatorial care. Secondary type specimens (paratypes, etc.) may be curated with primary types or placed in the main collection. All type material
must be marked in such a way that they clearly stand out as types to
comply with the various codes.

Catalogues

Catalogues provide information on specimens. There are two ways of cataloguing specimens, by lot or by specimen. Catalogues provide a means of recording information for each species or individual from a locality without having this information directly attached to the specimen. Cross-catalogues facilitate information retrieval.

Computer Cataloguing

Many curators have turned to computer cataloguing in an effort to automate part of the curatorial process. This has met with varying success depending on the commitment of the curator, the size of the existing collection, the training of those who input the data, and the money available.

Computer cataloguing is not a panacea for curatorial ills. In some cases, collections are so large that the cost of retrospective cataloguing (inputing specimens on lots that are already catalogued) is simply prohibitive. Further, when one looks at the cost of cataloguing compared to the total cost of collecting and preserving a particular specimen, the supposed savings are initially insignificant. But, there are some definite advantages. Name changes can be made by simply substituting the new name for the old. The next print-out will show the change. Also, range maps, print-outs of specimens held, and other types of information may be recovered quickly. Finally, some computer cataloguing systems permit the storage of numerical data with the specimens, facilitating data retrieval.

Fish Division, KU Museum of Natural History—An Example[1]

The following system of catalogues is utilized at the Fish Division of the University of Kansas Museum of Natural History. At Kansas we are currently implementing a SELGEM computer system for our collection but we are maintaining all of our other cross-cataloguing systems to ensure that the transition does not interrupt our ability to make material available, and to ensure that data is not lost. Once we have the system fully operational (i.e., once it produces a main catalogue, geographic file, and species sheets), we will dispense with our manual system except for the main catalogue (which will serve as the primary input source).

1. *Main catalogues.* The main catalogues are massive books composed of high quality paper and bound for permanence. Lots of fishes

[1] Credit for the system goes to Dr. Frank B. Cross, Curator–in–charge.

(all of the specimens of one species at one locality) are assigned a number and the following information is recorded:

a. Museum number
b. Number of specimens
c. Species or taxon name
d. Locality
e. Date collected
f. Collector

In some groups it is traditional to assign a number to each specimen (or parts of specimens). The data recorded are generally similar. In addition there is a "remarks" section which can be used to note exchanges or special handling of subsamples of the sample (as when we make skeletal preparations).

2. *Geographic file.* The geographic file consists of species cards arranged in geographic sequence from largest political unit to smallest political unit and arranges the species in alphabetical order by genus name within that political unit. The purpose of this file is geographic information retrieval, so a phylogenetic order is not maintained. For countries or states with few specimens, everything is arranged alphabetically by genus name because most requests are for particular species. For areas of extensive coverage (such as Kansas and Missouri in this collection), the file is subdivided to county. The geographic card on each species has extended sample information:

a. Group name (family)
b. Species
c. Catalogue number
d. Locality
e. Date of collection
f. Habitat
g. Collector
h. Collector's field number
i. Method of capture
j. Number of specimens
k. Size range
l. Original catalogue number (if an exchange)
m. Determinor (who identified the sample)
n. Cataloguer
o. Citation of published record
p. Remarks

The extent to which the blanks are filled in depends upon the quality of the field notes.

3. *Species sheets.* To facilitate requests for lists of holdings species sheets are maintained in a series of $8\frac{1}{2} \times 11''$ loose-leaf binders. Each

species has a separate sheet with multiple pages for species heavily represented. The file is arranged alphabetically by family and then by genus within the family. (Note, again, that no phylogenetic order is maintained.) The lots of each sample are entered as they are catalogued with the information contained in the main catalogue. If the museum is queried about a particular species, the species sheets can be copied and sent to the interested systematist.

4. *Curation of field notes.* We keep either the original or duplicate copies of all field notes associated with the specimens collected by our staff. These are arranged by name, year, and date. Access to the field notes is provided by the geographic file.

Curation as a Responsibility

The systematist lucky enough to have a research base in the form of an established systematic collection has the responsibility to see that this collection is properly maintained. This calls for a certain amount of knowledge about the specimens, their permanent preservation, upkeep, and storage. The modern curator tends to be a working systematist first and a curator second. This is accomplished without detriment to the collection by the presence of a supporting staff of professionals whose major functions involve the day-to-day curatorial activities involved in keeping any collection in excellent condition. It is, however, the curator's responsibility to see that this work is accomplished and to keep abreast of new curatorial techniques and innovations in collection management. It is the curator (or curators) who determines the overall quality and coverage of the collection and the strategy of collection growth and management.

Chapter 10

Characters and Quantitative Character Analysis

Intrinsic characters have been considered from a theoretical standpoint in Chapter 5. This chapter is devoted to various classes of intrinsic characters and the analysis of quantitative characters. In earlier chapters we have been concerned largely with qualitative morphological characters. This is more a product of the examples given and the need to present relatively simple situations than it is a judgement on the utility of other types of characters. Because morphological characters were used extensively in Chapter 5, I will not give additional examples here. Practical examples are given to illustrate the usefulness of other classes of intrinsic characters (such as biochemistry and karyotype) both to the process of phylogenetic hypothesis testing and to the elucidation of basic natural taxa (i.e., species). I will then briefly discuss quantitative character analysis, including a sample of multivariate techniques.

SOME CLASSES OF CHARACTERS

There are several ways of classifying intrinsic characters for discussion. If one is statistically oriented, the basic division might be made between qualitative and quantitative characters. That is, between characters that we describe by words or figures and those that are measured or counted. One might also recognize various characters on the basis of whether they are morphologic, behavioral, ecological, and so on. The latter is the method of presentation I will utilize, realizing that the divisions are frequently arbitrary and that each of the basic classes of characters may be expressed qualitatively or quantitatively. The subject of quantitative and qualitative characters will be discussed later in the chapter.

318

I will discuss most of the following classes of characters:

✓1. **Morphological characters.** Structural attributes of organisms at the cellular level or above.

2. **Karyological characters.** The structure of chromosomes.

3. **Biochemical characters.** The structure or other physical attributes of molecules, including DNA and the products of DNA.

4. **Physiological characters.** Nonstructural characters that result from metabolic activities.

5. **Behavioral characters.** Nonstructural characters that are the result of actions taken by organisms. (Not discussed.)

6. **Ecological characters.** Nonstructural characters that are a product of the interaction of an organism with its environment (including other organisms). (Not discussed.)

7. **Biogeographic characters.** Geographic range data.

I stress that this division of characters into classes is largely arbitrary. It is done purely for heuristic reasons. I do not doubt that chromosome structure and biochemical characters are, in fact, morphological. But, their interpretation and use differ from the usual morphological characters and thus they are discussed separately.

Morphological Characters

A morphological character is a structural attribute of an organism. Morphological characters are the primary source of characters in most groups (bacteria, many protists, and viruses excepted). The diversity of organisms precludes a discussion of particular kinds of morphological characters found in various groups. However, most of the examples used throughout the book utilize morphology and give an idea of the diversity of morphological evidence. Morphological characters usually comprise series of morphological complexes. A **morphological character complex** is an array of ontogenetically linked morphological characters. The flower of an angiosperm or the caudal fin of a fish are examples of such complexes. Each morphological character complex is composed of a number of more-or-less interrelated characters.

The utility of particular character complexes or of particular characters varies from group to group and at particular levels of universality within groups. For a particular set of characters to be useful they must (1) vary between taxa, (2) vary in a correlated, coherent manner, and (3) not be ecophenotypic. Such characters have been termed **taxonomic characters** by Mayr (1969). Characters that do not vary or that vary randomly between groups are of no use to unraveling phylogenetic relationships at that particular level of analysis. However, a character that is invariant may very well have a homologue at another level of universality. Systematists interested in particular taxa must familiarize themselves with the biology of these taxa and the past work done by other investigators. This knowledge will permit the investigator to (1) evaluate the usefulness of characters used by past investigators, and (2) point the way to previously unutilized characters and character complexes. For example, in my own work on killifishes of the *Fundulus nottii* species complex (Wiley and Hall, 1975; Wiley, 1977b), I had the advantage of having available the excellent previous study by Jerriam L. Brown (1958). From this work I could see that the character "number of dorsal fin rays" showed little variation between nominal taxa (species by my reckoning, subspecies by Brown's). In assessing variation close to the geographic limits of a species, "dorsal fin rays" was one character I did not have to take into account. In contrast, Brown's study indicated that body color pattern was important, and I found this character complex to be of great value in studying the species group. In the same studies, Hall and I turned to sensory pores on the head largely because Gosline's (1949) work showed the potential for this character complex. The pattern of supraorbital sensory pores happened to be useful for the *F. nottii* species group. But it is not useful for other species of *Fundulus* (as I found when I examined the other 26-odd species in the genus). One conclusion can be drawn from this example. The systematist cannot predict that a particular character or character complex will be useful for one taxon just because it is useful for another group or subgroup.

"New" or Underutilized Characters

Systematic progress is frequently accelerated when "new" or underutilized characters are explored. Frequently a new preparation technique or the availability of an established technique to the average systematist will cause a flurry of application to various groups of organisms. Various staining techniques for ciliate protozoans and the scanning electron microscope are examples. Often, previously unexplored morphological systems will yield new insights, especially when variation in more traditionally used characters is low.

External Morphology

External attributes are the main sources of characters in many groups. They may be observed easily, unless microscopic, with simple optical aids or by sight. External characters may be relatively simple or very complex and they have proven useful in distinguishing taxa at varying levels, from phyla to species. Being relatively easy to observe, external characters are the predominant characters in keys and diagnoses. Some types of external characters are:

1. *Size and shape.* Size and shape characters include overall estimates of body form and particular descriptions of various body features. The analysis of shape is usually considered more taxonomically important than size, although size may influence shape if allometry exists. Techniques for quantifying size and shape are presented in later sections of this chapter.

2. *Color and color pattern.* Species are frequently characterized by having particular coloration or color patterns. Color and color pattern have been useful in analysis of many insects, birds, mammals, fishes, angiosperms, and fungi (to name a few). The extent to which color can be relied on depends on a knowledge of the influences of ontogenetic and environmental factors. Color pattern, when it exists, may be more useful than color itself and can frequently be discerned from preserved specimens.

Color may be quantified by wavelength or by reference to an international color standard. Color pattern is frequently described rather than quantified, although quantification is possible in some circumstances.

3. *Counts of various repeated or comparable structures.* Many external features may be counted. Examples are: (1) counts of flower parts, (2) numbers of various setae or bristles, or (3) numbers of fin rays. Such characters are frequently termed **meristic characters**. In many groups standardized methods of counting such characters are employed. So long as the methods are useful, they should be used to produce conformity and to make comparisons between studies possible. Quantitative techniques for count data are discussed later in this chapter.

Internal Morphology

Internal anatomy includes any features not found on the surface of the organism. A transparent organism may be examined directly. Opaque specimens require some kind of preparation. This may involve dissection, sectioning, clearing and staining, skeletal preparations, or other manipulations. In some groups (protists, many algae, flatworms, nematodes), internal anatomy provides most of the phylogenetic characters, even at the species level. In that vertebrate skeletons are also internal, such characters are also useful at many levels in most vertebrate groups. Frequently, external morphology will fail to provide adequate characters to reconstruct phylogenies at higher levels of analysis but internal characters will be found useful.

Sexual Dimorphism and Castes

Sexual dimorphism occurs when a species displays two different sexually mature epiphenotypes, one male and one female (see Fig. 10.1). The

Figure 10.1   Sexual dimorphism in male (left) and female (right) Northern Shovelers (*Anas clypeata*). (From Palmer, 1976. Copyright Ralph S. Palmer, 1976. Used with permission. Original drawing by Robert Mengel, University of Kansas.)

phenomenon occurs in both plants and animals, and species displaying sexual dimorphism are frequently termed **dioecious species**. Sexual dimorphism may occur only during the breeding season, or it may occur throughout the year. The characters may be directly or indirectly linked to the primary reproductive organs and may involve specialized sex organs, body size and shape, color, color pattern or any number of characters. When sexual dimorphism is suspected, males and females of different species must be examined separately for those characters involved, and the holomorphology of the species is more complicated.

The holomorphology of certain species is further complicated by the presence of castes as well as dimorphic sexes (see Fig. 10.2). Castes are common in ants, termites, and bees. The bulk of a particular colony is made of sterile workers and/or soldiers. Ross (1974) characterizes such colonies as being equivalent to a bisexual nonsocial pair of individuals.

Embryonic Characters and Ontogeny

The characters exhibited by multicellular organisms are the result of a sequence of characters which change along specific ontogenetic pathways during the growth of the organism. This may be accomplished by more-or-less continuous growth or by a succession of discrete body forms which undergo one or more metamorphoses. Because the characters of an organism may change with growth and other factors such as seasonal variation or reproductive status, comparisons between individuals must take these phenomena into account. Hennig (1966) has termed an individual organism at a particular stage in its life history a semaphoront (see Chapter 5). He suggests, correctly, that critical comparisons involve analysis of comparable semaphoronts (i.e., instar 1 compared with instar 1; adult male in reproductive condition against adult male in reproductive condition, etc.).

Embryonic characters are among the most important phylogenetic characters used in reconstructing phylogenetic relationships among higher groups. Indeed, embryology provided most of the evidence for phylogenetic relationships in the nineteenth century (see Haeckel, 1866) and remains among the best class of characters for justifying phylogenetic relationships at higher taxonomic levels. As mentioned in Chapter 6, ontogenetic character transformations may also be useful in elucidating

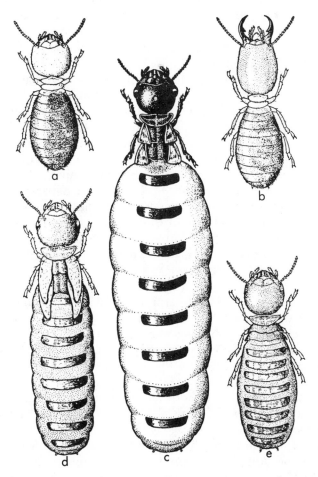

Figure 10.2 Castes in insects; five castes in the termite *Amitermes hastatus*: (*a*) worker, (*b*) soldier, (*c*) primary queen, (*d*) secondary queen, (*e*) tertiary queen. (From Wilson, 1971. Copyright Belknap Press of Harvard University Press. Used with permission.)

phylogenetic relationships within a higher taxon when the sister group relationships of that taxon are not well known.

Chromosomes

Speciation is often accompanied by rearrangements in the chromosomes of one of the two species involved (Grant, 1971; White, 1978a). Particular chromosome rearrangements may actually promote or insure lineage independence by rendering hybrids between the chromosome types sterile, of reduced fecundity, or inviable. Other types of chromosome rearrangements may vary polymorphically in a single lineage or they may be fixed in different lineages as a by-product of the speciation process (rather than its major cause). Chromosome data provide a different array of characters for the systematist to analyze and may provide the basis for species delineation and phylogenetic analysis in species where morphological similarity makes analysis on that level difficult.

In many groups (especially plants) chromosome data have become as much a part of the basic data to be reported in a standard systematic work as morphological data. Botanical systematists regularly report at least the number of chromosomes and frequently do in-depth analysis (see Davis and Heywood, 1965; Grant, 1971; Radford et al., 1974). In other organisms (such as fishes), the average systematist may report chromosome data if he or she has access to the skills and equipment, but the regular reporting of chromosome data in many animals has not yet become standard practice. The kinds of chromosome data reported depend on two factors: (1) the level of analysis that has been applied to the group, and (2) the amenability of the chromosomes themselves to the various techniques available. White (1978a) distinguishes no less than six levels of chromosome analysis. Three are commonly utilized in systematics:

1. **Alpha karyology.** The determination of chromosome number and chromosome size (the state of the art in many protists, fungi, and algae).

2. **Beta karyology.** Chromosome numbers and the location of the centromere are known, making morphological analysis of the chromosome arms possible (the state of the art in most eukaryotes—fishes are an example).

3. **Gamma karyology.** In addition to chromosome number and centromere position, various techniques are employed to stain regions of the chromosomes. This makes it possible to distinguish between homologue pairs whose arm lengths are similar (employed on a regular basis in dipterans and other insects and in many mammal groups).

The other three levels, delta, epsilon, and zeta karyology, are concerned with finer (and different) levels of analysis. Except for polytene chromosome analysis (zeta) in dipterans, these fine-level analyses do not have a sufficient data base to be useful at the level employed by the typical systematist—they are reserved for specialists in karyological systematics and are usually reported and analyzed separately.

A detailed discussion of karyological techniques is beyond the scope of this book. However, the beginning systematist will benefit from a short discussion of the nature of chromosomal data and a few specific recommendations about handling and reporting chromosomal data.

Basic Chromosome Data

1. *Chromosome number.* The chromosome number is expressed as a haploid (n) or diploid $(2n)$ number, including any variation observed. An indication of the tissue(s) used and the particular technique should accompany a report of chromosome number if the author determined the count. In comparing the works of various investigators, the reader should pay particular attention to variations in techniques.

2. *Chromosome morphology.* Chromosome morphology is best expressed with a photograph, drawing, and/or ideogram (see Fig. 10.3). The number of metacentrics, acrocentrics, and telocentrics is usually reported. Attempts should be made to pair the chromosomes of a diploid count so far as is possible.

3. *Banding patterns.* Detailed banding patterns make detailed homologue determination possible and comparisons between species can be carried out on chromosomes whose gross morphology is identical. Graphic and/or quantitative reporting procedures are indicated.

Chromosome Phylogenies

Chromosome phylogenies have been produced using whatever chromosome data is available for the group analyzed. White (1973; 1978a)

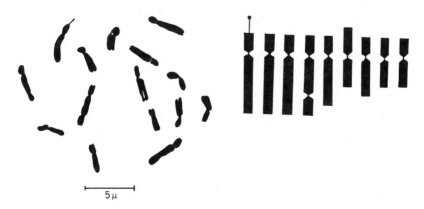

Figure 10.3 Chromosomes and ideogram of *Erigeron simplex* (From Spongberg *in* Radford et al., 1974. Used with the author's permission.)

has reviewed the data involving the number and shape of chromosomes for animals. Plant chromosome work has been discussed by Darlington (1958), Davis and Heywood (1963), Stebbins (1971) and Radford et al. (1974). More detailed work is possible by examining inversions, deletions, and translocations in banded chromosomes. Chromosome phylogenies are rarely cast by their proponents as phylogenetic in the sense of Hennig (1966). However, the assumption that observed differences in chromosome number or the presence of an inversion (especially an overlapping inversion) represents a unique historical event has a strong phylogenetic connotation.

Farris (1978a) has recently addressed the phylogenetic validity of inversion data. He found that the uniqueness postulate did not always result in a unique phylogenetic tree because the inversion (like any other phylogenetic character) must go through a polymorphic phase. This is shown in Fig. 10.4. A plesiomorphic chromosome banding sequence (p) exists in the ancestral species 1. An inversion occurs in species 1, making it heterozygous for p and the apomorphic a. Farris (1978a: 276) lists three possible fates for the apomorphic inversion, only one of which makes this inversion fixed for two sister species (see Fig. 10.4). Thus, whether the apomorphic inversion will show the sister group relationships between X and Y depends on certain probabilities which are similar in their properties to those of morphological data. Farris (1978a) suggests that phylogenetic trees that minimize the number of transitions from the heterozygous condition to the derived conditions provide the best solution to the problem under the principle of parsimony. Fortunately, this seems to be what most investigators have done with chromosome inversion data. This means that (1) chromosome data are handled in a manner similar to other characters, (2) that we can have some confidence that previously published phylogenetic hypotheses based on chromosome data will correlate strongly with phylogenetic analyses of other characters, and (3) that where morphological similarity prohibits phylogenetic analysis on that level, chromosome data can be

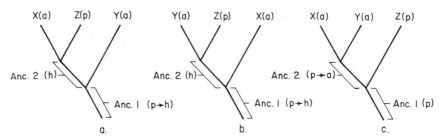

Figure 10.4 Three possible phylogenetic trees showing the outcome of fixation for a chromosome inversion (a) from the ancestral condition (p). "h" denotes a chromosome polymorphism. (Redrawn and modified from Farris, 1978a.)

used to produce genealogical hypotheses and provide a basis for phylogenetic classifications.

There are few chromosome phylogenies cast in an explicit framework of phylogenetic systematics. Yates et al. (1979), recently published such a study of three genera of peromyscine rodents. The study is interesting in that it utilizes out-group comparisons in attempting to establish the primitive chromosome bauplan for *Peromyscus* by examining the karyotypes of members of *Baiomys* and "*Neotomodon*". An interesting outcome of this study is that one of the "out-groups" (*Neotomodon-P. alstoni*, Fig. 10.5) seems to be more closely related to some *Peromyscus* than these are to other *Peromyscus*.

Chromosomes and Voucher Specimens

It is essential that voucher specimens from which chromosome data are taken (or samples of other individuals of the population taken at the specific site) be deposited in a collection where they will be available for study. Reference to the specimen(s) should be made in the original report and the specimen(s) should be catalogued and labeled as voucher specimens with an annotation that karyotypic data have been collected from them and where that data may be found. In the best of all possible worlds, the permanent slides containing the chromosomes would be deposited in the same collection for future inspection.

Biochemical Characters

Biochemical data includes estimates of the genetic diversity of structural enzymes and proteins, and direct assessment of the structure of DNA and RNA. Because I am not a biochemist I will not attempt to treat the various types of analyses in depth. Rather, I would like to discuss the potential of various classes of biochemical characters. One recent review of the various techniques is by Bush and Kitto (1978).

Alloenzyme Data—One Dimensional Electrophoresis

Alloenzymes are structural enzymes. They are usually studied by electrophoresis, an array of techniques which separates various enzymes by their molecular weights and/or electric charges (usually the latter). Its original application involved estimates of genetic variability in populations (Hubby and Lewontin, 1966; Lewontin and Hubby, 1966; Harris, 1966: see summary by Lewontin, 1974). This proved a great boon to population geneticists, even though the technique underestimates the actual genetic variability of structural genetic loci by as much as a factor of two or three (Bernstein et al., 1973) and even though it became ap-

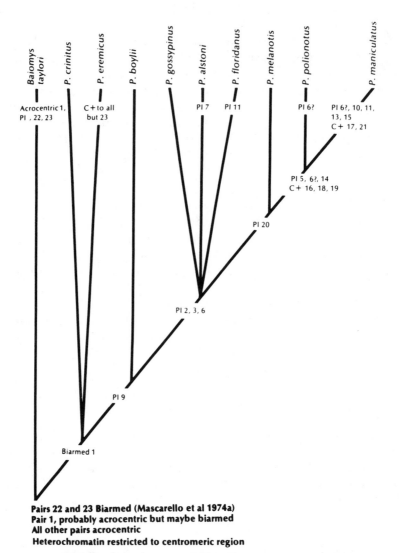

Figure 10.5   A phylogenetic hypothesis of relationships among some peromyscine rodents as evidenced by chromosome data. Inferred plesiomorphic characters are shown at the base of the tree, apomorphic characters are placed on the tree at their hypothesized origins. (From Yates, Barker, and Barnett, 1979. Used with permission of the editor, *Systematic Zoology*.)

parent that certain enzymes are less variable than others (Gillespie and Kojima, 1968; Kojima et al., 1970).

The study of populational variation properly belongs to the population geneticists. However, the phenomenon of geographic variation may be viewed as a proper subject for systematists since it is one end of a spectrum of structural gene variation between populations within species and between species themselves. No causal effect can be drawn from structural gene allele differences (Lewontin, 1974), but there is no *a priori* reason to think that true alloenzyme variation cannot be as valuable in systematics as other types of data. However, electrophoretic data are subject to restrictions, which have been summarized by Avise (1974). They are:

1. Two different gene loci may code for enzymes of identical electrophoretic mobility. The larger the number of species dealt with and the closer their phylogenetic relation, the greater the chances for spurious homology. The most direct way of detecting spurious homology is by amino acid sequencing, a technique not generally available. Thus, the tacit assumption is made that "electromorphs" are homologous.

2. Electrophoresis detects only those amino acid substitutions which effect electrophoretic mobility. The technique consistently underestimates actual differences between species. Thus, we must make the tacit assumption that two populations are equally likely to have mutations which affect electrophoretic mobility *and* that these mutations are a random sample of the total genetic changes in the enzyme.

3. Electrophoretic techniques are generally restricted to water soluble proteins encoded by structural genes.

With these restrictions in mind and with the caution that much of the analyses of electrophoretic data have been overtly phenetic rather than phylogenetic, we may examine some conclusions that have been derived from electrophoretic studies.

1. *Populations within recognized species are generally more similar electrophoretically than populations of different species.* In *Drosophila*, intraspecific populations similarity coefficients averaged between .71 and unity (Nei's coefficient: $\overline{X}$ = .91)[1] whereas between species coefficients ranged between .27 and .67 ($\overline{X}$ = .47) (Ayala et al., 1974). In closely related members of the *Drosophila willistoni* species group, where some confidence can be expressed in the genealogical closeness of the species, local populations within species had an average coefficient of .97 while subspecies and species ranged from .352 to .795. Similar results were found between ten species of sunfishes of the genus *Lepomis* (Avise and Smith, 1974a,b). Intraspecific similarity, using the Rogers coefficient, varied from .78 to .99 ($\overline{X}$ = .92); interspecific similarity varied from .37 to .79 ($\overline{X}$ = .54). Results obtained from a number of mammals (Johnson and Selander, 1971; Johnson et al., 1972; Selander

---

[1] There are a number of genetic similarity coefficients. They are sufficiently different from each other to make direct comparisons between studies difficult.

et al., 1969; Avise et al., 1974; Patton et al., 1972), reptiles (Webster et al., 1972; Hall and Selander, 1973), and amphibians (Hedgecock and Ayala, 1974) fall within or near these ranges.

These studies indicate that electrophoretic data can contribute to decisions about the species level. However, there is evidence that there is a correlation between ease of morphological identification and electrophoretic similarity. Two semispecies each of the mammal genera *Thomomys* (Patton et al., 1972) and *Mus* (Selander et al., 1969) are more similar electrophoretically (.84 and .77) than species of the genera *Dipodomys* ($\overline{X}$ = .61; Johnson and Selander, 1971) and *Peromyscus* ($\overline{X}$ approximately .66; calculated by Avise, 1974, from data reported by Avise et al., 1974). Two similar results ($\overline{X}$ = .79, Hall and Selander, 1973) compared to average coefficients among *Anolis* species ($\overline{X}$ = .21; Webster et al., 1972). If we can equate the term "semispecies" with difficult decisions on the species level, electrophoretic studies do not provide an infallible yardstick for decision-making.

In other studies, electrophoretic similarity appears decoupled from morphological divergence. Electromorph similarity among easily recognizable species of Death Valley pupfishes (Cyprinodontidae, *Cyprinodon*) averages .90 (range .84 to .97; Turner, 1974), a similarity coefficient comparable to that between populations of the same species of *Drosophila*. Coefficients of several genera of minnows (Cyprinidae) vary between that expected of local populations or sibling species of the *D. willistoni* group (Avise and Ayala, 1975). Genetic variation between *Pan* and *Homo* is comparable to that of sibling species of the same species group (King and Wilson, 1975). This indicates that there is an asymmetry in the data. Low values of electromorph similarity tend to corroborate decisions that two different geographic populations are species, but high values do not necessarily indicate that they are the same species.

2. *Electromorph similarities and differences are no different in kind than other sorts of data in evaluating phylogenetic hypotheses.* Put simply, electromorph data can be analyzed by phylogenetic methods in the same manner as other morphological characters. This conclusion has been demonstrated empirically by Mickevich and Johnson (1976) with a comparison of electromorph and morphological data on the atherinid fishes of the genus *Menidia*. The study is worth some extended discussion because it points out the shortfalls of phenetics as well as the advantages of alloenzyme data when analyzed phylogenetically.

Mickevich and Johnson (1976) analyzed sixteen populations representing five nominal species of *Menidia*. Nineteen alloenzyme and six morphological character complexes were examined. When analyzed phenetically, the alloenzyme and morphological data produced rather different results (Fig. 10.6c,d). However, a cladistic analysis (via a minimum-length Wagner tree algorithm) produced largely concordant results (Fig. 10.6a,b). The results were not completely congruent because there

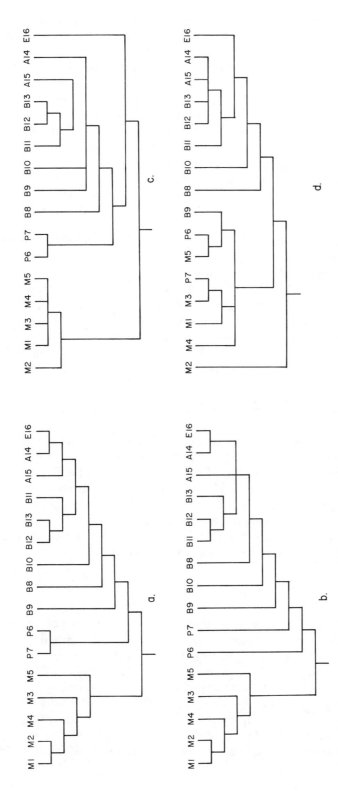

Figure 10.6 Phylogenetic and phenetic analyses of alloenzyme and morphological data for populations of the fish genus *Menidia*: (*a*, *b*) phylogenetic analyses; (*c*, *d*) phenetic analyses. (Redrawn from Mickevich and Johnson, 1976.)

331

were nonhomologous similarities (convergences, homoplasy) in the data set (see Farris, 1971). The study was summarized by Mickevich and Johnson (1976: 269) with three points. (1) The assumption that alloenzyme data is more valuable than morphological data in assessing phylogenetic relationships cannot be upheld. (2) The underlying biological differences in the genetic bases of alloenzyme and morphological data do not lead to incongruence in inferring genealogical relationships. In other words, there are apomorphic and plesiomorphic alloenzymes just as there are with morphological data. (3) Phylogenetic methods perform well when nonhomologous similarities are inadvertently included in the data.

While other investigators have claimed some consistency between alloenzyme and morphological data, these claims have been made in terms of phenetic analysis of the alloenzyme data compared to traditional classifications. Avise and Smith's (1974b) comparison with Branson and Moore's (1962) conclusion concerning *Lepomis* and Avise's (1974) comparison of electrophoretic and morphological classification of *Peromyscus* are not very convincing. Such comparisons would yield more interesting evolutionary results if studied phylogenetically because, in all probability, the contrasted similarities are plesiomorphic similarities.

Alloenzyme Data—Two-Dimensional Electrophoresis

A technique developed by O'Farrell (1975) permits more precise resolution of alloenzymes. It consists of separating proteins first by their electric charge (isoelectric focusing) and then by molecular weight in sodium dodecyl sulfate gels. The position of specific proteins can be determined by labeling. Because of its sensitivity, this technique can detect subtle differences which cannot be resolved by conventional electrophoresis. The usefulness of the data obtained for phylogenetic analysis has not, as yet, been examined (Bush and Kitto, 1978).

Amino Acids

Amino acid sequences may be compared between species of higher taxa for specific proteins. There are two basic ways of examining the changes in a protein amino acid composition, the indirect approach provided by immunology and the direct approach provided by amino acid sequencing. Immunological techniques are relatively rapid and inexpensive, but they do not provide the resolution of analysis provided by the more expensive sequencing techniques. Both are useful in assessing phylogenetic relationships. Before discussing these techniques, we must distinguish between types of homologous genes. Fitch (1970) recognized two different types of gene homologies:

1. Orthologous genes. Homologous genes that are the result of speciation so that the changes in these genes reflect the history of the group examined.
2. Paralogous genes. Homologous genes that are the result of gene duplication. Paralogous genes evolve independently during the history of speciation of the group examined.

The α hemoglobins of mice and men are orthologous. The α hemoglobin of mice and β hemoglobin of men are paralogous, the result of a gene duplication in the common ancestor of the two groups (and, of course, other groups). Fitch (1970: 113) emphasizes that "phylogenies require orthologous, not paralogous, genes." Thus, valid phylogenetic results involve the comparisons between proteins coded for by orthologous genes. This is not to say that the presence of duplicated genes cannot be interpreted phylogenetically—the gene duplication itself is a synapomorphy—as in the duplication of carbonic anhydrases in Mammalia (Maren, 1967).

Immunology—Microcomplement Fixation

Complement is a protein complex found in vertebrate blood serum. It has two special properties: (1) it binds irreversibly with antigen-antibody complexes, and (2) it can lyse (break open) blood cells that have been sensitized by a coating of anti-red blood cell antibody. Microcomplement fixation takes advantage of these properties to produce quantitative measurements of gross amino acid substitutions which have occurred between the orthologous proteins of two species. The steps of the method may be briefly summarized:

1. A highly purified protein is injected into a rabbit in multiple doses over several months to produce a serum antibody for the protein (albumin is a popular protein for tetrapods).
2. The antibody and complement are mixed with the protein. This is accomplished by serial dilution to insure that a maximum reactivity is obtained. The result is a complex of antibody, antigen, and complement bound together *plus* the excess complement that did not bind with the antigen-antibody complex.
3. The residual complement can then be reacted with sensitized red blood cells (RBC). The complement will bind to as many of the red blood cells as there is complement. Those red blood cells that are bound to the complement will lyse whereas those left will remain unlysed. The number of lysed red blood cells is directly proportional to the amount

of residual complement. The amount of residual complement, in turn, is related to how well the antibody and the antigen reacted. By measuring the amount of unlysed RBCs, a quantitative measure of the amount of antigen-antibody reaction can be gained.

4. The degree to which the antigen and antibody bind to each other and thus bind to complement is directly related to the similarity in amino acid sequence between the protein used to produce the antibody and the protein reacted with the antibody. In other words, an orthologous protein which is very similar in amino acid composition to the orthologous protein homologue used to produce the antibody will bind more complement than another orthologous protein which is less similar. Thus, by examining the relative amounts of free hemoglobin resulting from a series of tests using orthologous proteins from different species, a relative measure of immunological differences can be obtained.

The results are then used to compute dissimilarity indexes between species. Because there is a good correlation between immunological dissimilarity (distance) and the number of amino acid differences between two species' orthologous proteins (Prager and Wilson, 1971; Champion et al., 1974; Wilson, 1975), the method yields an indirect estimate of the gross number of amino acid substitutions which have occurred since the species diverged from their common ancestors. Given that these changes are stocastic, one may then construct a phylogenetic tree for the species compared.

Microcomplement fixation results do not lend themselves to classic phylogenetic analysis as it has been discussed in this book. This is because the analysis is, by its very nature, introspective. That is, it does not key on out-group comparisons for each discrete character complex (proteins) for formal recognition of primitive and derived orthologous proteins. However, the analysis may be intuitively cladistic in the sense that any changes which occur after separation from a common ancestor do represent unique evolutionary innovations and thus the overall similarity may be a measure of relative synapomorphic similarity. The tacit assumption that must be made is that nonhomologous similarities do not cloud the analysis. And this must always remain an assumption since the actual amino acid substitutions cannot be determined from the results. Another assumption is that the algorithm is grouping on the basis of synapomorphic similarity. However, it should be noted that the branching results and the determination of convergence are dependent on the algorithm used. Mickevich (1978) has pointed out that the branching sequences can be grouped on the basis of plesiomorphic similarity rather than apomorphic similarity. Thus, it is possible that some branching diagrams produced by some algorithms using microcomplement fixation are not phylogenetic. However, the close correspondence between cladograms produced using microcomplement fixation and recognized phylogenetic hypothesis (see Fig. 10.7) supports the view that microcomplement fixation, utilizing appropriate algorithms (e.g., Farris's, 1972) is a source of phylogenetic data.

For those interested in more information on the technique and its application, the following list of references divided into taxonomic groups is offered.

1. *General reviews.* Champion et al. (1974); Bush and Kitto (1978).
2. *Insects.* Brosemer et al. (1967); Marquardt et al. (1968); Fink et al (1970); Collier and MacIntire (1977).
3. *Lissamphibians.* Wallace et al. (1973); Maxson and Wilson (1974, 1975); Maxson et al. (1975); Maxson (1977); Bisbee et al. (1977).
4. *Reptiles.* Gorman et al. (1971).
5. *Birds.* Prager and Wilson (1971); Prager et al. (1974); Ho et al. (1976); Prager et al. (1976).
6. *Primates.* Sarich and Wilson (1966, 1967a,b); Wilson and Sarich (1969); Arnheim et al. (1969); Nonno et al. (1970); Sarich and Cronin (1974).

Amino Acid Sequencing

The most direct method of examining amino acid differences between orthologous proteins is to actually sequence the proteins and compare them. This operation is both expensive and time consuming and this is the reason why alternate indirect methods are still in wide usage. Unlike data from methods such as microcomplement fixation, sequence data is detailed enough to provide information on the actual changes at specific points in (ultimately) that section of DNA which codes for the protein examined.

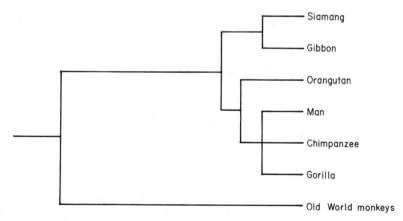

Figure 10.7 A phylogenetic tree of some anthropoid primates based on immunological differences in albumin proteins. (Modified from Sarich and Wilson, 1967.)

The usual procedure for sequencing begins by breaking up the protein into small peptide fragements. One run is done with one enzyme (usually trypsin), another with a second enzyme (usually chymotrypsin). The enzymes utilized hydrolize the protein at different points. The resultant peptides are separated by techniques such as column chromatography and each peptide is analyzed by a protein sequencer (or sequinator). Because each batch of proteins was hydrolized at different points, the overlap in amino acid sequences between the two groups of analyzed peptides may be compared to determine the complete amino acid sequence of the protein.

The datum resulting from this procedure is the sequence of amino acids which constitute the primary structure of the protein examined. In Table 10.1 the sequence 9 to 20 of cytochrome c are shown for humans, rhesus monkeys, and horses. Included are out-groups composed of "all other mammals" and "other metazoans" summarized by Dobzhansky et al. (1977: 296). The cytochrome c's are identical except for positions 19 and 20. Here humans and rhesus monkeys share isolucene (I) and methionine (M), whereas horses and the out-groups share valine (V) and glutamine (Q). The usual method of phylogenetic inference outlined in Chapter 5 can be applied to this problem. It is more parsimonious to consider the presence of isolucene and methionine, in positions 19 and 20, respectively, as synapomorphic for humans and rhesus monkeys (Primates) than to consider the presence of valine or glutamine as synapomorphic because a monophyletic Primates is more conqruent with other data than a monophyletic group composed of horses, grey whales,

Table 10.1 Amino acid sequences for positions 1–20 of three mammal species and two out-groups. Data from Dobzhansky et al. (1977). Because numbering is taken from domestic wheat, the first eight positions are missing. The out-group is composed of all other mammalian species, including such diverse mammals as sheep, dogs, the grey whale, and kangaroos.

Species		Amino Acid Position											
	1–8	9	10	11	12	13	14	15	16	17	18	19	20
Homo sapiens	—	G	D	V	E	K	G	K	K	I	F	I	M
Rhesus monkey	—	G	D	V	E	K	G	K	K	I	F	I	M
Equus	—	G	D	V	E	K	G	K	K	I	F	V	Q
Other mammals	—	G	D	V	E	K	G	K	K	I	F	V	Q
Other metazoans	—	G	D	V	E	K	G	K	K	I[a]	F	V	Q

[a] The "tuna" is an exception in having threonine at position 17 (an autapomorphy at this point of analysis but perhaps a synapomorphy somewhere within Actinopterygii).

and kangaroos (among other mammals). Further corroboration for this conclusion can be drawn from other out-groups. Four species of birds, one turtle, a bullfrog, and a tuna all have valine and glutamine at positions 19 and 20, respectively (Dobzhansky et al., loc. cit.).

The problem with manually working out the phylogenetic relationships of taxa as evidenced by protein sequences is the very large number of comparisons one must make. Cytochrome c varies from 103 to 112 amino acid positions in length (wheat is the longest, metazoans are uniformly 103 sequences). Thus, biochemical systematists have turned to computer algorithms built around the principle of maximum parsimony. That is, methods which minimize the number of amino acid substitutions needed to explain the sequence data. Some algorithms are better at doing this than others. Fitch (1971) points out that the early parsimony algorithms of Camin and Sokal (1965), Farris (1969, 1970), and Fitch and Margoliash (1967) provide hypotheses that would be counted among the relatively parsimonious solutions but that one cannot be sure that a more parsimonious hypothesis does not exist. Fitch's (1971) own algorithm attempts to provide for maximum parsimony solutions. The advantages of this algorithm are (1) it is phylogenetic sensu Hennig, (2) its assumptions are biologically reasonable, and (3) it has been mathematically proven (see Hartigan, 1973). Similar parsimony methods have been proposed and used by Dayhoff (1969) and Moore et al. (1973; also see Goodman et al., 1974). My apparent preference for Fitch's algorithm is due to my familiarity with its methods and objectives rather than any particular reason for doubting the validity of the other approaches. Certainly all three yield trees that correspond closely with the hypothesized phylogenetic relationships of the groups analyzed based on other characters (see, for example, Fig. 10.8).

The following compilation will provide some access to the literature. It is *not* comprehensive.

1. *General.* Fitch (1971, 1973, 1975, 1976); Fitch and Margoliash (1970); Boulter (1973b); Goodman (1976).
2. *Phylogenetic methods.* Fitch and Margoliash (1967); Dayhoff (1969); Fitch (1970, 1971, 1973, 1975); Moore et al. (1973); Fitch and Farris (1973).
3. *Applications* (not cited above). Tashian et al. (1972); Wooding and Doolittle (1972); Boulter et al. (1972); Boulter (1973a); Boulter and Peacock (1975).

DNA Hybridization

DNA hybridization techniques are used to measure how similar two single strands of DNA nucleotide chains are to each other. The techniques take advantage of three characteristics of DNA: (1) A double helix

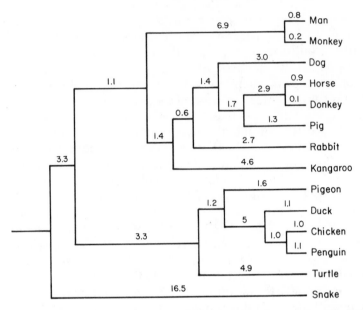

Figure 10.8 A phylogenetic tree of selected organisms based on nucleotide differences in cytochrome c. Numbers represent distances between branches. (From Fitch and Margoliash, 1967.)

may be denatured into a series of single strands of nucleotides by heating. (2) Parts of the double helix may be renatured by incubating a mixture of single strands. (3) The amount of renaturing, or annealing, is directly proportional to the similarity (complementarity) of the strands and this may be measured by heating the hybrid strand to determine its melting point.

Nontechnical summaries of DNA hybridization can be found in Dobzhansky et al. (1977) and Bush and Kitto (1977). There are many techniques; the following summary of steps is meant only to convey the basics of the analysis:

1.   Radioactive labeling the DNA of one species (this can now be accomplished *in vitro*: Commorford, 1971).

2.   Denaturing the DNA from each species and obtaining strands of approximately the same length (older techniques) or obtaining strands of "unique" DNA (the newest and best method for systematic work, see below).

3.   Producing hybrid DNA by mixing the strands of one species with those of a second species.

4.   Determining the melting point of the hybrid DNA by raising the temperature of the solution and counting the amount of radioactivity given off by eluding the mixture during heating.

The higher the temperature needed to denature the hybrid DNA strand, the more similar the two species' DNA. In other words, the more thermally stable the hybrid DNA, the more similar were the strands.

Early DNA hybridization work was accomplished without separating the DNA into repetitive and unique positions. The problems, as Nei (1975) and Throckmorton (1978) pointed out, is that one was never really sure what one was working with—thus the results were suspect. Bush and Kitto (1977) are more hopeful for the results obtained using only unique segments. Certainly the phylogenetic hypotheses produced by such studies as Kohne et al. (1972) are similar enough to well established phylogenetic relationships based on other characters to give one some confidence in the results obtained from such analyses.

Biochemical Techniques—A Summary

Biochemical techniques are not the panacea for phylogenetic reconstruction. But they do provide useful data which may be incorporated into phylogenetic studies when it is realized that the data may be analyzed in the same manner as other sorts of phylogenetic data. The old adage "the closer to the gene the closer to the truth" simply does not wash (Throckmorton, 1978). The closer to the gene, the more chances there are to pick up nonhomologues that are structurally identical and thus provide no chemical or morphological features that may lead to their identification as structural but not phylogenetic homologues. Given these problems, biochemical data must be analyzed carefully and by methods which are relatively robust to the problems of nonhomology. Such techniques as the Wagner tree algorithms used by Mickevich and Johnson (1976) or the maximum parsimony method of Fitch (1971) do seem to provide phylogenetic solutions (sensu Hennig) to complex biochemical data. Phenetic methods are too subject to the problems of nonhomology to give satisfactory answers.

QUANTITATIVE CHARACTER ANALYSIS

Quantitative characters are features of organisms which are either measured, counted, or scored. In contrast, qualitative characters are described in words or figures. A particular character is usually quantified for one of two purposes. (1) The investigator may feel that the character is more adequately described by quantifying it rather than trying to verbalize or illustrate it. Quantification in these cases elminates subjective descriptions which may not be understood by others. (2) The character or suite of characters may vary in such a way that the investigator may wish to use statistical techniques to study it. There are two basic types of quantitative characters.

1. **Continuous characters.** Characters normally described by the method of measuring them. In systematic work these usually comprise measures of length, width, breadth, volume, or shape of a feature of an organism. Shape is usually described by a combination of other continuous characters.

2. **Discontinuous characters.** Characters that describe the number of certain components which contribute to making a feature. Examples include the number of fin rays in a particular fin or the number of carpels in a flower, and numbers of chromosomes in a genome.

The following discussion is divided into two sections, univariate data analysis and multivariate data analysis. For the univariate section I assume that the reader has a working knowledge of some basic statistical procedures. Those mentioned are only a subsample of an array of procedures designed to answer specific questions. They are not necessarily the best for every application. For the multivariate section I do not assume a working knowledge of the techniques described. The purpose of this section is to describe two basic methods of multivariate analysis which I have found useful for certain systematic applications. Readers familiar with multivariate techniques are no doubt familiar with other techniques and many other applications. My objective is to introduce these basic procedures in the hope that those who have not used multivariate techniques will be interested enough to learn more about other techniques and their application. Before turning to these sections, I would like to briefly discuss quantitative character selection.

Selecting Quantitative Characters

The kinds of quantitative characters an investigator will select depend on the purpose of the study. The investigator's purpose may be to provide a basic description of a species or group of species or the investigator may wish to contrast character variation of individuals, species, or groups of species.

If the purpose of the investigator is to provide a basic description, then characters will be selected that describe size, shape, the number of repeated structures, and important anatomical and morphological features require quantitative description. The result of such a description is an account that "typifies" the taxon and documents estimates of variation found for the taxon for each character studied. The purpose of such a description is not comparative *per se*. However, basic descriptive data are the basis for comparative analysis between taxa and the investigator should make every effort to standarize counts, measurements, and other quantitative data with the methods of other investigators. Workers in some groups have developed standard methods. These should be followed when possible and reference should be made to the original paper outlining these methods.

If the purpose of the study is to compare taxa, then those characters which facilitate such comparisons should be selected. For example, the investigator may observe that the shape of a particular appendage differs from group to group. The investigator might select a series of measurements to describe the shape of the appendage, measure samples from each taxon, and apply statistical procedures to test the hypothesis that the appendages differ in one or several respects. Comparative studies are often facilitated by reviewing the results of previous investigator's studies. A series of counts or measurements useful in comparing species within one genus, for example, might also be useful in comparing species in a closely related genus. Previous analyses for certain characters may have already been made for the group investigated. If the study was done correctly the investigator would not have to repeat the counts and measurements. If the study was done incorrectly (in the opinion of the investigator), then it might be possible to obtain the original data and specimens. In either case, the work load of the investigator will be decreased, allowing time to examine other characters. I have developed a certain strategy for selecting quantitative characters for between-species analysis which I have found helpful in my own work:

1.   Check the literature for studies of geographic variation in the species. If data are available for certain characters, then these can be checked to see if geographic variation is present. Those characters that show no significant geographic variation can be eliminated from the list as requiring many specimens from many parts of the range.

2.   Determine if the raw data are available and if the specimens from which the raw data were taken are individually tagged. If so, then determine if your counts and measurements are the same on the same specimens as those of the original author (they will rarely be exactly the same, but the means or modes may not differ significantly).

3.   If these specimens represent an adequate geographic sample, then they represent a reservoir of specimens on which additional counts and measurements can be made without having to repeat the work of others. It is easy to fill out the needed quantitative data matrix when part of the data matrix has been filled by previous work.

4.   If these specimens do not represent an adequate geographic sample, then borrow or collect other samples.

5.   For characters not previously analyzed I like to do the following.

a.   If the analysis is concerned with species differences, take two or three geographically separated samples of one species and see if it is different from the species you are comparing it to. If not, you have the wrong quantitative measurement for your purposes. If it is significantly different, then proceed to analyze more populations.

    b. If the analysis is concerned only with basic description and not statistical comparison, then take counts and measurements which describe the species.

    c. If the analysis concerns geographic variation then compare widely separated population samples to see if such variation exists for the character. Do not blindly charge in on the eastern part of the range and work west only to find that they all look the same.

Univariate (Single Character) Analysis

Univariate analysis may involve one of two types of data manipulation. Descriptive analysis presents simple summary statistics for single characters for one or more populations or taxa. Comparative analysis tests the proposition that the observed differences in certain characters are, or are not, statistically different. Because analyses will differ depending on whether one is working with continuous or discontinuous data, each is discussed separately.

Continuous Characters

    1. *Descriptive statistics.* Descriptive statistics comprise a series of summary statements about the central tendencies, variation and size of a sample of continuous data. This summary for any particular continuous character may include, but need not be limited to the following:

    a. Observed limits and range. Observed limits (OL) are the minimum and maximum observed values, taken from the raw data.

OL: 65 to 95 millimeters

The range is computed from the number of units between the highest and lowest measurement:

$$\text{Range} = (X_h - X_l) - 1 = (95\text{--}65) - 1 = 29$$

This statistic is useful if frequency distributions are also to be graphed. (Note that in many studies the observed limits are often termed the "range.")

    b. Number of specimens examined (N or n). This datum gives the reader a feel for the adequacy of the sampling, permits the reader to know exactly how much data were actually used, and permits other investigators to compute standard error from standard deviation or vice versa.

c. Mean $(\overline{X})$. A measure of central tendency for continuously distributed data. The mean is computed from the formula:

$$\overline{X}_i = \frac{\sum X_i}{n}$$

d. Measures of variation around the mean. Several different and related statistics may be presented to show the variation around the mean of raw data. These include (besides the range):

$$\text{Variance } (s^2): \frac{\sum(X_i - \overline{X})^2}{n-1}$$

$$\text{Standard deviation } (s, s_D, SD): \sqrt{\frac{\sum(X_i - \overline{X})^2}{n-1}}$$

$$\text{Coefficient of variation } (CV \text{ or } V): \frac{100s}{\overline{X}}$$

The variance is rarely reported because it is in squared units. Standard deviation is an estimate of the dispersal of raw data around the estimated mean of a single population. The coefficient of variation is used as a measure of relative variation for populations that have different absolute sizes (see Simpson et al., 1960, for discussion).

2. *Inferential statistics.* One might wish to report inferential statistics as well as descriptive statistics. Two of the more common inferencial statistics applied to estimates of the mean are:

$$\text{Standard error } (s_{\bar{x}}, S_e, SE): \frac{s}{\sqrt{n}}$$

$$\text{Confidence limits: } \overline{X} \pm t_n \frac{s}{\sqrt{n}}$$

Standard error is an estimate of the standard deviation of the sample mean $(\overline{X})$ for a given sample size and is usually reported with the mean, for example, 10.82 ± 0.46. Confidence limits give a range within which the true mean is predicted to fall where t_n specifies the value of t at the confidence level desired for n degrees of freedom. If is usually reported in the same manner as the standard error (i.e., 54.9 ± 2.65). It is valid to use t only if normality assumptions are accepted. When one encounters a mean value and an inferential statistic with no reference as to whether it is a standard error or a report of confidence limits, the statistic usually refers to the standard error (Simpson et al., 1960). To avoid confusion, the investigator should always specify whether the statistic is standard error or confidence limits.

3. *Presentation of descriptive statistics.* The clearest way to present descriptive statistics is in a table such as that shown in Table 10.2.

Table 10.2   Mean, standard error, and number of observations for
four measurements on five species of *Fundulus* illustrating typical
morphometric data to be collected during the project.[a]

Species	Measurement[b,c]			
	SL	BD	CPD	SntL
F. lineolatus	41.3(1.4)27	9.66(0.45)27	5.24(0.26)27	3.16(0.21)27
F. escambiae	35.6(1.7)31	7.87(0.47)31	4.44(0.24)31	2.67(0.14)31
F. nottii	37.0(1.8)26	8.09(0.55)26	4.45(0.28)26	2.83(0.18)26
F. dispar	33.3(1.1)31	8.15(0.39)31	4.62(0.20)31	2.25(0.09)31
F. blairae	34.5(0.81)82	7.85(0.27)63	4.48(0.13)63	2.21(0.07)71

[a] Unpublished data from the author.
[b] Measurement abbreviations are: SL, standard length; BD, body depth anterior
insertion of dorsal fin; CPD, depth of caudal peduncle; SntL, length of snout.
[c] The table reads: mean (standard error) number of observations.

Raw data are rarely reported in publication. However, if the paper contains descriptions or redescriptions of species, measurements of primary
types are frequently reported. If a table format is not used, some standardized arrangement of the characters and their statistics presented
should be followed throughout the paper.

   4. *Tests of significance.*   Various tests are available to determine if
the observed differences between two or more mean values can be reasonably considered real differences or whether they should be considered the result of sampling error. Many of the common techniques rely
on the normal distribution of raw data and assume that the variances of
the populations from which the samples are drawn are the same. For
comparing two populations or species the **Student t-test** is frequently
appropriate. For comparing three or more populations the investigator
turns to **analysis of variance** (**ANOVA**). ANOVA should be coupled with
one of an array of **multiple range tests** to determine which, if any, of the
samples are statistically nonsignificantly different and which are significantly different. In other words, finding out that 10 populations show
a significant difference for a particular character is not enough. It may
be that eight of the populations are not significantly different from each
other while the other two are significantly different from the first eight
and from each other.

   5. *Presentation of statistical tests.*   I prefer to see statistical tests
presented in a table. Included should be (a) an indication of the significance level selected, (b) the degrees of freedom (Df) available for the
test, and (c) the test statistic value.

   6. *Graphic displays.*   Trends in geographic variation or differences
between species are frequently shown in systematic literature by means

of a Dice-Leraas diagram such as that shown in Fig. 10.9. This graphic technique was introduced to the systematic community by Dice and Leraas (1936) and popularized by Hubbs and Hubbs (1953). The mean, range, and confidence intervals (or $\pm$ 2 $S_{\bar{x}}$) provide a summary of these descriptive statistics and fulfill nicely their original purpose of displaying geographic variation (or lack thereof). However, they do not always substitute for more formal statistical tests of significance. Simpson et al. (1960) provide an excellent discussion of the use of statistical inferences which may be made from Dice-Leraas diagrams.

Discontinuous Data

Discontinuous data are data which fall into discrete classes with no intermediates. Continuous data are frequently grouped into discrete classes for certain purposes, but the usual type of discontinuous data involves counts or qualitative characters given character states (for example black = 0, gray = 1, white = 2). With this type of data the systematist is usually interested in (1) basic descriptions and (2) comparing the frequency of occurrence of counts or classes between populations or species.

1. *Basic statistics.* Discontinuous data are usually displayed in frequency tables (Table 10.3). This may be accompanied by the mean or mode.

2. *Tests of significance.* Tests of significance are frequently carried out with tests usually reserved for continuous data. Such tests are usually

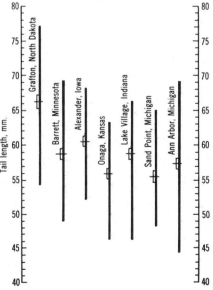

Figure 10.9 A Dice-Leraas diagram of variation in tail length of the rodent *Peromyscus maniculatus bairdii* from seven localities. (From Simpson, Roe, and Lewontin, 1960. Copyright G. G. Simpson. Used with permission.)

Table 10.3 Dorsal fin rays in nine species of *Fundulus* illustrating differences in meristic counts between subgenera in the genus

Subgenus/Species	7	8	9	10	11	12	13	14	15	16	17	n
Number of Dorsal Fin Rays												
Zygonectes[a]												
F. chrysotus			2	25	4							31
F. cingulatus			2	26	2							30
F. lineolatus	21	244	28									293
F. nottii	3	45										48
F. escambiae	11	92										103
F. dispar	11	41										52
F. blairae	29	179	19									227
Xenisma[b]												
F. catenatus						1		52	326	419	83	9	890
F. stellifer							2	121	220	38	5		386

[a] Unpublished data from the author plus published data from Brown (1958).
[b] Data from Thomerson (1969).

valid so long as there are enough classes. The investigator may also use chi-square or log-likelihood (G) tests for testing the significance of observed differences in the frequency distributions between counts or scores of two populations (or species).

3. *Graphic displays.*   A good frequency table is its own graphic display. Summary data may be put in several forms, such as the bar diagram shown in Fig. 10.10.

Bi- and Trivariate Character Analysis

The systematist can frequently sort out two or more populations or species when two or three characters are used together when single characters have failed to discriminate the populations. Bivariate analysis may be carried out simply by plotting scattergrams (Fig. 10.11), or by using scattergrams in concert with regression analysis and correlation coefficients. Statistical tests are available for (1) determining whether the slopes of the regression lines are significantly different from zero, and (2) determining if the slopes of two or more populations are significantly different. Correlation between characters may also be used to explore problems of allometry.

Three characters may be considered if they are plotted in a triangle (Fig. 10.12). Such plots may reveal species differences not fully apparent when only one or two characters are considered. They graphically display the ability of the investigator to differentiate the species involved. Trivariate scattergrams (like bivariate scattergrams) are actually quick-and-dirty multivariate analyses that require no computer but that do not yield the information on covariation that multivariate analyses can reveal.

Multivariate Character Analysis

Before systematists had ready access to electronic computers they were largely restricted to performing statistical routines on one or two characters. Multivariate clustering was limited to clusters formed by projecting individual specimens on axes of two or three raw characters. Thanks mostly to the availability of modern computers and user-oriented software in the form of package programs, systematists now have ready access to a vast array of multivariate analyses.

An excellent discussion of multivariate techniques as applied to systematics can be found in Neff and Marcus (1980). (Hopefully this reference will appear in a form more easily obtainable.) The more common

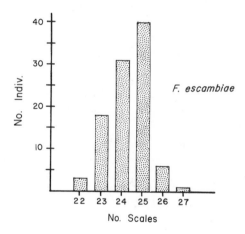

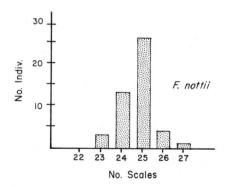

Figure 10.10 A bar diagram of the frequencies of individuals for the number of scales around the body in two species of *Fundulus* killifishes. Data based on Table 9.3.

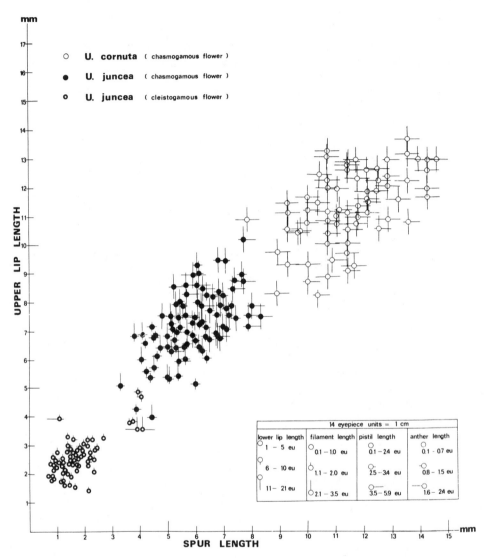

Figure 10.11 A scattergram of spur length versus upper lip length in the plants *Utricularia cornuta* and *U. juncea*. (From Kondo in Radford et al., 1974. Used with the author's permission.)

classes of techniques are:

1. Principal components analysis. A multivariate technique used to find structural relationships among variables and, in taxonomy, among specimens without *a priori* subdivision of the samples into discrete populations.

2. Linear discriminant analysis. An array of multivariate techniques used for comparing two or more populations. Character variability among populations is contrasted with that within populations. Specimens are assigned to populations *a priori*. Discriminant analysis per se refers to techniques for identifying new specimens as to their proper populations. Canonical variates analysis is used to find combinations of characters which discriminate overall and can be used for plotting scores to contrast within and among population variability in a few dimensions. Canonical variates analysis is also termed "canonical analysis," "multiple discriminant analysis," and "discriminant functions analysis."

3. cluster analysis. An array of multivariate techniques used to cluster populations or individuals into more inclusive groupings.

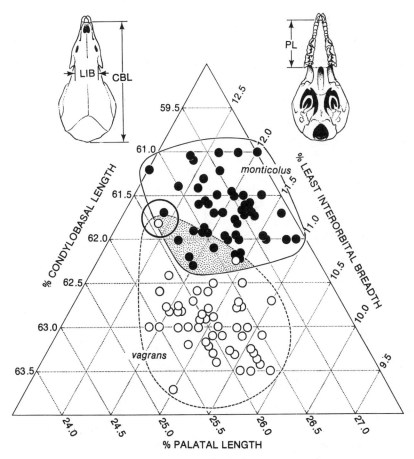

Figure 10.12 Discrimination of the shrews *Sorex monticolus* and *S. vagrans* using a trivariate plot of three measurements (shown above the plot) expressed as proportions. (From Hennings and Hoffman, 1977. Used through the courtesy of the University of Kansas and the authors.)

Principal Components Analysis

Reduces data, relat'
Reveals char.

Principal components analysis (PCA) is a method of data reduction designed to clarify the relationships between two or more characters and to divide the total variance of all characters into a limited number of uncorrelated new variables. These are termed "principal components." The first principal component summarizes more of the variability than any other variable. The second summarizes more of the variability not summarized by the first and uncorrelated with the first, and so on. This results in a significant benefit. Because the new variables are not correlated they may be interpreted independently. Thus, the total variation of a population may be broken down into *components*, each of which may say something about the size, shape, or other quantitative aspect of the members of the population. Because PCA is a method for sorting out factors contributing to size and shape of organisms, the systematist must select those characters that describe the feature studied. The number of components depends on the number of original variables. If we measure six original variables, the maximum number of components is also six.

The method is best illustrated by considering the graphic relationships between two variables, X_1 and X_2 as shown in Fig. 10.13. In Fig. 10.13a the original scattergram of individuals for each of the two variables has been summarized by a 95 percent concentration ellipse of variation. As a first step, the original axes are translated into deviation axes, calculated by substracting $\overline{X}_1$ and $\overline{X}_2$ from each original measurement. Thus, the intersection of the axes is shifted to $\overline{X}_1$ and $\overline{X}_2$, as shown in Fig. 10.13b. The deviation axes are then rotated into the principal component (PC) axes Y_1 and Y_2, as shown in Fig. 10.13c. The original data have separate variances for $\overline{X}_1$ and $\overline{X}_2$. But, there is also a total variance, a measure of the variability of all specimens for both characters. This is commonly expressed as the sum of the variances of all original variables. PC axes partition this total variance into fractions which depend on the covariation of the original variables. Because the axes are computed in such a way that they are orthogonal (at right angles) to each other and independent of each other, these fractions of total variation are additive. The **eigenvector** defines the relation of the PC axis to the original data axes. It is composed of **coefficients** associated with the original characters. If these values are scaled appropriately, they are termed **character loadings**, with each loading showing the relative contribution of each original variable to the PC axis. The **eigenvalue** defines the amount of total variation that is displayed (arrayed) on the PC axis. Frequently the proportion of variation accounted for by each principal component is expressed as the eigenvalue divided by the sum of the eigenvalues. This proportion is termed the **percent of trace** of the PC axis.

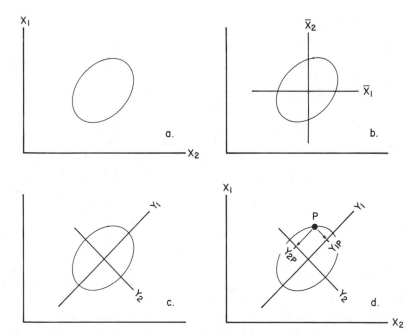

Figure 10.13 Bivariate example of transformation of raw variables to principal component axes: (*a*) raw data axes; (*b*) transformation of original data axes to deviation axes; (*c*) transformation of deviation axes ($\bar{X}_1$, $\bar{X}_2$) to the component axes Y_1 and Y_2; (*d*) the relationship of a particular individual value in raw score (P) to the component values of PC scores (Y_{1P} and Y_{2P}). The elipses represent a measure of the dispersal of the individuals in the original bivariate scattergram.

The first axis is oriented to the original specimen points in such a way as to account for the maximum amount of variation. This is simply shown by comparing the elipse of original variation and PC1 in Fig. 10.13*c*. The second PC axis is that axis oriented orthogonally to PC1 that explains the next greatest amount of the total variation (PC2, Fig. 10.13*c*). In two-space this is easily seen. Finally, note that although axes have been converted and transformed, the relative positions of individual specimens have not changed (as shown by the position of P). The position of individuals in relation to the PC axes is determined by projecting them onto the axes orthogonally. These positions are termed **PC scores**. In Fig. 10.13*d*, the PC scores for P are Y_{1P} and Y_{2P} for PC axes Y_1 and Y_2 respectively.

Matrices for PCA

Principal components analyses can be performed on one of two types of data matrices, a variance-covariance matrix or a correlation matrix.

Pimentel (1979) suggests the following guidelines:

1. If the data matrix is composed of data drawn from different kinds of measurements (length, counts, etc.), then a correlation matrix is probably preferred.

2. If the data matrix is composed of data of the same kind of measurements (for example, only lengths), then judgement must be used to determine which of the two matrices should be used. If the variance-covariance matrix is used, then absolute changes in morphology can be studied. If the correlation matrix is used, then changes relative to standardized data can be interpreted. Frequently both analyses are worthwhile.

Interpreting the Component Eigenvectors

Because eigenvectors are calculated based on the amount of total variation explained, every succeeding component will contain less and less of the total variability. Frequently the first axis will represent general size and it may explain up to as much as 50 to 95 percent of the observed variation. Such a size axis is indicated when all values of the eigenvectors have the same sign and nearly equal values. Other axes may describe shape. The last axis may explain as little as 1 percent or less of the total variation. Systematists are inclined not to attach much significance to components that explain little of the total variation. However, there is no *a priori* reason to disregard any component—some variable relationships may simply be subtle rather than blatant. In other words, each component must stand on its merits and none can be assumed *a priori* to be better than others.

Interpreting PCA—An Example

Johnston and Selander (1971) and Johnston (1973) studied 16 skeletal characters in the house sparrow, *Passer domesticus*. Individuals from several areas of North America and from western Europe were included in the study. We shall only be concerned with a portion of the original study, a data matrix composed of measurements from 644 males from North America. The principal objective was to study the covariation of the characters whereas Johnston's (1973) objective was to see if different data sets give replicate results (which they do). A part of the results are shown in Table 10.4. Three principal components (I, II, and II) are reported. The character loadings for each of the 16 original characters are given. The percent of trace is found under each component. A total of 68.2 percent of the total variance is explained by these principal components. Interpretation of the three axes follows:

1. On PC axis I the character loadings are all positive and all but one (for character 2) are rather high. This indicates that the first component

Table 10.4   Principal components analysis. Character loadings from a variance-covariance matrix onto the first three principal components of variation in a 16-character set of skeletal measurements on 644 specimens of North American house sparrows (*Passer domesticus*). Character loadings were scaled by dividing the eigenvectors by the standard deviation of the original characters and multiplying this value by the square root of each eigenvalue. All data are from Johnston (1973) after Johnston and Selander (1971) and are used with the author's permission

| Variable | Principal Component | | |
	I	II	III
1. Premaxilla length	0.458	0.048	−0.146
2. Narial width	0.067	0.070	−0.003
3. Cranial width	0.502	0.070	−0.074
4. Cranial length	0.644	0.012	−0.681
5. Dentary length	0.421	0.019	−0.117
6. Mandible length	0.513	0.016	−0.559
7. Coracoid length	0.758	0.027	0.084
8. Sternum length	0.750	−0.503	0.151
9. Keel length	0.717	−0.623	0.190
10. Sternum depth	0.454	−0.149	0.031
11. Humerus length	0.760	0.192	0.091
12. Ulna length	0.842	0.219	0.060
13. Femur length	0.741	0.294	0.060
14. Femur width	0.387	−0.033	−0.003
15. Tibiotarsus length	0.807	0.406	0.254
16. Tarsometatarsus length	0.683	0.444	0.170
Percent of trace	49.6	9.5	9.1

may be interpreted as a general size axis (Johnston, 1973; for discussion see Jolicoeur and Mosimann, 1960, and Roa, 1964).

2.   The second PC axis is a contrast between sternal characters (with negative character loadings) and limb characters (with positive loadings). Johnston (1973) interpreted this component as an expression of Bergmann's and Allen's rules which state that body size should increase and relative limb length decrease in colder climates compared to warmer climates. That is, those males living in colder parts of the sampled area have bigger bodies and relatively shorter limbs than those males living in warmer regions.

3.   PC axis III contrasts skull and limb characters. Johnston (1973)

interpreted this axis as reflecting some unspecified dimension of feeding adaptation.

Note that in each the interpretation depends on a judgement as to whether a particular original character contributes a significant amount of variation to the PC axis. In the case of Johnston's analysis, a method outlined by Blackith and Reyment (1971) was used to access significant contribution (at the 95-percent level) of each of the original variables. For example, characters 1 through 7 and 14 were found to be not significantly loaded on PC axis II (loading values from 0.012 to 0.070 and -0.033) whereas the other characters were found to be significant (loading values from 0.192 to 0.444 and from -0.603 to -0.149).

PCA as a Clustering Technique

PCA is not really a clustering technique nor is it designed to discriminate groups. Nevertheless, PCA can provide a representation of the data in which one can distinguish groups that were less than obvious in the original data. Further, there are certain circumstances where PCA may be superior to techniques that seek to separate preidentified groups, such as linear discriminant analysis or canonical variates analysis.

The utility of PCA in identifying less than obvious groups is illustrated by Temple's (1968) analysis of the nominal Silurian brachiopod *Toxorthis proteus*. Temple took several measurements of the pedical value, performed a PCA and projected individual specimens onto the PC axes. He found two different morphotypes (morphs A and B, Fig. 10.14). PCA should work in this manner whenever differences between groups occurs along the PC axes of variation.

PCA and Hybrid Analysis

Another useful application of PCA is in the analysis of possible hybrid individuals. Smith (1973) and Neff and Smith (1979) have discussed the relative advantages of PCA over linear discriminant analysis in dealing with hybrids. Discriminant analysis, as applied to hybrid analysis, assumes that (1) the original data are classified correctly and (2) the unknowns belong to one of the *a priori* known groups (Neff and Smith, 1979). The second assumption "may diminish the usefulness of discriminant function analysis for hybrid identification without an independently known hybrid population" (ibid.: 177). The problem, of course, is that the morphotype of suspected wild hybrids logically cannot be used as a standard by which future suspected hybrids can be tested if one wishes to evaluate whether any of the specimens are really hybrids.

Smith (1973) and Neff and Smith (1979) suggest PCA as a method of investigating suspected hybrids. An example is their analysis of two

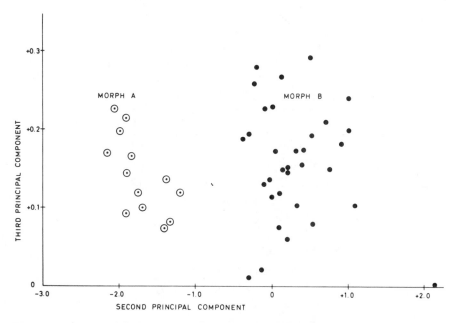

Figure 10.14 Temple's (1968) PCA of two morphs of the Silurian brachiopod *Toxorthis proteus*. [With permission from *Multivariate Morphometrics* by R. E. Blackith and R. A. Reyment. Copyright 1971 by Academic Press Inc. (London) Ltd.]

species of sunfishes (Centrarchidae) and their hybrids. Neff and Smith used 20 specimens each of *Lepomis macrochirus* and *L. cyanellus* and 40 known (pond raised) hybrids representing both reciprocal crosses. Seventeen characters were used including some that described basic shape and meristic attributes. The characters and their loading on the first two component axes are shown in Table 10.5. Plots of individual specimens projected on the first two axes are shown in Fig. 10.15. The ellipses are the 95 percent confidence ellipses for each population. Note that the hybrids fall between the known species and that the majority fall outside the 95 percent confidence ellipses of both parental species. Although Neff and Smith (1979) did not discuss introgression, I would suspect that PCA is also applicable to this phenomenon. Those interested in a fuller discussion of various aspects of PCA of hybrids should consult Neff and Smith (1979).

Discriminant Analysis

Most systematists analyze problems involving two or more populations. In such cases PCA is restricted to basic data exploration and specialized problems involving hybrid analysis. The investigator may then turn to linear discriminant analysis and canonical variates analysis to maximally

Table 10.5 Characters used in the principal components analysis (based on a correlation matrix of characters) of *Lepomis cyanellus, L. macrichirus*, and hybrids, and the loadings of each character on principal components I and II. The loadings are correlation coefficients.[a]

Character	PC I	PC II
1. Standard length	.62	.75
2. Length of head	.84	.51
3. Least depth of caudal peduncle	.70	.65
4. Dorsal origin to pelvic insertion	.52	.82
5. Length of pectoral fin	−.21	.84
6. Length of lower jaw	.88	.36
7. Length of upper jaw	.94	.23
8. Length of fifth dorsal spine	−.66	.71
9. Length of third dorsal spine	−.59	.74
10. Length of first dorsal spine	−.20	.55
11. Length of third anal spine	−.56	.72
12. Length of supramaxilla	.92	−.12
13. Number of lateral-line scales	.47	−.23
14. Number of dorsal spines	.04	−.12
15. Number of dorsal rays	−.46	.34
16. Number of anal rays	−.76	.31
17. Number of palatine teeth	.60	−.16

[a] From Neff and Smith (1979). Used with the authors' permission.

separate *a priori* designated groups, groups to which the investigator has assigned specimens before the analysis begins. To accomplish this, the technique is designed to produce new data axes that minimize variation within designated groups and maximize variation between designated groups. For statistical tests, the analysis depends on the following assumptions:

1. Each population sample is drawn randomly.
2. The populations exhibit multivariate normality (a dubious assumption, but the techniques are robust).
3. The dispersion of individuals around the population centroid (the "multivariate mean") are the same in each population.

These assumptions are the basic assumptions required for ANOVA extended to the multivariate case, MANOVA. Both techniques may be used for a variety of basic applications. The following are the most common applications:

1. Multivariate analysis of variance (MANOVA) may be used to test the equality of the centroids of two or more populations.

2. We may look at the distribution of individuals in relation to their original group assignments. Generally an individual specimen is assigned to the group whose centroid is closest to the individual. Thus although individual 1 may have been *a priori* assigned to group A, it does not necessarily follow that the analysis will reassign 1 to group A. It might be assigned to group B. If many individuals are misassigned such that no discrete clusters of individuals are apparent then we may doubt that two groups are present even if there is a significant difference between the centroids. However, a clear separation of individuals into two groups would lend support to the conclusion that two groups were indeed present.

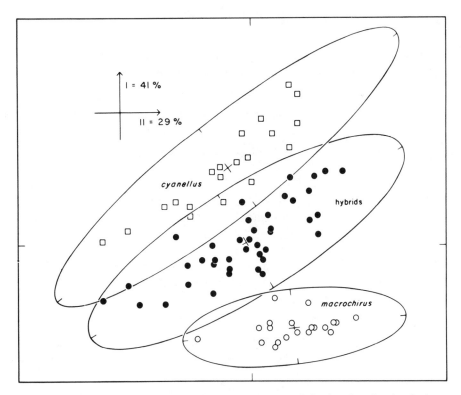

Figure 10.15 PCA of two species of sunfishes and their hybrids: plot of individuals on the first two PC axes. Squares are *Lepomis cyanellus*, open cirles are *L. macrochirus*, solid circles are hybrids of the two species. PCA data shown in Table 9.6. (From Neff and Smith, 1978. Used with the authors' permission.)

3.   Discriminant analysis may be used to diagnose unknown specimens in reference to two known populations.

Selection of Variables

The systematist selects those variables which are thought to measure group differences. Generally this is accomplished by taking enough measurements to describe the shape and form of the organism or some part of the organism. Because these analyses are designed to maximize the differences between populations, they will seek out those characters that provide the best separation. If two or more populations can be discriminated completely on the basis of one or more invariate characters then there is no reason to utilize these characters in a matrix (and one might question whether there is any reason to use discriminant or canonical variates analysis at all if the purpose is group discrimination). Pimentel (1979: 191) lists several criteria for the selection of variables. Some of those criteria are:

1.   *Theoretical approach.*   Knowledge of the literature should suggest measurements that might be correlated with group differences. Measurements of previous studies should be considered.
2.   *Accuracy, precision, and cost.*   One should obtain measurements that are accurate and precise. One must judge this in terms of cost, equipment, facilities, time, and effort.
3.   *Nature of variation.*   A good variable will vary more between groups than within any group.
4.   *Directness.*   Avoid taking a single measurement for an entire sample and applying it to all individuals of that sample.
5.   *Consistent variables.*   Variables should not change in meaning from specimen to specimen or taxon to taxon. Variables that do so are termed "inconsistent." Inconsistent variables are frequently nonhomologous. They may also be serial homologues.
6.   *Complete observations.*   Each individual must consist of a complete set of observations. If there are only a few missing observations there are techniques for estimating these data reasonably. If there are many missing data, the individual should be discarded from the analysis.

We may now consider two applications of discriminant analysis (one involving two groups, the other involving many groups) with examples of their use.

Two Group Linear Discriminant Analysis

Linear discriminant analysis can be applied to several interesting systematic questions. I will detail two of these questions. First, we may ask if two *a priori* identified groups are actually distinguished from each other using a particular set of characters. Such a question might be appropriate when considering two different geographic samples or two suspected species. Such questions concern *classification* (nonhierarchical classification of individuals). Second, we may ask if certain individual specimens belong to one or the other of two distinguishable groups. For example, we may have two species and a group of unknowns. Such questions are questions of *identity*, or *diagnosis*.

Discriminant Classification. In univariate analysis we may test the null hypothesis that two means $(\overline{X}_1, \overline{X}_2)$ of two populations are identical. Further, we may assign individual specimens to one population or the other based on their distance from these means and the variability of the two samples. Analogous procedures are used in discriminant analysis.

In Fig. 10.16 bivariate ellipses are presented for two populations, M_1 and M_2. The bivariate means, **centroids**, for each population are plotted (C_1 and C_2). The ellipses around the centroids represent the dispersal of individual specimens around their respective centroid at the 95 percent concentration ellipse. Discriminant algorithms standardize the original raw data in relation to the centroids and variation within groups. This produces circular dispersions. A single discriminant axis is then computed which best separates the two populations. This axis is based

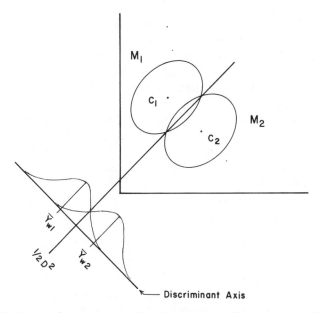

Figure 10.16 Linear discriminant analysis I, geometry. The projection of bivariate raw data (whose dispersals are represented by the two elipses) of two populations (M_1 and M_2) and their centroids (C_1 and C_2) onto a single discriminant axis.

on the original raw data weighted toward those characters that provide the best separation of the two groups. In other words, if we use a total of ten characters but only three of these characters provide good discrimination, then these three characters will have greatest influence in determining locations along the discriminant axis. This discriminant axis and its relationship to M_1 and M_2 are shown in Fig. 10.16. The contribution of each original variable to the score on this axis is usually termed a **discriminant coefficient**. The midpoint between the two original centroids is projected at right angles onto the discriminant axis. The distance between the centroids is an estimate of the distance, D, between parametric group means. The grand mean of all individuals, regardless of group membership is at the midpoint on the discriminant axis (i.e., the sum of deviations from this score equals zero). The original specimens are then projected at right angles onto the discriminant axis. The place on the axis where a specimen intersects is known as that specimen's **discriminant score** or **canonical variate score**. (In the figure the distribution of these scores is shown by the two normal curves.) If the grand mean is subtracted from an individual score, the difference will be either positive or negative and the sign will indicate to which group an unknown with that score will be assigned. Individual specimens are then assigned to the group whose mean discriminant score ($\overline{Y}_w$) is closest to the individual specimen's score. If the analysis fully separates the two groups, then each individual will be classified into the group to which it was originally assigned. The number of misclassifications (or "misses") provides a measure of discrimination for classificatory purposes.

The technique is so powerful that the simple fact that two populations can be separated cannot be used as convincing evidence that they represent (for example) two species. However, when two forms are known to be good species but are hard to separate using single characters, discriminant analysis may be used to attempt to find a multivariate character axis that can separate them. For example, Choate et al (1979) used linear discriminant analysis to separate and identify specimens of two species of white-footed mice (*Peromyscus*), which are hard to distinguish at all life stages and occur sympatrically in Kansas. Fourteen characters are used (Table 10.6). Some of these are fair univariate discriminators for certain age groups. A histogram of discriminant scores shows that all but a few subadults were discriminated (Fig. 10.17). These subadults represent less than 1 percent of the total number of each species (1 of 249 specimens of *P. maniculatus* and 2 of 189 *P. leucops*).

Identification. A more common use of linear discriminant analysis is the identification of unknown specimens in reference to two known populations. In the above example, Choate et al. (1979) are now in a position to identify any unknown specimens which might belong to one of the two species by entering these new specimens into the analysis as

Table 10.6 Discriminant coefficients for 14 characters of *Peromyscus leucopus* and *P. maniculatus* from Kansas (Choate et al., 1979)

Character	Discriminant Coefficient
1. Total length	0.032
2. Tail length	−0.097
3. Hind foot length	−0.341
4. Greatest length of skull	−1.426
5. Zygomatic breadth	−0.464
6. Interorbital constriction	−0.912
7. Greatest rostral breadth	0.321
8. Basonasal length	1.212
9. Length of bony palate	−0.977
10. Total length of toothrow	−0.650
11. Length of maxillary toothrow	−2.138
12. Breadth across molars	1.112
13. Pterygoid breadth	−0.391
14. Cranial depth	−0.334

unknowns. Their initial analysis indicates that about 99 percent of these unknowns will be correctly identified to species.

Canonical Variates Analysis (Multiple Group Discriminant Analysis)

Canonical variates analysis is an extension of linear discriminant analysis that considers the cases of three or more *a priori* designated groups. The procedures are similar to linear discriminant analysis. However, two points should be mentioned. First, because there are more than two groups to be discriminated, there can be more than one axis. In fact, the number of discriminant axes possible for any one analysis is one less than the number of *a priori* groups designated or number of variables used, which ever is less. For three groups there are two axes, and so on. This does not mean that all axes are good discriminators. As we shall see in an example later, a 21-group analysis produced only a single significant axis in spite of the fact that 20 axes were produced. Second, the scores are uncorrelated.

As an application of canonical variates analysis, we might be interested in seeing if a particular array of characters can distinguish three or more hard-to-identify species living in the same area. Another application of the technique is in exploring data sets to reach decisions on the species level using local populations as the *a priori* groupings. This approach

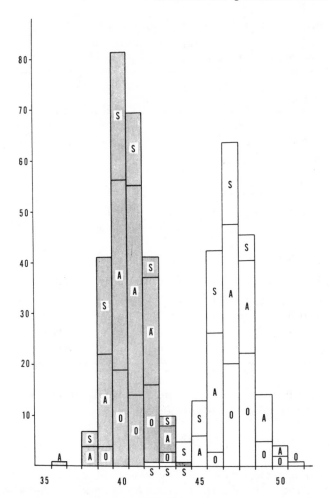

Figure 10.17 Linnear discriminant analysis II, discrimination of the rodents *Peromyscus maniculatus* (shaded) and *P. leucopus* (unshaded). Abbreviations are: S, subadults; A, adults; O, old adults. Note that the midpoint between the two species is approximately 44 rather than zero because the grand mean score was not subtracted from each individual score (a feature of some computer programs). (From Choate, Dowler, and Krause, 1979. Used with the authors' permission.)

was used by Heaney (1979) in a study of pocket gophers. The null hypothesis tested is that there are no significant differences between local population centroids. With a single species we should expect a large percentage of misclassifications and a reflection of this in the graphic display by large overlap between population plots. If two species are present in the array of population samples we might expect that two groups of populations would emerge from the analysis with little or no overlap between these two larger groups but much overlap among populations within each of the larger groups.

Heaney (1979) used this application of canonical analysis to examine populations of pocket gophers of the *Geomys bursarius* species complex. Two forms, *G. bursarius* and *G. lutescens*, have been considered by some as subspecies of the same species and by others as distinct species

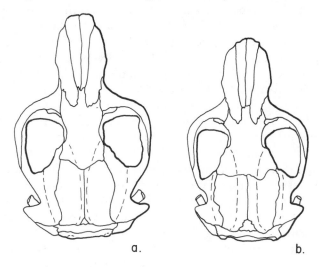

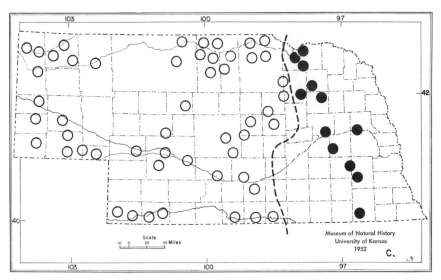

Figure 10.18 Skulls of the pocket gophers *Geomys lutescens* (*a*) and *G. bursarius* (*b*) and the geographic localities examined of the two forms in Nebraska. Skulls are shown in dorsal view. Map symbols are: open circles, *G. lutescens*; closed circles, *G. bursarius*. (From Heaney, 1979. Used with the author's permission.)

(see Heaney, 1979, for taxonomic history). The ranges of the two forms in Nebraska and a drawing of their skulls in dorsal view are shown in Fig. 10.18. Heaney (1979) analyzed 21 populations along the border between the two forms in Nebraska. Twelve characters were selected that he thought would best discriminate between the two. Each local populations was designated as an *a priori* group for the analysis. Table 10.7 shows the loadings for each character on the first two discriminant axes. Note that the first axis provides high loadings for some characters but the other two axes do not, making the first axis much easier to interpret. A plot of each individual on the first two axes is shown in Fig. 10.19*a*. These individuals are then grouped into their *a priori* local populations in Fig. 10.19*b*. Note that the 21 populations fall into two major groups, one composed of the three populations of *G. bursarius*, the other composed of the 18 populations of *G. lutescens*. If we inspect the loadings in Table 10.7, we can see that seven characters seem to provide most of the discrimination shown in Fig. 10.19: condylobasal length, length of the hind foot, and mastoid breadth are negatively loaded; total length of the skull, frontal square length, nasal breadth, and zygomatic breadth are postively loaded. Also note that the first axis eigenvalue is high whereas the second axis eigenvalue is low, and in this case so low as to make it uninterpretable. The first axis contrasts not the original *a priori* populations, but the two presumptive species. Compared to *G. lutescens*, *G. bursarius* has a shorter skull with broad mastoids and a narrow zygomatic arch and nasals, a longer condylobasal length, and a longer hind foot. An indication of some of these features can be seen in the skull drawings in Fig. 10.18. Heaney (1979) concluded that the nat-

Table 10.7 Character loadings and eigenvectors for 10 of 12 variables used by Heaney to analyze 21 populations of pocket gophers

Variable	Axis 1	Axis 2
Total length of skull	.209	.281
Length of hind foot	−.377	−.419
Condylobasal length	−.946	−.599
Zygomatic breadth	.136	.522
Mastoid breadth	−.198	.374
Nasal breadth	.148	−.082
Frontal square length	.180	.587
Frontal square breadth	.049	.315
Orbital length	.015	.249
Maxillary visibility	−.052	−.216
Eigenvalue	5.096	.595

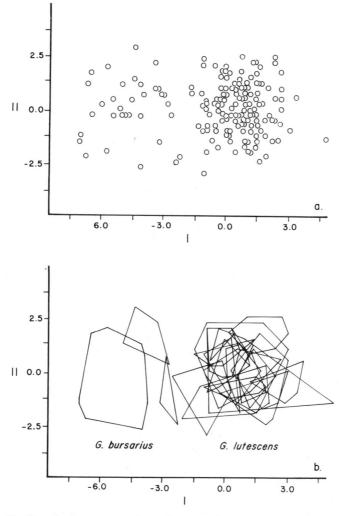

Figure 10.19 Canonical variates analysis of *Geomys lutescens* and *G. bursarius*: (*a*) plots of individuals on the first two canonical axes; (*b*) plots of population ranges for the 21 *a priori* indentified populations designated in the study. Note that in (*b*) the only discrimination is between the two presumptive species, not between *a priori* populations. (From Heaney, 1979. Used with the author's permission.)

ural clustering of *G. lutescens* and *G. bursarius* into distinct morphotypes without *a priori* identification of any population as to its presumed species was additional evidence that the two forms were good species. Naturally this cannot be considered proof positive. But it is another technique that is fruitful in making decisions and evaluating the conclusions of other workers. I believe that this approach to discriminant analysis will find wide application in the future.

Chapter 11

Publication and Rules of Nomenclature

Systematics is not simply the activity of collecting data or organisms and interpreting their historical relationships. Systematists must also be historical and taxonomic scholars. Historical scholarship enters at the level of tracing the history of names and in finding and interpreting those data and ideas presented by earlier workers. Such scholarship requires a knowledge of the kinds of systematic literature and where that literature may be found. A brief survey of these topics is dealt with in the first part of this chapter. Scholarship also involves the writing and publication of results. If there is an element of art in systematics it enters at the stage where the investigator prepares his or her paper for publication. The clear exposition of ideas is almost an artistic talent which many of us lack to a greater or lesser extent. To help those of us not blessed with such talent, various aids are available. A brief survey of the contents of a systematic paper forms the second part of this chapter. Finally, one mark of the phylogenetic systematist is that the investigator fulfills both the function of phylogenetic analyzer *and* taxonomist. To accomplish both, the investigator must be familiar with the rules of nomenclature applicable to the groups worked on. The mark of a taxonomic scholar *cum* phylogeneticist is the combination of critical phylogenetic analyses and correct taxonomic niceties. Thus, the last section of this chapter presents an overview of the three major nomenclatural codes.

SYSTEMATIC LITERATURE

We may now consider the various forms in which systematic literature may appear. The following is a survey of the various bibliographic aids that can be used to find the literature available for various groups of organisms. The listings are not meant to be exhaustive.

Types of Systematic Publications

1. *Descriptions of new species or higher taxa.* Descriptions of new species are among the most common systematic publications. Such a work may consist of the description of an isolated species or the descriptions of many species without much of an attempt to synthesize the study. The value of such publications varies with the purposes and experience of the author. The preparation of an isolated species account that is a real contribution to science (rather than a vehicle for recognition of the author as the discoverer of a new form) is a demanding exercise involving:

   a. A thorough knowledge of related species, including all synonyms.
   b. Careful examination of specimens of related species.
   c. Examination of type material to ascertain that the new species is, in fact, new.

Such isolated species descriptions can be made less isolated if comments are made regarding its presumed relationships and careful comparisons to closely related species are included. Incorporation of the new species into an existing key is also helpful. All of this helps put the new species into the context of what we know about the relationships of the group it belongs to.

Isolated species descriptions are often appropriate in the wake of a recently completed revision or monograph where the species-level entanglements one often encounters are cleared by one familiar with the group.

2. *Revisionary studies.* Revisionary studies range from synopses and reviews to monographs. Synopses and reviews are shorter papers that summarize current knowledge of a particular group. At their best, they provide a valuable source work in which all of the scattered references to a group are brought together. Revisions and monographs are more complete and usually include:

   a. The complete taxonomic history of a group.
   b. A classification (which may or may not be phylogenetic).
   c. Diagnoses and/or descriptions of each species.
   d. A key or series of keys for identification.
   e. All pertinent literature references.

Revisions and monographs are the most demanding of all systematic works for the phylogeneticist because the investigator must combine the traditional scholarship demanded of all taxonomists with the analytic techniques of the phylogenetic system.

3. *Classifications.* Classifications may be considered a type of synopsis since they present a synopsis of the author's views on the taxonomic relationships among groups (no matter how the term "relationship" is used). Examples include Berg (1947: fishes), Simpson (1945: mammals), Wetmore (1960: birds), Takhtajan (1969: angiosperms). Phylogeneticists usually present their classifications as summaries of revisionary or comparative work (for example, Patterson, 1973; Rosen, 1973; McKenna, 1975; Wiley, 1979c).

4. *Keys.* Keys are usually found in revisionary publications. However, keys may be published separately, usually for a specific region. Examples include the useful "How to Know" series (which is more than an elementary guide for identification). (The series comprises "The Pictured Key Nature Series," Wm. C. Brown Company Publishers.)

5. *Faunistic and floristic studies.* These may be termed monographs of specific areas rather than of particular groups. The results vary widely and are frequently limited to specific groups (e.g., vertebrates, angiosperms, bryophytes, etc.). Such a study includes:

a. The species which inhabit the area (and perhaps descriptions of any new species).
b. The distributional patterns of these species within the area.
c. Keys to identify the species covered.
d. The history of work in the area.

A faunistic or floristic work is a valuable systematic device when such things as the synonyms of species and species variation are reported. Radford et al. (1974: 378–379) list the following topics necessary for a complete floristic work in addition to the basic systematic data on the plants themselves:

a. Location and geography of the study area.
b. The history of botanical exploration.
c. A survey of applicable physiography and topology.
d. A summary of major biogeographic patterns placed in a historical context.
e. A summary of present ecology.
f. A resume of pertinent pedologic and geologic data.
g. A summary of pertinent climatological data.
h. A review of previous works.
i. A description of present land use or abuse.
j. A list of cited references.

6. *Atlases.* An atlas may include illustrations of the species of a particular taxonomic group, geographic area, and so on, and function as an aid for identification; it may consist only of illustrations, or, as is frequently the case, it may include distributional maps.

7. *Catalogues.* A catalogue is a tabulation of species detailing a varying amount of information. Included may be references to types, the original description, synonyms, and range.

8. *Checklists.* A checklist is a list of species for a particular group or a particular group in a specified geographic region. Checklists are frequently published for butterflies, beetles, and birds—groups popular with laypersons as well as specialists. Checklists are also commonly found in field guides.

9. *Handbooks.* Handbooks are designed to enable nonspecialists or laypersons to identify major groups and/or species occupying a particular political or geographic region.

10. *Taxonomic scholarship.* There are many publications that deal with nomenclatural problems and the rules of nomenclature themselves. Papers concerned with nomenclatural problems are usually published in journals appropriate to the problem. In zoology, papers potentially effecting nomenclature are published in *Zoological Nomenclature*, a journal sponsored by the International Commission in Zoological Nomenclature. For plants, *Taxon*, sponsored by the International Association of Plant Taxonomists, is the major forum for nomenclatural discussion.

11. *Phylogenetic analyses.* An increasing number of papers are not so much concerned with naming new taxa or taxonomic scholarship as with discovering the phylogenetic relationships among previously existing groups and perhaps suggesting changes in higher classifications to reflect this new knowledge.

Use of Literature

Systematics is a highly historical enterprise. Currently used taxonomies and current hypotheses of phylogenetic relationships result from the efforts of many investigators over the years and the systematist must know how to find and use this previous literature.

Access to the Literature

An exhaustive literature search is usually in order when a taxonomic revision is contemplated. The revisor must account for all the names ever used for the taxa revised and the nomenclatural history of these names. The best beginning is to consult the latest revision of the group. If catalogues exist, they should also be consulted. Once all pertinent references from these sources are accumulated, a literature search must then be conducted.

Literature Searches—Zoological Bibliographies

1. *Zoological Record.* The *Zoological Record* is published by the Zoological Society of London with the cooperation of the British Mu-

seum (Natural History). The *Record* summarizes the systematic literature from 1864 to the present (although it is now several years behind). The early volumes are spotty in coverage, the later fairly complete. New scientific names are given with reference to the paper in which they were described. The arrangement of families and higher categories follows the accepted taxonomy of the year of publication. The genera and species are listed alphabetically within families.

   2. *Archiv für Naturgeschichte* (now *Zeitschrift für Wissenschaftliche Zoologie*, Abteilung B. This bibliography covers the systematic literature from 1832 to.the present.

   3. *Biological Abstracts. Biological Abstracts* covers the entire biological sicences from 1926 to the present. Although more up-to-date than the *Zoological Record*, it is much less complete in its systematic coverage.

   4. *Berichte über die gesamte Biologie, Abt. A, Berichte über die Wissenschaftliche Biologie.* A German equivalent to *Biological Abstracts* covering the same period, 1926 to the present.

   5. *Bulletin Signaletique (CNRS) (several different fields).* Up-to-date and especially valuable for paleontological subjects, *Bulletin Signaletique* is unexcelled for filling in the blanks from the most recent *Zoological Record* to the present year.

Zoological Name Searches—Nomenclators

In addition to the various bibliographies just listed, direct access to names, authors, and dates for genera and species may be gained from various nomenclators. *Those proposing new generic names should always consult all nomenclators to avoid homonomy.* The various nomenclators, their coverage, and their applicability are shown in Table 11.1.

Literature Search—Botanical Bibliographies and Abstracts

I am much less familiar with botanical bibliographies. The following list includes references from Davis and Heywood (1965) and Radford et al. (1974). Additional extensive listings for angiosperm literature are listed in Radford et al. (1974).

   1. *Biological Abstracts* (1926 to the present).

   2. *Excerpta Botanica* (1959 to the present). *Section* A concerns taxonomic literature. A useful but incomplete bibliography.

   3. *Horticultural Abstracts* (1931 to the present). Good coverage of taxonomic works concerning cultivated plants.

   4. *Plant Breeding Abstracts.* Davis and Heywood (1965) list this work as a good reference to cultivated plants.

Table 11.1 Nomenclators—animals. Blackwelder (1967: 234) has conveniently ordered these by period of coverage. Other references, with emphasis on vertebrata, are found in Blackwelder (1972)

Period	Reference	Coverage
1758–1800; 1801–1850	Sherborn, C. D. (1902), (1922–1933). Supplement by Poche (1938)	Genera and species
1758–1842	Agassiz, L. (1846)	Genera
1758–1842	Agassiz, L. (1848)	Genera
1758–1873	Marschall (1873)	Genera
1758–1882	Scudder (1882)	Genera
1758–1926	Schulze, et al. (1926–1954)	Genera
1758–1935; 1936–1945	Neave (1939–1940), (1950)	Genera
1891–1900; 1901–1910	Waterhouse (1902, 1912)	Genera

5. *Bulletin Signaletique* (CNRS) (1961 to the present).

6. *Referativnyi Zhurnal, seriya Botanika.* Davis and Heywood (1965) are impressed with the coverage of this series. I have not seen it.

7. *Index Holmensis* (Tralau, 1969 and onward). A world phytogeographic index to distribution maps.

8. *Index Londinensis* (1929, supplement 1941). An index to illustrations of angiosperms, ferns, and fern allies.

9. *Taxonomic Literature* (Stafleu, 1967; Stafleu and Cowan, 2nd Ed., greatly enlarged, 1976 and onward). A selective guide to literature between 1753 and 1940.

10. *Royal Botanic Gardens, Kew: Current Awareness List* (1973 and onward). A monthly mimeographed summary of current literature.

11. *Regional bibliographies.* In addition to the international series listed, there are several regional indexes. A sample of these include:

a. *Index to American Botanical Literature (Bulletin of the Torrey Botanical Club,* 1886 and onward). ·This now includes the *Taxonomic Index* previously published in *Brittonia* (coverage: 1938–1967).

b. *Flora Maleniana* (1947 and onward). Indo-Malaysian region.

c. *A.E.T.F.A.T. Index* (Association pour l'Estude Taxononique de la Flore d'Afrique Tropicale). Sub-Sahara Africa and Madagascar.

d. *Index to European Literature* (1965–1969). This series was superseded by the *Kew Record of Taxonomic Literature* listed later.

Literature Search—Botanical Names

Botanists do not have the inclusive nomenclators available to zoologists.

The following list of references provides major access to botanical names. Additional references are found in Stafleu (1967) and Radford et al. (1974).

1. Steudel (1821–1841). *Nomenclator Botanicus.* Covers plant names from 1753–1840.

2. Pfeiffer (1873–1874). *Nomenclator Botanicus.* A valuable reference for suprageneric names.

3. *Index Kewensis.* This index covers generic and specific names for seed plants from 1753 to the latest supplement. It is the basic source for generic and specific names, but it does not include information on infraspecific names.

4. *The Kew Record of Taxonomic Literature* (1974 and onward). This work covers, to date, the years between 1971 and 1974 and includes literature concerning infraspecific names.

5. *Grey Herbarium Card Index* (1866 and onward). This index covers all new names and name combinations for extant vascular plants of the Western Hemisphere.

6. Willis (8th Edition, 1973, revised by Shaw). This work is a dictionary of family and genus names of vascular plants.

7. *Index Filicum.* (1905–1965). This index covers genera and species of ferns. Post-1964 names are covered in supplements.

8. *Index Muscorum* (Wijk et al., 1959–1969). This index covers the genera and species of mosses from 1801 to the present. Biennial supplements are published in *Taxon.*

9. Crosby and Magill (1978). A dictionary of moss family and genera names.

10. *Index nominum genericorum (Plantarum)* (1979, compiled by Farr, Leussink, and Stafleu). This work is a generic nomenclator and includes names for all plants, including algae, fungi, and fossils, with the type species indicated. More editions and fuller coverage are expected to follow.

PUBLICATION OF SYSTEMATIC STUDIES

The results of a systematic study must be published if they are to become part of the general knowledge of science. In many respects, a study that is not published might just as well never been done. Thus, the systematist must think of publication as the final step in the scientific process. The format of systematic papers varies with the type of study and the journal to which the systematist submits the manuscript. Most journals contain a section in each issue which instructs prospective authors on matters of style. The *CBE Style Manual* (now in its third edition) is an invaluable source reference for the prospective author.

The usual systematic paper contains the following parts:

1. *Title.* The title must be informative without being overly long. Many workers review the literature by scanning titles (for example, by scanning *Current Contents*) and an obscure title can produce an obscure paper no matter what the value of its contents may be. If the name(s) or group(s) dealt with are not immediately recognizable, the name of the major group and family should also appear in the title.

2. *Author's Name.* It is best for bibliographic purposes for the investigator to use the same form of his or her name. The address of the author(s) usually follows the name(s). This address is usually the address where the work was done. The mailing address, if different, follows.

3. *Introduction.* The scope of the paper should be outlined. A historical review of the problem may also be appropriately included.

4. *Acknowledgments.* Those who have contributed to the author's efforts should be credited. This may include institutions and staffs who assisted in loaning specimens, persons who helped with techniques, suggested problems, reviewed the manuscript, typed, prepared illustrations, or any number of other things which helped the author(s).

5. *Methods and materials.* This section is reserved for outlining the way the author proceeded, what techniques were used, and what specimens were examined. In revisionary work the specimens examined are frequently listed for each species in the body of the text. In this case, this section is appropriately used for institutional abbreviations.

6. *Body of the text.* The body of the text may include the results, conclusions, and discussion, as appropriate. Various aspects of more formal taxonomic sections of the body will be detailed later.

7. *Summary or abstract.* A particular journal may require a summary or an abstract but usually not both. In many respects this section is of critical importance *because it may be the only part that is actually read by many workers.* A good abstract or summary will interest the prospective reader to read the entire paper.

8. *References cited or bibliography.* Most journals which publish systematic papers require a compilation of references at the end of the paper. The alternative is to list references in footnotes (an expensive alternative). Consult the *CBE Style Manual* and instructions for style in the journal to which the paper will be submitted.

9. *Appendices.* Appendices are frequently useful for including information not readily placed in the main body of the paper. This might include a list of specimens examined, tables of statistics, lists of localities, or techniques.

10. *Footnotes.* Footnotes should be used only when the information contained in them would detract from the word flow of the main body of the text and yet is essential for full meaning. Footnotes should be a last resort.

Major Features of the Formal Taxonomic Work

Certain features of a species description or group revision have a more formal manner of presentation than do papers that do not have to conform to the various aspects of nomenclatural codes. These include: (1) name presentation, (2) diagnosis and/or description, (3) synonomies, and (4) keys. Additional subjects that deserve comment are: (5) comparisons, (6) the presentation of distributional data, (7) etymology, and (8) illustrative material. Comments on these topics will be detailed later.

A format for the formal taxonomy of a species or other taxon is listed briefly. This is but one of several possible formats and is presented only for discussion:

1. Presentation of the valid/correct name.
2. Reference to figures.
3. Synonomy.
4. Type material examined.
5. Other material examined.
6. Diagnosis.
7. Description.
8. Comparisons.
9. Discussion.
10. Distribution.
11. Etymology.
12. Keys.

Name Presentation and Figure References

I prefer to present the full name of the taxon, including the year of description. This is immediately followed by figure references or common name and figure references. Some examples are:

Tinthiini LeCerf, 1917[1]

(Figure 11a; male genitalia, Figure 6)

Bidyanus bidyanus (Mitchell, 1838)[2]

Silver perch

Figure 77

---

[1] A subtribe of sesiid moths, from Naumann (1977).
[2] A species of teraponid teleost, from Vari (1978).

Synonomies

Synonyms are various names that have been validly published and applied to the same taxon. The rules governing synonyms are discussed in a later section. For now we are concerned with presenting synonomies in taxonomic publications.

A major part of the revision of a taxon or the naming of a species is a presentation of the history of names that have been applied to the taxon. Such a historical compilation requires a thorough literature search. A synonomy fulfills this function and can be used to convey other useful information. Synonomies come in one of two basic forms. **Complete synonomies** purport to give every reference to every name ever applied to the taxon. The publication of a complete synonomy is a relatively rare event, except for small infrequently cited taxa. **Abbreviated synonomies** purport to list only those references that directly affect the taxonomy of the taxon and additional references that might directly involve the taxonomy of the group and provide entry into pertinent taxonomic literature (e.g., well known field guides, faunal or floral works, etc.).

Although only valid synonyms legally affect the history of a name, species synonomies frequently contain reference to unacceptable names and mistakes in identity as well as synonyms. Such names should be annotated to call attention to their status.

Species synonomies minimally include: (1) the original form of the scientific name, (2) the author, (3) date of publication, (4) reference, and (5) page number. In addition, they may include (6) type locality, (7) where the type is deposited, and (8) any other annotations the revisor feels are necessary. There are two ways to present the reference. First, the reference may be cited in the synonomy itself. Second, it may be cited in the bibliography. I prefer the latter when the same references are cited in several synonomies. However, the former is traditional in many groups and demanded by some journals.

There are two basic formats. The synonomy may be arranged by date of publication or it may be arranged by date of name appearance. Particular journals may require one or the other. Several examples are presented.

Example 1. Arrangement by date of publication, reference in bibliography (from Bartram, 1977).

Proterus elongatus Wagner 1863

1863 *Proterus elongatus* Wagner: 645
1881 *Notagogus macropterus* Vetter: 46

1895 *Proterus speciosus* Wagner; Woodward: 184, pl. 3, fig. 5
1914a *Proterus speciosus* Wagner; Eastman: 407, pl. 13, fig. 1.

Example 2. Arrangement by date of name, reference in bibliography (from Rindge, 1976).

Plataea calcaria (Pearsall)

Apricrena calcaria Pearsall, 1911, p. 205. Barnes and McDunnough, 1917, p. 122.
Plataea calcaria: Barnes and McDunnough, 1918, p. 151.
Plataea triangularia Barnes and McDunnough, 1916, p. 27, pl. 3, fig. 18 (holotype male); 1917, p. 115, 1918, p. 151 (placed as synonym of *calcaria*).
Plataea dulcinia Dyar, 1923, p. 23.
Plataea dulcinea [sic]: McDunnough, 1938, p. 170 (listed as synonym of *calcaria*).

Example 3. Arrangement by date of name, reference included in synonomy (from Smith, 1972).

Pteronotus suapurensis (J. A. Allen)

Dermonotus suapurensis J. A. Allen, 1911, Bull. Amer. Mus. Nat. Hist., 20: 229, 20 June.
Pteronotus suapurensis J. A. Allen, 1911, Bull. Amer. Mus. Nat. Hist., 30: 265, 20 December.
Pteronotus suapurensis centralis Goodwin, 1942, Journ. Mamm., 28: 88, 16 February.

Example 4. Arrangement by date, reference included, synonymy in paragraph style (from Koponen, 1968).

7. *Genus Orthomnion* Wils. 1857
    Orthomnion Wilson in Mitten, Kew Journ. Bot. 9: 368. 1857.—*Mnium* *Orthomnion* (Wils., Mitten, Journ. Linn. Soc. London, Suppl. Bot. 1: 142. 1859.—Typus: *Orthomnion bryoides* (W. Griff.) Norkett (cf. Norkett 1958).

Generic synonomies differ from specific synonomies principally in that (1) only available synonyms are listed and (2) the type species and the reason it is the type species are given (particulars about type species are found later in the chapter). For example, a generic synonomy of *Lepisosteus* given by Wiley (1976) is shown.

Lepisosteus Lacépède

Lepisosteus Lacépède, 1803: 331 (type species *L. gavialis* by subsequent designation, Jordan and Evermann, 1896: 109).
Sarchirus Rafinesque, 1818a: 418 (type species *S. vittatus* by subsequent designation, Jordan, 1877: 9).
Cylindrosteus Rafinesque, 1820: 72: (type species *C. plastostomus* by subsequent designation, Jordan, 1877: 11).
Lepidosteus (Lacépède): Koenig, 1825: 12; Agassiz, 1843: 2 [amended spelling of *Lepisosteus*].

Material Examined

The material examined may be listed in a variety of places. I like to list at least the type specimens I examined immediately after the synonomy and before the diagnosis. If the list of other specimens examined is extensive, it is probably better to list them at the back of each species' section or in an appendix. The style of the section should be as abbreviated as possible. Some authors quote the collectors notes, others give only localities, and yet others do not list the specimens examined but reference where such a list can be obtained. The way that specimens are cited frequently varies depending on the scope of the article and the level of previous knowledge of the group.

The Diagnosis

A **diagnosis** is a brief listing of those characters which differentiate a taxon from related and/or similar taxa. If the diagnosis contrasts these characters with other taxa, it is a **differential diagnosis**. Differential diagnoses are usually longer but more informative than straight diagnoses.

The purpose of a diagnosis is to specify in the most concise manner possible those characters that distinguish a taxon. A good diagnosis will aid other investigators to identify members of the taxon without having to compare specimens with the more extensive description. When there is no key, the diagnosis is essential. Even when a key is provided, diagnoses assist other investigators to determine whether they have a new taxon. New species frequently key to their closest relative(s) and a proper diagnosis of these relatives may assist the investigator to determine the status of the possibly new form before comparing it with the more extensive description.

There is a tendency for some to write a diagnosis as if it were a description. This should be avoided because it obscures the usefulness of the diagnosis.

The length of a diagnosis may vary from a single sentence to a rather long paragraph. There is no need to repeat characters shared by members of the inclusive taxon. The number of distinguishing characters and the nature of the diagnosis (differential or not) usually determines the length. When several taxa are diagnosed, the diagnoses should be arranged so that comparisons are facilitated.

Botanists must write at least the diagnosis of a new taxon in Latin. Stearn (1973) allows almost anyone to write comprehensible Latin and also provides a wealth of valuable information about botanical Latin and a dictionary. A full description should be written in Latin only if the author thinks it will benefit the general botanist as well as the specialist.

The Description

A **description** is an account of the characters, or of certain characters, of a taxon. Descriptions may be exhaustive or they may be concise. I prefer descriptions that concentrate on the important aspects of the morphology of the species as it relates to other species. Such a description concentrates on what Mayr et al. (1953) and Mayr (1969) term "taxonomic characters." A good description will convey an overall picture of the organism without inundating the reader with minutiae. Characters found in all members of a family do not need repeating in the description of a genus. The same goes for species within genera. If several taxonomic levels are included in the work, descriptions can be structured in such a way that there is little repetition. Descriptions may be telegraphic or in standard prose. The wishes of the editor and the policies of the journal usually determine if regular prose is allowed. If the investigator is producing a monograph or revision, the order or descriptive characters should be standardized so that each description is directly comparable; the utility of the work will be reduced if this is not done. **Redescriptions** should be cast in the same format. In describing a single species, an effort should be made to order the description in a manner similar to good descriptions of other members of the genus. This will make the work of a revisor much easier (suggestions for standard plant descriptions are given by Radford et al., 1974).

Descriptions in Atlases and Floral and Faunal Works

The emphasis of the description in a floral or faunal work is to supplement the key and/or illustration. These description are necessarily short and discuss only those characters that aid in identifying the specimen to species.

Illustrations and Graphics

Illustrations of an organism and graphic displays of its characteristics may be considered part of the description of a taxon. There are several methods of illustrating the features of an organism. Some techniques are superior for certain features but ineffective for other features. No one technique is suitable for every situation. The systematist should be familiar with these various methods and how different printing techniques affect the original illustration. Several excellent books cover this subject rather fully. These include Staniland (1953), Zweifel (1961), and Papp (1976). The following are a few of the many techniques:

1. *Line drawings.* Good line drawings are deceptively simple but require skill to render correctly. A good line drawing is a prelude to other types of drawings. Fortunately, various mechanical aids are available to help render the drawing. These include the camera lucida, the "lazy lucy" (a large platform lucida), and tracing from photographs.

2. *Stippled and shaded drawings.* The details of shape and depth can be accentuated with stippling or shading. Stipple drawings are slightly cheaper to reproduce, shaded drawings faster to complete, once the skill is acquired. The two can be combined by using stipple board which has a textured surface.

3. *Photographs.* Photographs are frequently superior to drawings in illustrating whole specimens. The lighting, exposure, and printing are critical. Journals frequently reproduce photos at reduced contrast, which means that the investigator should work for higher contrast in the original. Good scientific photographs are an art unto themselves and the beginner should be braced for many disappointing prints before learning the art. Blaker (1965) is a good reference for those who have mastered the essentials.

Most systematists must produce their own illustrations. With practice and the use of various mechanical aids even those who have no innate artistic skills can learn to do at least acceptable diagrammatic line drawings. Good photographs can be obtained with patience. If the systematist is lucky enough to have the services of a professional illustrator, so much the better. The results must be carefully checked by the systematist to ensure that they are scientifically accurate as well as artistic. On occasion the systematist may discover that the artist has observed characters previously overlooked, much to the embarrassment of the systematist.

Comparisons

A comparison section may be used to contrast the characteristics of various closely related species. These comparisons might concentrate on finer details and more characters than would be suitable for a differential diagnosis. A comparison section is also appropriate for discussing characters as possibly apomorphic or plesiomorphic.

Discussion

Within the formal descriptive format, a discussion section may be necessary. It may be used to: (1) discuss reasons for circumscribing a taxon in the manner adopted, (2) bring to the reader's attention certain fragmentary specimens that have been wrongly identified in the past, (3) point out geographic areas in need of additional collecting, or (4) call attention to new or previously poorly known characters. For papers involving the description of a new species, the discussion section provides a format for discussing the relationship of the species to other taxa.

Distributional Data

The range of the taxon should be specified in words. Additionally, a distributional map showing the localities of collections examined should be included. Such maps are the most straightforward way of presenting range data and they provide basic data for biogeographers. To be of value the maps must be accurate. It is better to underestimate the range of a taxon than to publish range data for populations not personally examined by the investigator or checked by another competent systematist. There are several different formats for mapping.

   1. *Spot maps.* This is the most accurate type of map and is the preferred format for taxonomic publication. Each locality is shown by a symbol on the map (Fig. 11.1). Each species on the same map has a different symbol.
   2. *Boundary-line maps.* The entire range of the taxon is shown by a range line (a track). If two or more taxa are shown on the same map, they may be hatched or differentially shaded. Boundary-line maps are suitable for higher taxa where the emphasis is on the cumulative ranges of species. They are also suitable for less formal works such as field guides.
   3. *Pictorial maps.* This type of map combines both geographic and morphological information. They are popular in botanical works and effectively portray geographic variation for selected characters.

Etymology

The origin of names is a point of scholarship. Descriptions of new taxa should include the origin (and gender if a genus) of the proposed name. I also prefer to include the etymology of previously described species in a revision. Two currently available books that give etymological as-

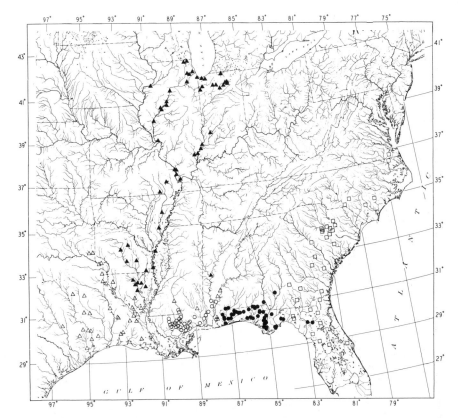

Figure 11.1   A spot map of the distributions of five members of the *Fundulus nottii* species group. Open triangles, *F. blairae*; black triangles, *F. dispar*; open circles, *F. nottii*; black circles, *F. escambiae*; open squares, *F. lineolatus*. (From Wiley, 1977.)

sistance are Stearn (1973) and Brown (1978).

Keys

Keys are devices used to identify specimens. Clarity and convenience are the most important qualities of a good key. Although these qualities can be incorporated into a **natural key** (i.e., one that follows a natural classification), they are more frequently met by an **artificial key**. *A simple artificial key is to be preferred over a more complex natural key.* Classifications, not keys, are the vehicle for presenting hypotheses of relationship.

The most convenient keys are arranged in dichotomous **couplets**. Each member of the couplet is a **lead**. Two types of keys are usually encountered, **indented keys** and **bracket keys**. The following is an example of each using the original key for kingbirds from Brodkorb (1968):

Indented Key

    A.   Belly white
        B.   Tail with white tip ....*T. tyrannus*
        B.B.   Tail without white tip ....*T. dominicensis*
   A.A.   Belly yellow
        C.   Exposed culmen equal to or greater than
        tarsus ....*T. melancholicus*
        C.C.   Exposed culmen shorter than tarsus...
           D.   Outer primary shorter than sixth, outer
           web of lateral rectix brownish with narrow
           gray edging ....*T. vociferans*
           D.D.   Outer primary longer than sixth,
           outer web of lateral rectix white to shaft ...*T. verticalis*

Bracket Key

   1a. Belly white 2
     b. Belly yellow 3
   2a. Tail with white tip ....*T. tyrannus*
     b. Tail without white tip ....*T. dominicensis*
   3a. Exposed culmen equal to or longer than tarsus *T. melancholicus*
     b. Exposed culmen shorter than tarsus 4
   4a. Outer primary shorter than sixth ....*T. voliferans*
     b. Outer primary longer than sixth ....*T. verticalis*

The indented format is suitable for artificial synoptic keys and natural keys. The characters distinguishing the taxa are presented in a readily appreciated, synoptic form. However, indented keys have two practical difficulties: (1) the contrasting leads are frequently separated from each other and (2) they take up a larger amount of page space. The second difficulty may become acute when publishing in a double-column format. I prefer the bracket key format.

Several other types of keys may be encountered. Pictorial keys are appropriate when they are intended for use by nonspecialists. Multiple entry keys are frequently used by botanists. These keys may consist of stacks of cards, representing taxa, with holes punched in them, representing characters. Sorting is done by inserting a rod into the stack and pulling out all taxa that have certain characters, and repeating this process until the specimen is keyed out. The advantage of such a system is that a specimen in any condition (such as a vegetative specimen) can either be keyed out or its possible identity severely circumscribed. A disadvantage, shared with any single character key, is that if the specimen deviates in a single character in the sequence it may be wrongly keyed.

A good key has the following characteristics:

1. Each couplet is composed of contrasting leads. To confirm an identification two or three characters should be used. If too many characters are used the efficiency of the key will be reduced.

2. The style is telegraphic.

3. The characters used are as readily observable as possible.

4. The character choices should be absolute and should not depend on a judgment of quality (for example, darker versus lighter).

5. The characters should apply to all ages and both sexes or (a) multiple keys should be provided, or (b) it should be specified that the key works only for a particular sex or age group.

6. Reference should be made to an illustration if the contrasting characters are confusing or call for expert judgment.

Constructing a good key is a demanding task. A good way to start is to construct a table of characters and construct the key from the table. Once the key is made, it should be tested against a random selection of specimens *and* by asking others to use it to identify specimens. Additional discussion concerning keys include Lawrence (1951), Voss (1952), Mayr et al. (1953), Metcalf (1954), Stearn (1956), Davis and Heywood (1965), Leenhouts (1966), Blackwelder (1967), Mayr (1969), and Radford et al. (1974). An excellent review of the principles of key construction and an introduction to computerized keys is given by Pankhurst (1978).

THE RULES OF NOMENCLATURE

Linnaeus provided the systematic community with a binomial nomenclature and a limited set of higher inclusive categories. He also provided a set of rules for botanists to follow in naming plants. No such rules were available for animals. The rules proposed by Linnaeus were gradually forgotten. Ross (1974: 262) states:

Later there developed a trend for making names descriptive. If, for example, a species had been named *flavus* because it was yellow and someone later said it was brown, he might change the name to *brunneus*. If it were discovered that several names applied to one species or one genus, the most appropriate name might be used, or the name proposed by the most eminent of the respective authors. In these instances, different investigators often chose different names for the same taxon.

The general situation became unstable as each authority adopted their own nomenclature. This prompted de Candolle (1813) and Strickland (1842) to propose codes for botany and zoology, respectively. Strickland's codes were prepared for the British Association for the Advancement of Science. This was followed by codes for the United States (1877), France (1881), Germany (1894), and codes for groups (the Douvillé Code, 1881; the A.O.U. Code, 1885). The zoologists decided to formulate an international code and established the First International Congress of Zoology in 1889 to prepare a draft. After several Congresses and an expansion of the membership to include international representatives, *Règles internationales de la Nomenclature zoologique* was accepted by the Fifth Congress (1901) and published under the auspices of the Sixth Congress in 1905. The code was rewritten in 1958 as the International Code of Zoological Nomenclature and has been revised and updated since that time; the most current is the code published in 1963. An International Code of Botanical Nomenclature was first adopted in 1905 and has been revised several times since. The most recent revisions were adopted in 1975 and published in 1978.

Purposes of the Codes

The three codes are designed to provide for the universal application of the names of organisms in such a manner that continuity of names and stability of taxonomic nomenclature are promoted to the maximum possible that is compatible with the free exercize of taxonomic judgment. Groups of organisms are given latinized names and the codes provide that these names are universally applicable. All codes are based on the principle of priority, that the oldest name for a taxon should (except under special circumstances) be the name that is universally applied to the taxon in question. Further, only this one name is the valid name for that taxon. Names for certain groups are objectively defined by types. Thus, each code provides a set of rules that promotes a single, universally accepted system of names for the organisms governed by each code. The authority of the individual taxonomist is replaced by the authorities of the various codes. This authority exists as a service to taxonomists and should be viewed as such.

The Codes

Zoologists are governed by the *International Code of Zoological No-*

menclature, adopted by the Fifteenth International Congress of Zoology in 1963. The published version and a supplement covering changes since 1963 are available from the International Trust for Zoological Nomenclature (14 Belgrave Square, London S.W. 2, England, U.K.). A new zoological code should be forthcoming in the near future. Botanists are governed by the *International Code of Botanical Nomenclature*, adopted by the Twelfth International Botanical Congress in 1975. It is available from the International Bureau for Plant Taxonomy and Nomenclature (106 Lange Nieuwstraat, Utrecht, Netherlands). Bacteriologists are governed by the *International Code of Nomenclature for Bacteria* (1969 and 1971, *Internat. J. Syst. Bacteriol.* 16(4):459–490 and 21(1):111–118). An invaluable guide to the various codes is Jeffrey (1973).

General Principles

1. Each code is independent of the others.

2. Taxa are given Latin or latinized names, a binomial for species and a uninomial for supraspecific taxa.

3. Taxa assigned to certain categories may have only one correct name.

4. In general, the correct name for a taxon is the earliest proposed name that is correctly published (the principle of priority).

5. In general, two taxa can not have the same name. This principle applies within, not between, the codes. No two genera of animals may have the same generic name. But since the codes are independent, it is possible to have the same genus name for a plant and an animal.

6. Taxa assigned to certain categories are objectively defined in reference to types.

7. The most current codes are retroactive unless specifically stated otherwise.

Basic Nomenclatural Concepts

Each code is built around certain basic concepts and utilizes certain basic principles. Among these are the concepts of priority, effective publication, typification, synonomy, and homonymy. These will be discussed in separate sections.

Priority

Priority pertains to date of publication. A validly published name has priority over another validly published name if it was published earlier. In most circumstances the valid (correct) name of a taxon is the oldest name of that taxon and it has precedence over names published later.

Example: *Lepisosteus* Lacépède 1803 has priority over *Sarchirus* Rafinesque 1818 because *Lepisosteus* was validly published earlier.

Priority under the zoological code does not extend to taxa ranked as superfamilies and lower categories. In botany, priority extends to family, while in bacteriology it extends to order. However, all codes encourage the stabilization of higher taxon names, even if priority is not followed.

Priority is not meant to replace names which have been firmly established and extensively used with older names simply because these older names exist. Each code has provisions which permit the **conservation** of at least some of these younger and well established names. In zoological nomenclature the International Commission has the power to suppress an older name and make a younger name the valid name of a taxon. In botany, a list of **conserved names** of families and genera are found in appendices II and III of the International Code. In all codes there are provisions for formal suspension of the law of priority. In zoology, the International Commission must act to suppress senior synonyms. In botany, the commission can act to suppress a senior synonym if there is a controversy surrounding its use.

The Limits of Priority

Priority extends back to certain specified taxonomic works specific for each group of organisms. Names applied to taxa before these specified works are not valid (correct) names.

Zoological names. Zoological nomenclature begins with Linnaeus's *Systema Naturae*, edition 10, considered published January 1, 1758. Other works published in 1758 are considered published after *Systema Naturae*.

Botanical names. Botanical names begin with various works published from 1753 to 1900. The nomenclature of fossil plants begins with Sternberg's *Flora der Vorwelt, Versuch I*, assigned the publication date December 21, 1820. Selected Recent groups are listed (the full list is found in Article 13 of the botanical code).

1. Spermatophyta, Pteridophyta. May 1, 1753. Linnaeus, *Species Plantarum*, ed. 1.
2. Musci (except Sphagnaceae). January 1, 1801. Hedwig. *Species Muscorum*.
3. Sphagnaceae and Hepaticae. May 1, 1753. Linnaeus, *Species Plantarum*, ed. 1.
4. Fungi (Uredinales, Ustilaginales, Gasteromycetes). December

31, 1801. Persoon, *Synopsis Methodica Fungorum.*

5. Fungi caeteri. January 1, 1821. Fries, *Systema Mycologicum* (including *Elenchus Fungorum*, 1828).

6. Most algae. May 1, 1753. Linnaeus, *Species Plantarum*, ed. 1 [exceptions are listed in the code, Art. 13.1(g)].

7. Myxomycetes. May 1, 1753. Linnaeus, *Species Plantarum*, ed. 1.

Bacteriological names. Bacteriological names begin May 1, 1753, with Linnaeus's *Species Plantarum*, ed. 1.

Names

The names used in our systems of nomenclature are based on Latin. This is a natural outgrowth of the extensive use of Latin by earlier Western scholars. It provides a nomenclature that transcends language barriers and promotes the universality of biological nomenclature.

Specific rules are contained in the codes for the formation of taxon names and they should be consulted directly for this purpose. There are certain general principles:

1. The names of species are binomial names, consisting of a genus name and a species name (epithet). The generic name begins with a capital letter, the species name with a lower case letter. (Exceptions to the latter are allowable, but not mandatory, in botany). Specific names are adjectival and either adjectives, genitives, or nouns are used as adjectives.

2. The names of genera, subgenera, and (in botany) sections are treated as singular nouns in the nominative case. In botany, certain subgeneric names are formed as plurals of a specific epithet.

3. The names of suprageneric taxa are treated as plural nouns in the nominative case.

Name Endings

Certain name endings are specified by the various codes. The specified endings in botanical and bacteriological codes are more extensive than in the zoological code. The ranks and associated specified endings are as follows.

Taxon Rank	Botany	Zoology
Phylum (Division)	-phyta, mycota (fungi)	None
Subphylum (Subdivision)	-phytina, mycotina (fungi)	None

Taxon Rank	Botany	Zoology
Class	-opsida (higher plants) -phyceae (algae) -mycetes (fungi)	None
Subclass	-opsida (higher plants) -phycidae (algae) -mycetidae (fungi)	None
Order	-ales	None
Suborder	-ineae	None
Superfamily	None	-oidea
Family	-aceae	-idae
Subfamily	-oideae	-inae
Tribe	-eae	-ini
Subtribe	-inae	-ina

Available and Legitimate Names

The codes specify certain conditions that must be met for a name to enter biological nomenclature. A name that is correctly formed and validly and effectively published is termed an **available name** (zoology) or a **legitimate name** (botany and bacteriology). Such names have standing in biological nomenclature and are available under certain specific conditions for use by taxonomists. Names that do not satisfy the conditions specified in the codes are termed **unavailable** and **illegitimate names**. The three codes specify certain conditions that determine whether a name can be accepted for usage. In certain cases the year or period during which the name was proposed may determine whether the name is available or legitimate. For a taxon name proposed under the present codes to be available/legitimate, it must meet the following general conditions:

1. It must be effectively published.
2. It must be in such a form as to comply with the rules of word formation.
3. It must be accompanied by a description or diagnosis (which in zoology differentiates it from other taxa) or (in botany) it must give reference to such a description or diagnosis.
4. In botany, the description or diagnosis (or both) must be published in Latin.

There are specific articles in each of the codes specifying additional criteria, including rules for replacement names.

Synonyms

Although a name may be available or legitimate, it may not be the valid or correct name of the taxon in question. There may be several names for the same taxon. Such names are termed **synonyms**. A **senior synonym** is the oldest available/legitimate name of a taxon. **Junior synonyms** are younger names for the same taxon. A **valid name** (zoology) or **correct name** (botany and bacteriology) is the only name which may be used for the taxon in question. Following the law of priority, the senior synonym is usually the valid/correct name of a taxon. Names continue to be available/legitimate even if they are junior synonyms. Junior synonyms may be the correct name of a botanical taxon of certain ranks if the name is **conserved**. Likewise, a junior synonym may be the correct name of a zoological taxon if the senior synonym is **suppressed**.

Example. *Vexillifer* Ducke 1922 is a legitimate generic name as is *Dussia* Krung et Urban 1892. Both are the *correct* names of their respective genera. However, if these genera are united into a single genus, *Dussia* Krug et Urban becomes the correct name because it is the senior synonym. *Vexillifer* Ducke would be a junior synonym and its usage would be incorrect.

Types and Synonomy

If two taxon names are based on the same type they are considered **objective synonyms** (zoology and bacteriology) or **nomenclatural synonyms** (botany). Such synonyms are junior synonyms and are neither available nor correct unless they are conserved or their senior objective synonym is suppressed. If two taxa are based on different types they are considered **subjective synonyms** (zoology) or **taxonomic synonyms** (botany). Such names are available and may be correct or valid depending on the taxonomy favored by the investigator.

Example. *Hydragira* Lacépède 1803 is based on the type species *H. swampina*. *Hydragira swampina* is a synonym of *Fundulus heteroclitus* Linnaeus. *Fundulus heteroclitus* is the type species of *Fundulus*. Thus, *Hydragira* is an objective synonym of *Fundulus* and is not available as a valid generic name (from Bailey and Wiley, 1976).

Example. *Zygonectes* Agassiz is based on the type species *Zygonectes olivaceus* Agassiz. Both type species are considered to belong to *Fundulus*. *Fundulus* and *Zygonectes* are subjective synonyms. If the genus is split and the type species of *Zygonectes* is the correct name of that genus given that it is the oldest generic name. Thus, *Zygonectes* may be validly used depending on the taxonomic scheme of killifishes used by the investigator.

Homonyms

Two names applied to taxa of the same rank (or grouping of ranks) are **homonyms**. In such cases the **senior homonym** is the earlier name whereas the **junior homonym** is the later name. No code permits junior homonyms to be valid names. In zoology, the law of homonymy states: "Any name that is a junior homonym of an available name must be rejected and replaced" (Art. 53).

The zoological code specifies that homonyms must be identical in spelling (although the suffix may differ in the case of family-group names). The botanical code specifies that orthographic variants of the same name are to be treated as homonyms.

Example. *Lepisosteus* Lacépède and *Lepidosteus* Koeing are not homonyms under the zoological code because they are not identical in spelling. If these names were applied to plants, the botanical code would consider *Lepidosteus* a junior homonym of *Lepisosteus*.

Example. *Clastes* Cope was proposed for a fossil genus of gars. Whitley found that *Clastes* was preoccupied by a genus of spiders and he proposed the replacement name *Clastichthytes*.

Other provisions for homonyms apply to particular taxon ranks. These will be discussed in later sections.

The Type Concept

Certain taxa of particular ranks are established in reference to a **type**. The type may be a specimen or it may be another taxon of lower rank. There are three basic types.

1. **Type specimen(s).** A single specimen or series of specimens that comprise the type of a taxon of the species category or below. Since all other types eventually refer back to a type species, they ultimately refer back to a type specimen (or a type series). Under certain circumstances, a type specimen may be an illustration of a specimen or the "work of an animal." There are various kinds of type specimens. These will be discussed later.

2. **Type species.** The named species that is the type of a taxon in the genus group. The type species is a nominal species—that is, it must

be an available/legitimate name, but not necessarily a valid/correct name.

3. **Type genus.** A nominal genus that is the type of a family group taxon.

The value of the type concept lies in the fact that typification pinpoints a particular taxon to a specimen or a lower taxon which thereafter serves as a reference point for further taxonomic changes. On the species level, the specimen or specimens named as types (with the exception of neotypes) serve to provide specimens actually examined by the describer and determined by the describer to pertain to the nominal species. On the supraspecific level, type species or genera serve to anchor names in reference to particular restricted groups of organisms. If a series of species is removed from a genus, then the law of priority and available generic synonyms in reference to type species will determine the valid name of the newly recognized genus. Considered this way, the type concept is simply a tool in taxonomy and not a relic of typology. Types are not meant to be typical of any taxon they represent, nor are they supposed to represent variation within the taxon they represent.

Publication

Each code provides criteria for publishing names. Names that do not meet the criteria for publication are unavailable (illegitimate) and have no standing in biological nomenclature. The various codes specify similar criteria for names appearing in current literature. However, whether a name published in the earlier literature was published validly depends on when and under what circumstances it was published. In zoology, a taxon described before 1950 without the describer's name is validly published but one described after 1950 is not validly published [Art. 9(7)]. As another example, a botanical work published by indelible autograph before 1953 is effectively published, while one published in the same manner on or after January 1, 1953 is not effectively published. The taxonomist should consult the various codes if doubt exists about the validity of previously published names. The following criteria are collected to serve as a guide in assuring that names published under the present codes are effectively published.

1. The work must be printed and available for distribution. The zoological code specifies that it must be a process that places ink on paper in such a way that numerous identical copies are produced. In botany, mimeograph copies are specifically prohibited; in zoology, mimeographing is discouraged but not specifically prohibited.

2. The paper must be available for distribution through sale or as a

gift or exchange.

3.   The names must meet various criteria to become available. These criteria will be discussed in later sections.

Each code also specifies forbidden methods of publication. Works "published" in any of these manners have no validity in biological nomenclature:

1.   Distribution of microfilm, microcards, or other similar reproductions made from unpublished manuscripts. Thus, "publication" of U.S. Ph.D. theses via University Microfilms have no validity in biological nomenclature.

2.   A mention of the name at a scientific meeting.

3.   Labeling of a specimen or specimens in a collection or garden.

4.   The distribution of proof sheets (zoology).

5.   The deposition of a document (such as a thesis) in a library.

6.   Private distribution of a note or illustration, even if printed (zoology).

The date of publication is the date it becomes available. For new works, this is generally established by receipt of the publication by certain institutions. In the absence of proof otherwise, the date appearing on the publication is considered the date of publication.

Rules Affecting Taxa

Each of the codes provides rules for certain taxon groups, specifically those to which the laws of priority and the principle of typification apply. In bacteriology the law of priority extends to the rank of order. In botany and zoology, the law of priority extends to the rank of family or the family group (including superfamilies). In botany, it is recommended that priority be followed above the family level and the code provides criteria for forming such names (Arts. 16 and 17).

In the following discussion, I will cover the major features of the codes concerning family, genus, and species group names. All facets of the rules are not covered and I assume the reader will become familiar with the details of the rules for the group of interest.

Taxa of the Family Group

Scope.   In zoology the family group includes taxa ranked from superfamilies to tribes. In botany, it includes taxa ranked from families to tribes.

Names.   The name of a family group taxon is formed from the root of its type genus and an appropriate suffix. Exceptions to this include conserved names and, in zoology, certain situations of synonymy.

Type Basis. A taxon of the family group is objectively based on a type genus. The family name is formed on the basis of the name of the type genus according to provisions of the codes (Rosaceae from *Rosa* L.; Poeciliidae from *Poecilia*, etc.). A family name based on an illegitimate or invalid genus is invalid unless specifically conserved. In some cases, a family name may be conserved and the type genus differs from the family name. In such cases in botany, an alternate family name based on the type genus is acceptable.

Example. Palmae is a conserved family name considered validly published. It is so conserved because of long usage (botanical code, Art. 18.5). The type genus is *Areca* L., so the alternate acceptable name is Arecaceae, based on the type genus.

In zoology the family name is unaffected by the synonymy of the type genus with another genus so long as that taxonomic synonymy occurred after 1960.

Example. The genus *A-us* Doe 1882 is the type genus for A-idae Doe 1882. *A-us* is a junior synonym of *B-us* Smith 1880. If A-idae has not been replaced by B-idae by 1961, or if B-idae has not won general acceptance, then A-idae Doe 1882 continues to be the family name. However, if B-idae has already replaced A-idae and if B-idae has won general acceptance, then B-idae is considered the correct name and takes on the date of publication of A-idae (i.e., 1882). (From the *International Code of Zoological Nomenclature*, Art. 40a–b.)

Controversies concerning the term "general acceptance" must be dealt with by the Commission.

Priority. Names for the family group are subject to strict application of the law of priority. Names conserved for stability must compete with other subjective (taxonomic) synonyms.

Homonymy. Taxa of the family group are subject to the law of homonymy. Because homonyms in zoology must be exactly alike, homonymy on the family level occasionally occurs when two genus names have the same stem. An example cited in the zoological code (Art. 39) concerns *Merope* (Insecta) and *Merops* (Aves). Correct family names in such situations are determined by the Commission.

Higher Taxon	Type Genus	Stem	Family Name	Opinion
Insecta	*Merope*	Merop-	Meropidae	Meropeidae
Aves	*Meropes*	Merop-	Meropidae	Meropidae

Also in zoology, homonymy does not exist if a corrected or misspelled name is introduced (Art. 55). For example, *Lepisosteus* Lacépède was corrected by Koeing to *Lepidosteus*, with a subsequent spelling of the family name of Lepidosteidae. Although Lepidosteidae differs in only a single letter from Lepisosteidae, it is not a homynym and remains available.

In botany such problems are avoided by considering orthographic variations as homonyms and, in the case of *Merope* and *Merops*, by Article 75.1 which states that names so similar as to cause confusion may be treated as homonyms when they are based on different types.

Subtaxa.   Subtaxa of the family group are based on nomenclatural type genera, as with family names. Names are formed from the root of the type genus and a suffix specified by the various codes. **Nominate subtaxa** are subtaxa which include the type genus of the family. For example:

Taxon	Type Genus
Family Tipulidae	*Tipula* Linnaeus
Subfamily Tipulinae	*Tipula* Linnaeus

If a subtaxon is established that is not a nominate subtaxon because its type genus is not the type genus of the family, this causes the automatic formation of coordinate nominate subtaxa.

Taxa of the Genus Group

Scope.   The genus group includes taxa ranked between genus and species. The most commonly used taxa in the genus group are:

Zoology	Botany
Genus	Genus
Subgenus	Subgenus
"Collective group"	Section
	Subsection
	Series
	Subseries

The zoological code also allows for such terms as "division" and "section" which are nomenclaturally treated as subgenera.

Names.   In all codes, genus group names are Latin nouns in the singular case or words treated as such. The name may be based on a previously existing word or even an arbitrary combination of letters. All names are treated as Latin even if their endings are not latinized (e.g., *Abudefduf*, a genus of pomacentrid fishes). Nonlatin letters are acceptable.

Subtaxon Name Citation. In zoology, the subgenus name or the uninominal primary subdivision of a genus (such as a section) is placed in parentheses between the generic and specific names. It is not considered part of the binomial.

Example. *Fundulus (Zygonectes) nottii*

In botany, the name of a subtaxon of a genus is a combination of the genus name and a subtaxon name connected by a specific term denoting the rank of the subtaxon.

Example. *Euphorbia* sect. *Tithymalus.*

Name Formation. The correct formation of names for the genus group are specified in the codes. The reader should consult the codes to determine such matters as gender and latinization.

Types. Taxa of the genus groups are defined with reference to type species. The exception is the "collective group" which is treated as an assemblage of species of uncertain generic position and which is permitted to exist without reference to a type species.

In zoology (Art. 13) and botany (Art. 37) the publication of a new genus group name must be accompanied by the designation of a type species. In zoology, genus group names published after 1930 without designation of a type species are invalid. The same holds for botanical genus group names after January 1, 1958.

In botany, the type should be placed immediately following the Latin diagnosis or description and is indicated by tagging it with the word "*typus.*"

In zoology, the mention of a species as an example of, or typical of, the genus group taxon is not sufficient to establish a type species. Designation of zoological type species may be handled in one of four ways (Art. 68):

1. *Original designation.* The ·investigator designates a single species as the type species.
2. *Epithet designation.* The investigator names one species "*typicus*" or "*typus.*"
3. *Monotypy.* The investigator establishes a new genus or subgenus containing a single species.
4. *Tautonymy.* The investigator establishes a new genus and includes a species having the generic or subgeneric name for the species epithet (e.g., *Ratus ratus*). This applies even if the tautonym is a junior

synonym.

For those genus group names validly established without type species, any subsequent biologist may designate a type species from among the nominal species included in the taxon.

Types of Hybrid Plant Genus Group Names.   Hybrid plant taxa of the genus group are effectively published merely if their parent taxa are indicated.

Example (from Art. H.9).   × *Philageria* Masters was published with a statement of its parentage (*Lapageria* × *Philesia*) and is, therefore, validly published.

Priority.   Names of the genus group are subject to strict application of the law of priority. Conserved names are not excused from priority because they are conserved names when they compete for priority with names against which they have been explicitly conserved.

Example (botanical code, Art. 14.3).   If *Weihea* Spreng. (1825) is united with *Cassipourea* Aubl. (1775), the correct genus name is *Cassipourea* despite the fact that *Weihea* is a conserved name.

Synonymy.   In most circumstances, an available/legitimate name for a genus group taxon published after the valid/correct name of that taxon becomes a junior synonym. If the junior synonym is based on the same type species it is an objective/nomenclatural junior synonym. If based on a different type species, it is a subjective/taxonomic junior synonym. In later subdivisions of the genus into two or more genera, the valid/correct names of the genera are determined by reference to available names and type species.

Occasionally, one junior synonym of an array of synonyms becomes an established name. In such a case, the correct procedure in botany is to petition the International Commission for conservation of the name. In zoology, the Commission should be petitioned to suppress the senior synonym. (In many cases, the controversy is simply forgotten, no action is taken, and the junior synonym remains in use.)

Homonymy.   Taxa of the genus group are governed by the law of homonymy. In botany, homonymy may result from orthographic errors or variant spelling. Further, homonymy may result when name spellings are so similar as to cause confusion (Art. 75.1).

Examples.   *Asterostemma* and *Astrostemma*
                 Bradlea, *Bradleja*, and *Braddleya*.

In zoology, even a single letter difference is not sufficient to place a name in homonymy. Neither of the following examples are homonyms.

Examples. *Lepisosteus, Lepidosteus*
  Merope, Merops.

However, if certain fossil species of a valid genus are given a different ending (e.g., *-ites, -ytes,* or *-ithes*), these names are considered junior homonyms. And, when two homonymous genus group names have identical dates of publication, the name of highest rank is considered the senior homonym.

If a senior synonym of a genus group taxon is found to be a junior homonym, then next available synonym becomes the valid/correct name. If no synonym exists, the revisor should form a **replacement name**.

Taxa of the Species Group

Scope. In zoology, the species group includes the species and the subspecies categories only. In botany, it includes the species, subspecies, and a variety of infraspecific categories such as form (forma), variety, and so on.

Names of Species. The species name is a binomial formed from the genus name and the species name. Species group names that are used as adjectives in the nominative singular must agree in gender with the generic name. Genetive epithet and nouns used in apposition do not have to agree in gender with the generic name. Note that in zoology the specific epithet may be a **tautonym** (e.g., *Ratus ratus*) and that in botany tautonyms are not permitted (e.g., *Linaria linaria* is not permitted).

Names of Infraspecific Groups. In zoology, the subspecies is a trinomial. In botany the rank of the taxon is mentioned. Infraspecific epithets agree grammatically with the generic name except when they are used as substantives.

Examples. *Trifolium stellatum* forma *nanum*
  Andropogon ternatus subsp. *macrothrix.*

Name Formation. Specific rules and recommendations for name formation and for determining gender are found in the various codes.

Types. Names of the species group are objectively defined in reference to type specimens. Article 7 of the botanical code and Articles 71 to 75 of the zoological code govern the designation of type specimens for the species group. In addition, the botanical code provides a guide for the determination of types. Species group names published under the present codes must be accompanied by designation of type material (preferably a holotype) to be validly published. The kinds of types and notes on their usage are listed:

1. Holotype. A single specimen or part of an organism used by the author or specifically designated by the author as the type of the species.

2. Isotype. Any specimen that is a duplicate of the holotype (botany).

3. Syntype. Any two or more specimens cited by the author when no holotype is designated or when all are simultaneously designated as types. "Co-types" or simply "types" are considered syntypes under the zoological code.

4. Paratypes. Any specimens of the type series that were not designated as the holotype. Paratypes exist when a holotype was designated unless (in botany) two or more specimens were designated as types. In this case, the remaining specimens are paratypes.

5. Lectotype. One of several syntypes designated as the type specimen of a species after the original description. Lectotypes *may* be designated by any biologist but they *should* be designated only by specialists familiar with the group. In zoology, the remaining syntypes become **paralectotypes**. The codes give specific recommendations for designating lectotypes.

6. Neotype. A specimen or part of a specimen selected by a revisor as the type specimen of a species when all of the original material on which the species was based is missing. In zoology, a neotype can only be designated in connection with a revisionary work and only when designation is necessary in the interest of nomenclatural stability. Also, a number of additional qualifying conditions are specified in the code (Art. 75c). In botany, restrictions are less stringent but the proposer of a neotype is admonished to exercise care in selecting a neotype. In zoology, the rediscovery of original type specimens after the designation of a neotype must be referred to the Commission. In botany, if original type specimens are rediscovered, they automatically supercede the neotype.

Priority. Names of the species group are subject to strict application of the law of priority. In zoology, species names and subspecies names compete in synonymy on equal footing. In botany, priority of species group names compete only with names of the same rank.

Example. *Aus bus* Doe (1880) and *Aus cus dus* Smith (1807) are synonomyzed but *Aus cus cus* (1807) is retained as a separate species. In zoology, *Aus dus* Smith and *Aus cus* are the correct names, since the subspecific epithet *dus* cannot compete for priority with the specific epithet *bus*.

Synonomy. Unless specified in one of the codes or by ruling of a Commission, the oldest specific epithet is the valid/correct epithet for a species. All names applied to the species after the valid/correct epithet are junior synonyms. If a junior synonym is an established name, it may be conserved by applying to the Commission. In such a case, the senior synonym is suppressed.

Placing a species in another genus does not change the specific epithet except, where needed, in gender, or when homonymy occurs. In zoology, if the specific name is applied to a genus other than that originally named by the describer, the describer's name is placed in parentheses.

Example. *Cobitis heteroclita* Linnaeus becomes *Fundulus heteroclitus* (Linnaeus).

In botany the same practice is followed except that the person who made the new combination is added.

Example. *Alsophila Kalkbreyi* Baker becomes *Trichipteris Kalkbreyi* (Baker) Tryon.

Homonymy. The law of homonymy also applies for taxa of the species group. In zoology, subspecies are coordinated with species under the code. Thus, homonyms may exist between a species and a subspecies of another species. In botany, each infraspecific level competes only at its own hierarchical level.

Example (zoology). *A-us x-us* and *A-us yus x-us* are homonyms.

Example (botany). *Z-us aus* ssp. *bus* and *Z-us bus* are not homonyms, but if *Z-us aus* ssp. *bus* is raised to species level, then *bus* is a homonym and must be replaced.

In zoology and botany, species epithets do not need to be exactly identical to be considered homonyms. This departure for the zoological code is specified in Article 58 and concerns the origin and meaning of the specific epithet.

In zoology, there are two types of species group homonymy. **Primary homonyms** are two or more identical specific names applied to the same genus at the times of their publications.

Example. *Taenia ovilla* Rivolta and *Taenia ovilla* Gmelin are primary homonyms because they were both originally described in the same nominal genus, *Taenia* Linnaeus. Junior primary homonyms are permanently rejected.

Secondary homonyms are species that were originally described in different genera but were later placed in the same genus by the action of a revision.

Example. *Mius* contains *M. aus* and *Zus* contains *Z. aus*. If the genera are combined, homonymy between the two species names results and the junior homonym must be replaced by the next oldest synonym or with a replacement name.

The problem is that many workers may not agree that *Mius* and *Zius* should be combined (Mayr, 1969; Crowson, 1970). The zoological code specifies that such secondary homonyms may be used if the genera are again separated so long as the synonymy of the genera occurs after 1960 [Art. 59(c)].

Example. *A-us niger* Smith 1960 is believed by Jones (1970) to be congeneric with *B-us niger* Dupont 1950. He labels *A-us niger* Smith as a junior secondary homonym of *B-us niger* Dupont and proposes the replacement name *B-us ater* Jones. Subsequently, Doe (1979) splits *A-us* from *B-us*. *A-us niger* Smith is revived and *B-us ater* Jones becomes (at least to Doe) a junior objective synonym of *A-us niger* Smith.

Interspecific Hybrids (Botany). Hybrids or putative hybrids between two species (or two infraspecific taxa) of the same genus are designated by a formula, or, where it seems useful or necessary, by a name. These names are subject to the same rules as those for species.

Example. *Salix* × capreola Kevnes ex Anderson = *Salix avrita* L. × *S. caprea* L, or, alternately, *Salix aurita* × *caprea*.

However, polyploids may be treated as species and bear an epithet without a multiplication sign. Segrates and backcrosses of hybrids are termed **nothomorphs** (abbreviation, NM) and are equivalent in rank to variety.

Literature Cited

Adams, E. N. 1972. Consensus techniques and the comparison of taxonomic trees. Syst. Zool. 21:390–397.

Adamson, M. 1763. Familles des plants, Vol. I. Vincent, Paris.

Agassiz, L. 1833–1844. Recherches see les Poissons Fossiles. Neuchatel. 5 vols. with supplement.

Agassiz, L. 1846, 1848. *Nomenclatoris Zoologici.* Index Universalis, Soloduri.

Anderson, E. 1949. Introgressive hybridization. John Wiley and Sons, Inc., New York.

Anderson, E. 1953. Introgressive hybridization. Biol. Rev. 28:280–307.

Arnheim, N., E. M. Prager, and A. C. Wilson. 1969. Immunological prediction of sequence differences among proteins. J. Biol. Chem. 244:2085–2094.

Ashlock, P. H. 1971. Monophyly and associated terms. Syst. Zool. 20:63–69.

Ashlock, P. H. 1973. Monophyly again. Syst. Zool. 21:430–437.

Ashlock, P. H. 1974. The uses of cladistics. Annu. Rev. Syst. Ecol. 5:81–99.

Ashlock, P. H. 1980. An evolutionary taxonomist's view of classification. Syst. Zool. 28:441–450.

Ashlock, P. H., and D. J. Brothers. 1979. Systematization and higher classification in evolutionary systematics through cladistic and anagenetic analysis. Manuscript.

Avise, J. C. 1974. Systematic value of electrophoretic data. Syst. Zool. 23:465–481.

Avise, J. C., and F. J. Ayala. 1975. Genetic differentiation in speciose versus depauperate phylads: evidence from the California minnows. Evolution 29:411–426.

Avise, J. C., and M. H. Smith. 1974a. Biochemical genetics of sunfish. I. Geographic variation and subspecific intergradation in the bluegill, *Lepomis machrochirus.* Evolution 28:42–56.

Avise, J. C. and M. H. Smith. 1974b. Biochemical genetics of sunfish. II. Genetic similarity between hybridizing species. Amer. Natur. 108:458–472.

Avise, J. C., M. H. Smith, R. K. Selander, T. E. Lawlor, and P. R. Ramsey. 1974. Biochemical polymorphism and systematics in the genus *Peromyscus.* V. Insular and mainland species of the subgenus *Haplomylomys.* Syst. Zool. 23:226–238.

Ayala, F. J., M. L. Tracey, D. Hedgecock, and R. C. Richmond. 1974. Genetic differentiation during the speciation process in *Drosophila.* Evolution 28:576–592.

Bailey, R. M., and E. O. Wiley. 1976. Identification of the American cyprinodontid fish *Hydrargira swampina* Lacépède. Proc. Biol. Soc. Washington 89(41):477–480.

Balinsky, B. I. 1970. An introduction to embryology. W. B. Saunders Co., Philadelphia, Pennsylvania.

Ball, I. R. 1975. Nature and formulation of biogeographic hypotheses. Syst. Zool. 24:407–430.

Bartram, A. W. H. 1977. The Microsemiidae, a Mesozoic family of holostean fishes. Bull. British Mus. Natur. Hist. (Geol.) 2a(2):137–234.

Beatty, J., and W. L. Fink. 1980. Simplicity (a review of Sober). Syst. Zool. 28:643–651.

Beckner, M. 1959. The biological way of thought. Columbia University Press, New York.

Bell, M. A. 1979. Persistence of ancestral sister-species. Syst. Zool. 28:85–88.

Berg, L. 1947. Classification of fishes both recent and fossil. J. W. Edwards, Ann Arbor, Michigan.

Bernstein, S. C., L. H. Throckmorton, and J. L. Hubby. 1973. Still more genetic variability in natural populations. Proc. Natl. Acad. Sci. U.S.A. 70:3928–3931.

Bigelow, R. S. 1958. Classification and phylogeny. Syst. Zool. 7:49–59.

Bisbee, C. A., A. M. Baker, A. C. Wilson, I. Hadji-Azimi, and M. Fischberg. 1977. Albumin phylogeny for clawed frogs (*Xenopus*). Science 195:785–787.

Blackith, R. E., and R. A. Reyment. 1971. Multivariate morphometrics. Academic Press, London and New York.

Blackwelder, R. E. 1967. Taxonomy, a text and reference book. John Wiley and Sons, Inc., New York.

Blackwelder, R. E. 1972. Guide to the taxonomic literature of vertebrates. The Iowa State University Press, Ames.

Blaker, A. A. 1965. Photography for scientific publication—A handbook. W. H. Freeman and Company, San Francisco, California.

Blaney, R. M., and P. K. Blaney. 1979. The *Nerodia sipedon* complex of water snakes in Mississippi and Southeastern Louisiana. Herpetologica 35(4):350–359.

Bock, W. J. 1965. The role of adaptive mechanisms in the origin of higher levels of organization. Syst. Zool. 14:272–287.

Bock, W. J. 1969. The concept of homology. Ann. New York. Acad. Sci. 167:111–115.

Bock, W. J. 1974. Philosophical foundations of classical evolutionary taxonomy. Syst. Zool. 22:375–392.

Bock, W. J. 1977. Foundations and methods of evolutionary classification. *In*: Hecht, Goody, and Hecht (Eds.), Major patterns of vertebrate evolution. Plenum Press, New York. Pages 851–895.

Bonde, N. 1975. Origin of "higher groups." Viewpoints of phylogenetic systematics. Coll. Int. C.N.R.S. 218:293–324.

Bonde, N. 1977. Cladistic classification as applied to vertebrates. *In*: Hecht, Goody, and Hecht (Eds.), Major patterns in vertebrate evolution. Plenum Press, New York. Pages 741–804.

Borror, D. J., D. W. Delong, and C. A. Triplehorn. 1976. An introduction to the study of insects. Holt, Rinehart and Winston, New York.

Boulter, D. 1973a. Amino acid sequences of cytochrome *C* and platocyanins in phylogenetic studies of higher plants. Syst. Zool. 22:549–553.

Boulter, D. 1973b. The use of amino acid sequence data in the classification of higher plants. *In*: Bendz and Santesson (Eds.), Nobel Symposium 25, Chemistry in botanical classification. Academic Press, New York.

Boulter, D. and D. Peacock. 1975. Estimating evolutionary relationships in higher plants from cytochrome *C* and plastocyanin amino acid sequences. *In*: Estabrook (Ed.), Proc. Eighth Internat. Conf. on Numerical Taxonomy. W. H. Freeman and Co., San Francisco, California. Pages 176–188.

Boulter, D., J. A. M. Ramshaw, E. W. Thompson, M. Richardson, and R. H. Brown. 1972. A phylogeny of higher plants based on the amino acid sequences of cytochrome *C* and its biological implications. Proc. Roy. Soc. London, Ser. B, 181:441–455.

Branson, B. A., and G. A. Moore. 1962. The lateralis components of the acousitco-lateralis system in the sunfish family Centrarchidae. Copeia 1962:1–108.

Bremer, K. 1976a. The genus *Relhania* (Compositae). Opera Bot. 40:1–85.

Bremer, K. 1976b. The genus *Rosenia* (Compositae). Bot. Notiser 129:97–111.

Bremer, K. 1978a. The genus *Leysera* (Compositae). Bot. Notiser 131:369–383.

Bremer, K. 1978b. *Oreoleysera* and *Antithrixia*, new and old South African genera of the Compositae. Bot. Notiser 131:449–453.

Bremer, K., and H.-E. Wanntorp. 1978. Phylogenetic systematics in botany. Taxon 27(4):317–329.

Bremer, K., and H.-E. Wanntorp. 1979. Geographic populations or biological species in phylogeny reconstruction? Syst. Zool. 28:220–224.

Bretsky, S. S. 1975. Allopatry and ancestors: a response to Cracraft. Syst. Zool. 24:113–119.

Bridgman, P. W. 1945. Some general principles of operational analysis. Psychol. Rev. 52:246.

Briggs, J. C. 1974. Operation of zoogeographic barriers. Syst. Zool. 23:248–256.

Britten, R. J., and E. H. Davidson, 1971. Repetitive and non-repetitive DNA sequences and a speculation on the origins of evolutionary novelty. Quart. Rev. Biol. 46:111–138.

Brodkorb, P. 1968. Part Five - Birds. *In*: Blair, Blair, Brodkorb, Cagle, and Moore (Authors), Vertebrates of the United States. McGraw-Hill Book Co., New York. Pages 269–451.

Brooks, J. L. 1957. The species problem in freshwater animals. *In*: Mayr (Ed.), The species problem. Amer. Assoc. Adv. Sci. Publ. 50:81–123.

Brosemer, R. W., D. S. Grosso, G. Estes, and C. W. Carlson, 1967. Quantitative immunochemical and electrophoretic comparisons of glycerophosphate dehydrogenases in several insects. J. Insect Physiol. 13:1757–1767.

Brothers, D. J. 1975. Phylogeny and classification of the aculeate Hymenoptera, with special reference to Mutillidae. Univ. Kansas Sci. Bull. 50:483–648.

Brothers, D. J. 1978. How pure must cladistic classification be?—A response to Nelson on Michener. Syst. Zool. 27:118–122.

Brown, J. L. 1958. Geographic variation in southeastern populations of the cyprinodont fish *Fundulus notti* (Agassiz). Amer. Midl. Natur. 59(2):477–488.

Brown, R. W. 1978. Composition of scientific words. Smithsonian Institution Press, Washington, D. C. (reprint.)

Brundin, L. 1966. Transantartic relationships and their significance, as evidenced by chironimid midges. Kungl. Svenska Vetenskap. Handl. 11:1–472.

Brundin, L. 1968. Application of phylogenetic principles in systematics and evolutionary theory. *In*: Orvig (Ed.), Nobel Symposium 4, Current problems in lower vertebrate phylogeney. Alguist and Wiksell, Stockholm. Pages 473–495.

Brundin, L. 1972. Phylogenetics and biogeography. Syst. Zool. 21:69–79.

Buck, R. C., and D. L. Hull. 1966. The logical structure of the Linnean hierarchy. Syst. Zool. 15:97–111.

Bush, G. L. 1969. Sympatric host race formation and speciation in frugivorous flies of the genus *Rhagoletis*. Evolution 23:L237–251.

Bush, G. L. 1975a. Modes of animal speciation. Annu. Rev. Ecol. Syst. 6:339–364.

Bush, G. L. 1975b. Sympatric speciation in phytophagous parasitic insects. *In*: Price (Ed.), Strategies of parasetic insects. Plenum Press, London. Pages 187–206.

Bush, G. L., and G. B. Kitto. 1978. Application of genetics to insect systematics and analysis of species differences. *In*: Romberger (Gen. Ed.), Beltsville Symposia in Agricultural research (2): Biosystematics in agriculture. Pages 89–118.

Cain, A. J. 1954. Animal species and their evolution. Harper and Row, New York.

Cain, A. J., and G. A. Harrison. 1958. An analysis of the taxonomist's judgement of affinity. Proc. Zool. Soc. London 131:85–98.

Cain, S. A. 1944. Foundations of plant geography. Harper and Row, New York.

Camin, J. H., and R. R. Sokal. 1965. A method for deducing branching sequences in phylogeny. Evolution 19:311–326.

Camp, W. H. 1947. Distributional patterns in modern plants and the problems of ancient dispersals. Ecol. Monogr. 17:159–183.

Candolle, A. P. de. 1813. Théorie elémentaire de la Botanique. Detréville, Paris.

Candolle, A. P. de. 1820. Geographie botanique. In: Dictionnaire des sciences naturelles, vol. 18. Strasbourg and Paris.

Champion, A. B., E. M. Prager, D. Wachter, and A. C. Wilson. 1974. Microcomplement fixation. In: Wright (Ed.), Biochemical and immunological taxonomy of animals. Academic Press, London. Pages 397–416.

Choate, J. R., R. C. Dowler, and J. E. Krause. 1979. Mensural discrimination between Peromyscus leucopus and P. maniculatus (Rodentia) in Kansas. Southwest Natur. 24(2):249–258.

Colless, P. H. 1972. A note on Ashlock's definition of "monophyly." Syst. Zool. 21:126–128.

Collier, G. E., and R. J. MacIntire. 1977. Microcomplement fixation studies on the evolution of alpha-alycerophosphate within the genus Drosophila. Proc. U.S. Nat. Acad. Sci. 74:684–688.

Commorford, S. L. 1971. Iodination of nucleic acids in vitro. Biochemistry 10:1993–2000.

Coombs, M. C. 1975. Sexual dimorphism in chalicotheres (Mammalia: Perissodactyla). Syst. Zool. 24:55–62.

Cracraft, J. 1973. Continental drift, paleoclimatology, and the evolution and biogeography of birds. J. Zool. 169:455–545.

Cracraft, J. 1974a. Phylogenetic models and classification. Syst. Zool. 23:71–90.

Cracraft, J. 1974b. Phylogeny and evolution of the ratite birds. Ibis 116:494–521.

Cracraft, J. 1975. Historical biogeography and earth history: perspectives for a future synthesis. Ann. Missouri Bot. Garden 62:227–250.

Croizat, L. 1952. Manual of phytogeography. Vitgeverij Dr. W. Junk, The Hague.

Croizat, L. 1958. Panbiogeography. Published by the author, Caracas.

Croizat, L. 1964. Space, time and form; the biological synthesis. Published by the author, Caracas.

Croizat, L., G. J. Nelson, and D. E. Rosen. 1974. Centers of origin and related concepts. Syst. Zool. 23:265–287.

Crosby, M. R., and R. E. Magill, 1978. A dictionary of mosses. Monogr. in Syst. Bot., Missouri Bot. Garden 3:1–43.

Crowson, R. A. 1970. Classification and biology. Heineman Education Books Ltd., London.

Darlington, C. D. 1958. The evolution of genetic systems. Basic Books, Inc., New York.

Darlington, P. J., Jr. 1957. Zoogeography: the geographic distribution of animals. John Wiley and Sons, Inc., New York.

Darlington, P. J., Jr. 1959. Darwin and zoogeography. Proc. Amer. Phil. Soc. 103:307–319.

Darlington, P. J., Jr. 1970. A practical criticism of Hennig-Brundin "Phylogentic (sic) Systematics" and Antarctic biogeography. Syst. Zool. 19:1–18.

Darwin, C. 1859. On the origin of species by means of natural selection, or the preservation of favored races in the struggle for life. John Murry, London.

Davidson, E. H. and R. J. Britten. 1979. Regulation of gene expression: possible role of repetitive sequences. Science 204:1052–1059.

Davis, P. H., and V. H. Heywood. 1965. Principles of angiosperm taxonomy. P. Van Nostrand Co., Inc., Princeton and New York.

Day, A. 1965. The evolution of a pair of sibling allotetraploid species of Cobwebby Gilias (Polemoniaceae). Aliso 6:25–75.

Dayhoff, M. O. 1969. Computer analysis of protein evolution. Sci. Amer. 221:86–95.

Dice, L. R., and H. J. Leraas. 1936. A graphic method for comparing several sets of measurements. Contrib. Lab. Vert. Genet. Univ. Michigan, No. 3:1–3.

Dobzhansky, T. 1951. Genetics and the origin of species. Columbia University Press, New York.

Dobzhansky, T. 1970. Genetics of the evolutionary process. Columbia University Press, New York.

Dobzhansky, T. F. J. Ayala, G. L. Stebbins, and J. W. Valentine. 1977. Evolution. W. H. Freeman and Co., San Francisco, California.

Dobzhansky, T. H., and C. Epling. 1944. Contributions to the genetics, taxonomy, and ecology of *Drosophila pseudoobscura* and its relatives. Carnegie Inst. Washington Publ. No. 554:1–46.

Dodson, P. 1975. Taxonomic implications of relative growth in lambeosaurine hadrosaurs. Syst. Zool. 24:37–54.

Dupuis, C. 1978. Permanence et actualite de la Systematique: La "Systematique phylogenetique" de W. Hennig (Historique, duscussion, choix de references). Cahiers des Naturalistes 34:1–69.

Erlich, P., and R. W. Holm. 1963. The process of evolution. McGraw-Hill, Inc., New York.

Erlich, P., and P. H. Raven. 1969. Differentiation of populations. Science 165:1228–1232.

Eldredge, N., and S. J. Gould. 1972. Punctuated equilibria: an alternative to phyletic gradualism. In: Schopf (Ed.), Models in paleontology. Freeman, Cooper, and Co., San Francisco, California. Pages 82–115.

Endler, J. A. 1977. Geographic variations, speciation, and clines. Princeton University Press, Princeton, New Jersey.

Engelmann, G. F. and E. O. Wiley. 1977. The place of ancestor-descendant relationships in phylogeny reconstruction. Syst. Zool. 26:1–11.

Estabrook, G. F. 1977. Does common equal primitive? Syst. Bot. 2:36–42.

Estabrook, G. F. 1979. Some concepts for the estimation of evolutionary relationships in Systematic Botany. Syst. Bot. 3:146–158.

Estabrook, G. F., J. G. Strauch, and K. L. Fiala. 1977. An application of compatibility analysis to the Blackith's data on orthopteroid insects. Syst. Zool. 26:269–276.

Farr, E. R., J. A. Leussink, and F. A. Stafleu. 1979. *Index Nominum Genericorum (Plantarum)*. Regnum Veg. vols. 100–102. Utrecht.

Farris, J. S. 1969. A successive approximations approach to character weighting. Syst. Zool. 18:374–385.

Farris, J. S. 1970. Methods for computing Wagner trees. Syst. Zool. 19:83–92.

Farris, J. S. 1971. The hypothesis of nonspecificity and taxonomic congruence. Annu. Rev. Syst. Ecol. 2:277–302.

Farris, J. S. 1972. Estimating phylogenetic trees from distance matrices. Amer. Natur. 106:645–668.

Farris, J. S. 1973. A probability model for inferring evolutionary trees. Syst. Evol. 22:250–256.

Farris, J. S. 1974. Formal definitions of paraphyly and polyphyly. Syst. Zool. 23:548–554.

Farris, J. S. 1976. Phylogenetic classification of fossils with Recent species. Syst. Zool. 25:271–282.

Farris, J. S. 1977. On the phenetic approach to vertebrate classification. *In*: Hecht, Goody, and Hecht (Eds.), Major patterns in vertebrate evolution. Plenum Press, New York. Pages 823–850.

Farris, J. S. 1978a. Inferring phylogenetic trees from chromosome inversion data. Syst. Zool. 27:275–284.

Farris, J. S. 1978b. The 11th Annual Numerical Taxonomy Conference—and part of the 10th. Syst. Zool. 27:229–238.

Farris, J. S. 1980. The information content of the phylogenetic system. Syst. Zool. 28:483–519.

Farris, J. S., A. G. Kluge, and M. J. Eckardt. 1970. A numerical approach to phylogenetic systematics. Syst. Zool. 19:172–189.

Felsenstein, J. 1977. The number of evolutionary trees. Syst. Zool. 27:27–33.

Fink, S. C., C. W. Carlson, S. Gurasiddaiah, and R. W. Brosemer. 1970. Glycerol-3-phosphate dehydrogenases in social bees. J. Biol. Chem. 245:6525–6532.

Fitch, W. M. 1970. Distinguishing homologous from analogous proteins. Syst. Zool. 19:99–113.

Fitch, W. M. 1971. Toward defining the course of evolution: Minimum change for a specific tree topology. Syst. Zool. 20:406–416.

Fitch, W. M. 1973. Aspects of molecular evolution. Annu. Rev. Genet. 7:343–380.

Fitch, W. M. 1975. Toward finding the tree of maximum parsimony. *In*: Estabrook (Ed.), Proc. Eighth Internat. Conf. on Numerical Taxonomy. W. H. Freeman and Co., San Francisco, California. Pages 189–230.

Fitch, W. M. 1976. Molecular evolutionary clocks. *In*: Ayala (Ed.), Molecular evolution. Sinauer Assoc., Inc., Sunderland, Massachusetts. Pages 160–178.

Fitch, W. M., and E. Margoliash. 1967. The construction of phylogenetic trees. Science 155:279–284.

Fitch, W. M., and E. Margoliash. 1970. The usefulness of amino acid and nucleotide sequences in evolutionary studies. Evol. Biol. 4:67–109.

Gaffney, E. S. 1979. An introduction to the logic of phylogeny reconstruction. *In*: Cracraft and Eldredge (Eds.), Phylogenetic analysis and paleontology. Columbia University Press, New York. Pages 79–111.

Gardiner, B. G. 1960. A revision of certain actinopterygian and coelacanth fishes, chiefly from the lower Lias. Bull. British Mus. Natur. Hist. (Geol.), 4:239–284.

Gardiner, B. G. 1973. Interrelationships of teleostomes. *In*: Greenwood, Miles, and Patterson (Eds.), Interrelationships of fishes. Academic Press, London. Pages 105–135.

Gardner, R. C., and J. C. La Duke. 1979. An estimate of phylogenetic relationships within the Genus *Crusea* (Rubiaceae) using character compatibility analysis. Syst. Bot. 3(2):179–196.

Garstang, W. 1922. The theory of recapitulation: a critical re-statement of the biogenetic law. Zool. J. Linnean Soc. London 35:81–101.

Gegenbaur, C. 1873. Ueber das Archipterygium. Jenaische Z. Med. Nat. Wiss. 7:131–141.

Ghiselin, M. T. 1966. On psychologism in the logic of taxonomic controversies. Syst. Zool. 15:207–215.

Ghiselin, M. T. 1969. The triumph of the Darwinian Method. University of California Press, Berkeley.

Ghiselin, M. T. 1974. A radical solution to the species problem. Syst. Zool. 23:536–544.

Ghiselin, M. T. 1980. Natural kinds and literary accomplishments. The Michigan Quart. Rev. 19(1):73–88.

Gillespie, J. H., and K. Kojima. 1968. The degree of polymorphism in enzymes involved in energy production compared to that in nonspecific enzymes in two *Drosophila ananassae* populations. Proc. Nat. Acad. Sci. 61:582–585.

Gilmour, J. S. L. 1940. Taxonomy and philosophy. *In*: Huxley (Ed.), The new systematics. Clarendon Press, London. Pages 461–468.

Gilmour, J. S. L. 1961. Taxonomy. *In*: Macleod and Cobley (Eds.), Contemporary botanical thought. Quadrangle Books, Chicago, Illinois.

Gingerich, P. D. 1979. The stratophenetic approach to phylogeny reconstruction in vertebrate paleontology. *In*: Cracraft and Eldredge (Eds.), Phylogenetic analysis and paleontology. Columbia University Press, New York. Pages 40–70.

Goldschmidt, R. 1940. The material basis of evolution. Yale University Press, New Haven, Connecticut.

Goldschmidt, R. 1952. Evolution as viewed by one geneticist. Amer. Sci. 40:84–98.

Good, R. 1964. The geography of the flowering plants. Longmans, Green and Co. Ltd., London.

Goodman, M. 1976. Protein sequences in phylogeny. *In*: Ayala (Ed.), Molecular evolution. Sinauer Assoc., Inc., Sunderland, Massachusetts. Pages 141–159.

Goodman, M., and G. W. Moore. 1971. Immunodiffusion systematics of the primates I. The Catarrhini. Syst. Zool. 20:19–62.

Goodman, M., G. W. Moore, J. Barnabas, and G. Matsuda. 1974. The phylogeny of human globin genes investigated by the maximum method. J. Molec. Evol. 3:1–48.

Gorman, G. C., A. C. Wilson, and M. Nakanishi. 1971. A biochemical approach towards the study of reptilian phylogeny: Evolution of serum albumin and lactic dehydrogenase. Syst. Zool. 20:167–185.

Gosline, W. A. 1949. The sensory canals in the head of some cyprinodont fishes, with particular reference to the genus *Fundulus*. Occ. Pap. Mus. Zool. Univ. Michigan 519:1–17.

Gould, S. J. 1969. An evolutionary microcosm: Pleistocene and Recent history of the land snail *P. (Poecilozonites)* in Bermuda. Bull. Mus. Comp. Zool. 138:407–531.

Gould, S. J., and N. Eldredge. 1977. Punctuated equilibria: the tempo and mode of evolution reconsidered. Paleobiology 3(2):115–151.

Gould, S. J., and R. F. Johnston. 1972. Geographic variation. Annu. Rev. Syst. Ecol. 3:452–498.

Grant, P. R. 1975. The classical case of character displacement. Evol. Biol. 8:237–337.

Grant, V. 1963. The origin of adaptations. Columbia University Press, New York.

Grant, V. 1964. Genetic and taxonomic studies in *Gilia*. XII. Fertility relationships of the polyploid Cobwebby Gilias. Aliso 5:479–507.

Grant, V. 1971. Plant Speciation. Columbia University Press, New York.

Grant, V., and A. Grant. 1960. Genetic and taxonomic studies in *Gilia*. XI. Fertility relationships of the diploid Cobwebby Gilias. Aliso 4:435–481.

Griffiths, G. C. D. 1974. Some fundamental problems in biological classification. Syst. Zool. 22:338–343.

Griffiths, G. C. D. 1976. The future of Linnean nomenclature. Syst. Zool. 25:168–173.

Haeckel, E. 1866. Generelle Morpholigie der Organismen, II. Georg. Reiner, Berlin.

Hall, W. P., and R. K. Selander. 1973. Hybridization of karyotypically differentiated populations in the *Sceloporus grammicus* complex (Iguanidae). Evolution 27:226–242.

Hanson, E. D. 1977. The origin and early evolution of animals. Wesleyan University Press, Middletown, Connecticut.

Harper, C. W., Jr. 1976. Phylogenetic inference in paleontology. J. Paleontol. 50:180–193.

Harré, R. 1972. The philosophies of science. An introductory survey. Oxford University Press, Oxford.

Harris, H. 1966. Enzyme polymorphisms in man. Proc. Royal Soc., Ser. B, 164:298–310.

Hartigen, J. A. 1973. Minimum mutation fits to a given tree. Biometrics 29:53–65.

Hays, J. D. 1970. Stratigraphy and evolutionary trends of Radiolaria in North Pacific deep-sea sediments. Geol. Soc. Amer. Mem. 126:185–218..

Heaney, L. R. 1979. Taxonomy and hybridization of Great Plains pocket gophers: A study of mammalian speciation. Ph.D. Thesis, University of Kansas.

Hedgecock, D., and F. J.' Ayala. 1974. Evolutionary divergence in the genus *Taricha* (Salamandridae). Copeia 1974:738–747.

Hempel, C. G. 1965. Aspects of scientific explanation and other essays in the philosophy of science. The Free Press, New York.

Hempel, C. G. 1966. Philosphy of natural science. Prentice-Hall, Inc., Englewood Cliffs, New Jersey.

Hennig, W. 1950. Grundzüge einer Theorie der phylogenetischen Systematik. Deutscher Zentralverlag, Berlin.

Hennig, W. 1960. Die Dipteran-Fauna von Neuseeland als systematisches un tiergeographisches Problem. Beitr. Entomol. 10:221–329. (English translation, 1966, Pacific Insects Mon. 9:1–81).

Hennig, W. 1965. Phylogenetic systematics. Annu. Rev. Entomol. 10:97–116.

Hennig, W. 1966. Phylogenetic systematics. University of Illinois Press, Urbana.

Hennig, W. 1969. Die Stammesgeschichte der Insekten. Senchenberg Naturf. Ges., Frankfurt am Main.

Hennig, W. 1975. "Cladistic analysis or cladistic classification?": A reply to Ernst Mayr. Syst. Zool. 24:244–256.

Hennings, D., and R. S. Hoffmann. 1977. A review of the taxonomy of the *Sorex vagrans* species complex from Western North America. Occas. Pap. Mus. Natur. Hist. Univ. Kansas No. 68:1–35.

Heslop-Harrison, J. 1958. The unisexual flower—a reply to criticism. Phytomorphology 8:177–184.

Ho, C. Y-K., E. M. Prager, A. C. Wilson, D. T. Osuga, and R. E. Feeney. 1976. Penguin evolution: protein comparisons demonstrate phylogenetic relationship to flying aquatic birds. J. Molec. Evol. 8:271–282.

Holland, W. J., and O. A. Peterson. 1914. The osteology of the Chalicotheroidea with special reference to a mounted skeleton of *Moropus elatus* Marsh, now installed in the Carnegie Museum. Mem. Carnegie Mus. 3:189–406.

Holmes, S. J. 1944. Recapitulation and its supposed causes. Quart. Rev. Biol. 18:319–331.

Hooker, J. D. 1853. The botany of the Antarctic voyage of H. M. Ships Erebus and Terror in the years 1839–1843. Part II. Flora Novae-Zelandiae, Vol. I. Lovell Reeve, London.

Hooker, J. D. 1860. On the origin and distribution of species.—Introductory essay on the flora of Tasmania. Amer. J. Sci. Arts, ser 2, 29:1–25, 305–326.

Hubbs, C. L. 1955. Hybridization between fish species in nature. Syst. Zool. 4:1–20.

Hubbs, C. L. 1961. Isolating mechanisms in the speciation of fishes. *In*: Blair (ed.), Vertebrate speciation. University of Texas Press, Austin. Pages 5–23.

Hubbs, C. L., and C. Hubbs. 1953. An improved graphical analysis. Syst. Zool. 2:49–56, 92.

Hubbs, C. L., L. C. Hubbs, and R. E. Johnson. 1943. Hybridization in nature between species of catostomid fishes. Contrib. Lab. Vert. Biol., Univ. Michigan No. 22:1–76.

Hubbs, C. L. (ed.). 1958. Zoogeography. Publ. Amer. Assoc. Adv. Sci. 51:1–509.

Hubby, J. L., and R. C. Lewontin. 1966. A molecular approach to the study of genetic heterozygosity in natural populations. I. The number of alleles at different loci in *Drosophila pseudoobscura*. Genetics 54:577–594.

Hull, D. L. 1964. Consistency and monophyly. Syst. Zool. 13:1–11.

Hull, D. L. 1966. Phylogenetic numericlature. Syst. Zool. 15:14–17.

Hull, D. L. 1968. The operational imperative—sense and nonsense in operationalism. Syst. Zool. 17:438–457.

Hull, D. L. 1971. Contemporary systematic philosophies. Annu. Rev. Syst. Ecol. 1:19–54.

Hull, D. L. 1976. Are species really individuals? Syst. Zool. 25:174–191.

Hull, D. L. 1980. The limits of cladism. Syst. Zool. 28:416–440.

Humphries, C. J. 1979. A revision of the genus *Anacyclus* L. (Compositae: Anthemideae). Bull. British Mus. Natur. Hist. (Bot.) 7(3):83–142.

Hutton, F. W. 1873. On the geographical relationships of the New Zealand fauna. Trans. Proc. New Zealand Inst. 5:227–256.

Huxley, J. S. 1940. Towards the new systematics. *In*: Huxley (Ed.), The new systematics. Clarendon Press, London.

Huxley, J. S. 1958. Evolutionary process and taxonomy with special reference to grades. Uppsala Univ. Arsskr. 6:21–39.

Huxley, T. H. 1868. On the classification and distribution of the Alectoromorphae and Heteromorphae. Proc. Zool. Soc. London 1868:294–319.

Jardine, N. 1967. The concept of homology in biology. British J. Philo. Sci. 18:125–139.

Jardine, N. 1969. The observational and theoretical components of homology: a study based on the morphology of the dermal skull-roofs of rhipidistian fishes. Biol. J. Linnean Soc. London 1:327–361.

Jardine, N., and C. J. Jardine. 1969. Is there a concept of homology common to several sciences? Classification Soc. Bull. 2:12–18.

Jardine, N., and R. Sibson. 1971. Mathematical taxonomy. John Wiley and Sons, London.

Jarvik, E. 1977. The systematic position of acanthodian fishes. *In*: Andrews, Miles, and Walker (Eds.), Problems in vertebrate evolution. Academic Press, London. Pages 199–225.

Jeffrey, C. 1973. Biological nomenclature. Crane, Russak & Co., New York.

Jessen, H. 1972. Schultergurtel und Pectoralflosse bei Actinopterygiern. Fossils and Strata 1:1–101.

Jessen, H. 1973. Interrelationships of actinopterygians and brachiopterygians: evidence from pectoral anatomy: *In*: Greenwood, Miles, and Patterson (eds.), Interrelationships of fishes. Academic Press, London. Pages 227–232.

Johnson, W. E., and R. K. Selander. 1971. Protein variation and systematics in kangaroo rats (genus *Dipodomys*). Syst. Zool. 20:377–405.

Johnson, W. E., R. K. Selander, M. H. Smith, and U. J. Kim. 1972. Biochemical genetics of sibling species of the cotton rat (Sigmodon). Studies in genetics VII. Univ. Texas Publ. 7213:297–305.

Johnston, R. F. 1973. Evolution in the House Sparrow. IV. Replicate studies in phenetic covariation. Syst. Zool. 22:219–226.

Johnston, R. F., and R. K. Selander. 1971. Evolution in the House Sparrow. II. Adaptive differentiation in North American populations. Evolution 25(1):1–28.

Jolicoeur, P., and J. E. Mosimann. 1960. Size and shape variation in the painted turtle, a principal component analysis. Growth 24:339–354.

Jollie, M. 1973. Chordate morphology, Robert Krieger Publishing Co., Inc., Huntington, New York.

Jordan, D. S. 1905. The origin of species through isolation. Science 22:545–562.

Jordan, K. 1896. On mechanical selection and other problems. Novit. Zool. 3:426–525.

Jordan, K. 1905. Der Gegensatz zwichen geographischer und nicht-geographischer variation. Z. wiss. Zool. 83:151–210.

Jussieu, A. L. de. 1789. *Genera Plantarum*. Paris.

Kavanaugh, D. H. 1972. Hennig's principles and methods of phylogenetic systematics. The Biologist 54:115–127.

Keast, J. A. 1977. Zoogeography and phylogeny: The theoretical background and methodology to the analysis of mammal and bird faunas. *In*: Hecht, Goody, and Hecht (Eds.), Major patterns in vertebrate evolution. Plenum Press, New York. Pages 249–312.

Key, K. H. L. 1968. The concept of stasipatric speciation. Syst. Zool. 17:14–22.

King, M. C., and A. C. Wilson. 1975. Evolution at two levels in humans and chimpanzees. Science 188:107–116.

Kiriakoff, S. G. 1959. Phylogenetic systematics versus typology. Syst. Zool. 8:117–118.

Kluge, A. G., and J. S. Farris. 1969. Quantitative phyletics and the evolution of anurans. Syst. Zool. 18:1–32.

Kluge, A. G., and J. S. Farris. 1979. A botanical clique. Syst. Zool. 28:400–411.

Kohne, D. E., J. A. Chicson, and B. H. Hoyer. 1972. Evolution of primate DNA sequences. J. Human Evol. 1:627–644.

Kojima, K., J. Gillespie, and U. M. Tobari. 1970. A profile of *Drosophila* species' enzymes assayed by electrophoresis. I. Number of alleles, heterozygosities, and linkage disequilibrium in glucose-metabolizing systems and other enzymes. Biochem. Genet. 4:627–637.

Koponen, T. 1968. Generic revision of Mniaceae Mitt. (Bryophyta). Ann. Bot. Fenn. 5:117–151.

Koponen, T. 1973. *Rhizomnium* (Mniaceae) in North America. Ann. Bot. Fenn. 10:1–28.

Lawrence, G. J. M. 1951. Taxonomy of vascular plants. The Macmillan Co., New York.

Leenhouts, P. W. 1966. Keys in biology: a survey and a proposal of a new kind. Proc. Konin. Nederlandse Akad. Van Wetenschappen 69 (ser. C):571–596.

LeQuesne, W. J. 1972. Further studies based on the uniquely derived character concept. Syst. Zool. 21:281–288.

Lewontin, R. C. 1974. The genetic basis of evolutionary change. Columbia University Press, New York.

Lewontin, R. C., and J. L. Hubby. 1966. A molecular approach to the study of genetic heterozygosity in natural populations. II. Amount of variation and degree of heterozygosity in natural populations of *Drosophila pseudoobscura*. Genetics 54:595–609.

Løvtrup, S. 1974. Epigenetics, a treatise on theoretical biology. John Wiley and Sons, New York.

Løvtrup, S. 1977. Phylogeny of vertebrata. John Wiley and Sons, Inc. New York.

Løvtrup, S. 1978. On von Baerian and Haeckelian recapitulation. Syst. Zool. 27:348–352.

Løvtrup, S. 1979. The evolutionary species. Fact or Fiction? Syst. Zool. 28:386–392.

MacArthur, R. H. 1972. Geographical ecology. Harper and Row, New York.

MacArthur, R. H., and E. O. Wilson. 1967. The theory of island biogeography. Princeton University Press, Princeton, New Jersey.

McCoy, E. D., and K. L. Heck, Jr. 1976. Biogeography of corals, seagrasses, and mangroves: An alternate to the center of origin concept. Syst. Zool. 25:201–210.

McKenna, M. C. 1975. Toward a phylogenetic classification of the Mammalia. *In*: Luckett and Szalay (Eds), Phylogeny of the primates. Plenum Press, New York. Pages 21–46.

Maren, T. H. 1967. Carbonic anhydrase: chemistry, physiology, and inhibition. Physiol. Rev. 47:595–781.

Margulis, L. 1970. Origin of eukaryotic cells. Yale University Press, New Haven, Connecticut.

Marquardt, R. W., C. W. Carlson, and R. W. Brosemer. 1968. Glyceraldehyde phosphate dehydrogenase: crystallization from honeybees; quantitative immunochemical and electrophoretic comparisons of the enzyme in other insects. J. Insect Physiol. 14:317–333.

Marschall, A. 1873. *Nomenclator Zoologicus*. Vindobonal.

Matthew, W. D. 1915. Climate and evolution. Ann. New York Acad. Sci. 24:171–318.

Maxson, L. R. 1977. Immunological detection of convergent evolution in the frog *Anotheca spinosa* (Hylidae). Syst. Zool. 26:72–76.

Maxson, L. R., and A. C. Wilson. 1974. Convergent morphological evolution detected by studying proteins of tree frogs in the *Hyla eximia* species group. Science 185:66–68.

Maxson, L. R., and A. C. Wilson. 1975. Albumin evolution and organismal evolution in tree frogs (Hylidae). Syst. Zool. 24:1–15.

Maxson, L. R., V. M. Sarich, and A. C. Wilson. 1975. Continental drift and the use of albumin as an evolutionary clock. Nature 255:397–400.

Mayr, E. 1942. Systematics and the origin of species. Columbia University Press, New York.

Mayr, E. 1949. Speciation and evolution. *In*: Jepsen, Mayr, and Simpson (Eds.), Genetics, paleontology, and evolution. Princeton University Press, Princeton, New Jersey. Pages 281–298.

Mayr (Ed.). 1952. The problem of land connections across the South Atlantic, with special reference to the Mesozoic. Bull Amer. Mus. Natur. Hist. 99:79–258.

Mayr, E. 1957. Species concepts and definitions. *In*: Mayr (Ed.), The species problem. Amer. Assoc. Adv. Sci. Publ. 50:1–22.

Mayr, E. 1963. Animal species and evolution. Harvard University Press, Cambridge, Massachusetts.

Mayr, E. 1969. Principles of systematic zoology. McGraw-Hill, New York.

Mayr, E. 1970. Populations, species and evolution. Belknap Press of Harvard University, Cambridge, Massachusetts.

Mayr, E. 1974. Cladistic analysis or cladistic classification? Z. Zool. Syst. Evolut.-forsch. 12:94–128.

Mayr, E. 1976a. Is the species a class or an individual? Syst. Zool. 25:192.

Mayr, E. 1976b. Evolution and the diversity of life. Belknap Press of Harvard University, Cambridge, Massachusetts.

Mayr, E., and L. L. Short. 1970. Species taxa of North American birds—A contribution to comparative systematics. Publ. Nuttal Ornith. Club, No. 9:1–127.

Mayr, E., E. G. Linsley, and R. L. Usinger. 1953. Methods and principles of systematic zoology. McGraw-Hill Book Company, Inc., New York.

Mecham, J. S. 1960. Introgressive hybridization between two southeastern treefrogs. Evolution 14:445–457.

Mecham, J. S. 1975. Incidence and significance of introgressive hybridization in anuran amphibians. Amer. Zool. 15:831.

Meglitsch, P. A. 1954. On the nature of the species. Syst. Zool. 3:49–65.

Merxmüller, H., P. Leins, and H. Roessler. 1977. Inuleae—systematic review. *In*: Heywood et al. (Eds.), The biology and chemistry of the Compositae 1:577–602.

Metcalf, Z. P. 1954. The construction of keys. Syst. Zool. 3:38–45.

Michener, C. D. 1963. Some future developments in taxonomy. Syst. Zool. 12:151–172.

Michener, C. D. 1970. Diverse approaches to systematics. *In*: Dobzhansky, Hecht, and Steere (Eds.), Evolutionary biology, vol. 4. Appleton-Century-Crofts, New York. Pages 1–38.

Michener, C. D. 1977. Discordant evolution and the classification of allodapine bees. Syst. Zool. 26:32–56.

Michener, C. D. 1978. Dr. Nelson on taxonomic methods. Syst. Zool. 27:112–118.

Mickevich, M. F. 1978. Taxonomic congruence. Syst. Zool. 27:143–158.

Mickevich, M. F., and M. S. Johnson. 1976. Congruence between morphological and allozyme data in evolutionary inference and character evolution. Syst. Zool. 25:260–270.

Miles, R. S. 1977. Dipnoan (lungfish) skulls and the relationships of the group: a study based on new species from the Devonian of Australia. Zool. J. Linnaean Soc. 61:1–328.

Mill, J. S. 1879. A system of logic. Longmans Green, London.

Milne, M. J., and L. J. Milne. 1939. Evolutionary trends in caddis worm case construction. Ann. Entomol. Soc. Amer. 32:533–542.

Minkoff, E. C. 1965. The effects on classification of slight alterations in numerical technique. Syst. Zool. 14:196–213.

Moore, G. W., J. Barnabas, and M. Goodman. 1973. A method for constructing maximum parsimony ancestral amino acid sequences on a given network. J. Theor. Biol. 38:459–485.

Naef, A. 1931. Die Gestalt als Begriff und Idee. *In*: Bolk et al. (Eds.), Handbuch der vergleichender Anatomie der Wirbeltiere, Vol. 1. Urban and Schwarzenberg Verlag, Berlin. Pages 77–118.

Naumann, C. M. 1977. Studies on the systematics and phylogeny of Holarctic Sesiidae (Insecta, Lepidoptera). Amerind Publishing Co., New Delhi (translated from Bonner Zoolog. Monogr. 1, 1971).

Neave, S. A. 1939–1940, 1950. *Nomenclator Zoologicus*, 4 vol. + suppl. vol., London.

Neff, N. A., and L. F. Marcus. 1980. A survey of multivariate methods for systematics. Published by the authors, New York. (Inquiries for obtaining copies should be directed to Marcus, Dept. of Invertebrates American Museum of Nat. Hist., CPW at 79th street, New York, NY 10024).

Neff, N. A., and G. R. Smith. 1979. Multivariate analysis of hybrid fishes. Syst. Zool. 28:176–196.

Nei, M. 1975. Molecular population genetics and evolution. North-Holland, Amsterdam and Elsevier, Inc., New York.

Nelson, G. J. 1967. Branchial muscles in some generalized teleostean fishes. Acta Zool. 48:277–288.

Nelson, G. J. 1969. Gill arches and the phylogeny of fishes with notes on the classification of vertebrates. Bull. Amer. Mus. Natur. Hist. 141:475–552.

Nelson, G. J. 1970. Outline of a theory of comparative biology. Syst. Zool. 19:373–384.

Nelson, G. J. 1971. Paraphyly and polyphyly: redefinitions. Syst. Zool. 20:471–472.

Nelson, G. J. 1972a. Phylogenetic relationship and classification. Syst. Zool. 21:227–231.

Nelson, G. J. 1972b. Comments on Hennig's "Phylogenetic Systematics" and its influence on ichthyology. Syst. Zool. 21:364–374.

Nelson, G. J. 1973a. Notes on the structure and relationships of certain Cretaceous and Eocene teleostean fishes. Amer. Mus. Novit. 2524:1–31.

Nelson, G. J. 1973b. "Monophyly again?"—a reply to P. H. Ashlock. Syst. Zool. 22:310–312.

Nelson, G. J. 1973c. Comments on Leon Croizat's biogeography. Syst. Zool. 22:312–320.

Nelson, G. J. 1974. Classification as an expression of phylogenetic relationships. Syst. Zool. 22:344–359.

Nelson, G. J. 1975. Historical biogeography: An alternative formulation. Syst. Zool. 23:555–558.

Nelson, G. J. 1978a. Ontogeny, phylogeny, and the biogenetic law. Syst. Zool. 27:324–345.

Nelson, G. J. 1978b. From Candolle to Croizat: Comments on the history of biogeography. J. Hist. Biol. 11:293–329.

Nelson, G. J. 1979. Cladistic analysis and synthesis: Principles and definitions, with a historical note on Adanson's *Familles des Plantes* (1763–1764). Syst. Zool. 28:1–21.

Nelson, G. J. ms. Cladograms and trees.

Nonno, L. H., H. Hershman, and L. Levine. 1970. Serologic comparisons of the carbonic anhydrases of primate erythrocytes. Arch. Biochem. Biophys. 136:316–367.

O'Farrell, P. H. 1975. High resolution two-dimensional electrophoresis of proteins. J. Biol. Chem. 250:4007–4021.

Ohno, S. 1970. Evolution by gene duplication. Springer Verlag, Berlin.

Owen, R. 1848. Report on the archetype and homologies of the vertebrate skeleton. Reports 16th Meeting, British Assoc. Adv. Sci. Pages 169–340.

Palmer, R. S. (Ed.). 1976. Handbook of North American birds. Vol. 2. Yale University Press, New Haven, Connecticut.

Pankhurst, R. J. 1978. Biological identification. The principles and practice of identification methods in biology. University Park Press, Baltimore, Maryland.

Papp, C. S. 1976. Manual of scientific illustration. American Visual Aid Books, Sacramento, California.

Patterson, B. 1949. Rates of evolution in taeniodonts. *In*: Jepsen, Simpson, and Mayr (Ed.), Genetics, paleontology, and evolution. Princeton University Press, Princeton, New Jersey. Pages 243–278.

Patterson, C. 1973. Interrelationships of holosteans. *In*: Greenwood, Miles, and Patterson (Eds.), Interrelationships of fishes. Academic Press, London. Pages 233–305.

Patterson, C. 1977. The contribution of paleontology to teleostean phylogeny. *In*: Hecht, Goody, and Hecht (Eds.), Major patterns in vertebrate evolution. Plenum Press, New York. Pages 579–643.

Patterson, C. 1978. Verifiability in systematics. Syst. Zool. 27:218–222.

Patterson, C., and D. E. Rosen. 1977. Review of ichthyodectiform and other Mesozoic teleost fishes and the theory and practice of classifying fossils. Bull. Am. Mus. Natur. Hist. 158:172.

Patterson, J. T., and W. S. Stone. 1952. Evolution in the genus *Drosophila*. Macmillan, New York.

Patton, J. L., R. K. Selander, and M. H. Smith. 1972. Genetic variation in hybridizing populations of gophers (genus *Thomomys*). Syst. Zool. 21:263–270.

Pfeiffer, L. 1873–1874. *Nomenclator Botanicus*. Kassel.

Pimentel, R. A. 1979. Morphometrics, the multivariate analysis of biological data. Kendall/Hunt Publishing Co., Dubuque, Iowa.

Platnick, N. I. 1976a. Drifting spiders or continents? Vicariance biogeography of the spider subfamily Laroniinae (Araneae: Gnaphosidae). Syst. Zool. 25:101–109.

Platnick, N. I. 1976b. Are monotypic genera possible? Syst. Zool. 25:198–199.

Platnick, N. I. 1976c. Concepts of dispersal in historical biogeography. Syst. Zool. 25:294–295.

Platnick, N. I. 1977a. Paraphyletic and polyphyletic groups. Syst. Zool. 26:195–200.

Platnick, N. I. 1977b. Cladograms, phylogenetic trees, and hypothesis testing. Syst. Zool. 26:438–442.

Platnick, N. I. 1977c. The hypochiloid spiders: a cladistic analysis, with notes on the Atypoidea (Arachnida, Araneae). Amer. Mus. Novitates 2627:1–23.

Platnick, N. I. 1977d. Parallelism in phylogeny reconstruction. Syst. Zool. 26:93–96.

Platnick, N. I., and G. J. Nelson. 1978. A model of analysis for historical biogeography. Syst. Zool. 27:1–16.

Platnick, N. I., and M. U. Shadab. 1976. A revision of the mygalomorph spider genus *Neocteniza* (Araneae, Actinopodidae). Amer. Mus. Novit. 2603:1–19.

Poche, F. 1938. Supplement Zu C. D. Sherborn's *Index Animalium*. Festsch. 60. Geburtstage E. Strand 5:477–615.

Popper, K. R. 1968a. The logic of scientific discovery. Harper Torchbooks, New York.

Popper, K. R. 1968b. Conjectures and refutations. Harper Torchbooks, New York.

Prager, E. M., and A. C. Wilson. 1971. The dependence of immunological crossreactivity upon sequence resemblance among bird lysozymes: Microcomplement fixation studies. J. Biol. Chem. 246:5978–5989.

Prager, E. M., A. C. Wilson, D. T. Osuga, and R. E. Geeney. 1976. Evolution of flightless birds on southern continents. Transferrin comparisons shows monophyletic origin of ratites. J. Molec. Evol. 8:283–294.

Prager, E. M., A. H. Brush, R. A. Nolan, M. Nakanishi, and A. C. Wilson. 1974. Slow evolution of transferrins and albumin in birds according to microcomplement fixation analysis. J. Molec. Evol. 3:243–262.

Prothero, D. R., and D. B. Lazarus. 1980. Planktonic microfossils and the recognition of ancestors. Syst. Zool. 28:119–129.

Radford, A. E., W. C. Dickison, J. R. Massey, C. R. Bell. et al. 1974. Vascular plant systematics. Harper and Row, New York.

Ratti, J. T. 1979. Reproductive separation and isolating mechanisms between sympatric dark- and light-phase Western Grebes. The Auk 96:573–586.

Raven, P., and D. I. Axelrod. 1974. Angiosperm biogeography and past continental movements. Ann. Missouri Bot. Garden 61:539–673.

Rayner, D. H. 1948. The structure of certain Jurassic holostean fishes with special reference to their neurocrania. Philo. Trans. Royal Soc., 233B:287–345.

Read, D. W. 1975. Primate phylogeny, neutral mutations and "molecular clocks." Syst. Zool. 24:209–221.

Remane, A. 1956. Die Grundlagen des naturlichen Systems der vergleichenden Anatomie und Phylogenetik. 2. Geest und Portig K. G., Leipzig.

Richardson, R. H. 1974. Effects of dispersal, habitat selection, and competition on a speciation pattern of *Drosophila* endemic to Hawaii. *In*: White (Ed.), Genetic mechanisms of speciation in insects. Australia and New Zealand Book Co., Sydney. Pages 140–164.

Riedl, R. 1978. Order in living organisms. A systems analysis of evolution. John Wiley and Sons, New York.

Rindge, F. H. 1976. A revision of the moth genus *Plataea* (Lepidoptera, Geometridae). Amer. Mus. Novit. 2595:1–42.

Rivas, L. R. 1964. A reinterpretation of the concepts "sympatric" and "allopatric" with the proposal of the additional terms "syntopic" and "allotopic." Syst. Zool. 13:42–43.

Roa, C. R. 1964. The use and interpretation of principal component analysis in applied research. Sankhaya, ser. A, 26:329–358.

Romanes, G. J. 1897. Darwin and after Darwin, vol. 3. Open Court Publishing Co., Chicago, Illinois. Pages 41–100.

Romer, A. S. 1966. Vertebrate paleontology. University of Chicago Press, Chicago, Illinois.

Rosa, J. 1918. Ologenesi. Bemparad, Firenze.

Rosen, D. E. 1973. Interrelationships of higher euteleostean fishes. In: Greenwood, Miles, and Patterson (Eds.), Interrelationships of fishes. Academic Press, London. Pages 397–513.

Rosen, D. E. 1974. Phylogeny and zoogeography of salmoniform fishes. Bull. Amer. Mus. Natur. Hist. 153:265–326.

Rosen, D. E. 1975. A vicariance model of Caribbean biogeography. Syst. Zool. 24:431–464.

Rosen, D. E. 1978. Vicariant patterns and historical explanation in biogeography. Syst. Zool. 27:159–188.

Rosen, D. E. 1979. Fishes from the uplands and intermontane basins of Guatemala: revisionary studies and comparative biogeography. Bull. Amer. Mus. Natur. Hist. 162(5):267–376.

Ross, H. H. 1974. Biological systematics. Addison-Wesley Publication Co., Inc., Reading, Massachusetts.

Sarich, V. M., and J. E. Cronin. 1974. Primate evolution at higher taxonomic levels: a molecular view. Amer. J. Phys. Anthro. 41:502.

Sarich, V. M., and A. C. Wilson. 1966. Quantitive immunochemistry and the evolution of primate albumins: Microcomplement fixation. Science 154:1563–1566.

Sarich, V. M., and A. C. Wilson. 1967a. Immunological time scale for hominid evolution. Science 158:1200–1203.

Sarich, V. M., and A. C. Wilson. 1967b. Rates of albumin evolution in primates. Proc. Nat. Acad. Sci. 58:142–148.

Schaeffer, B. 1967. Osteichthyan vertebrae. J. Linnean Soc. London 47:185–195.

Schaeffer, B., M. K. Hecht, and N. Eldredge. 1972. Paleontology and phylogeny. Evol. Biol. 6:31–46.

Schindewolf, O. H. 1950. Grundfragen der Palaontologie. Schweizerbart, Stattgart.

Schlee, D. 1968. Hennig's principle of phylogenetic systematics, an "intuitive, statistico-phenetic taxonomy"? Syst. Zool. 18:127–134.

Schlee, D. 1971. Die Rekonstrucktion der Phylogenese mit Hennig's Prinzip. Waldemar Kramer, Frankfurt am Main.

Schuh, R. T. 1976. Pretarsal structure in the Miridae (Hemiptera) with a cladistic analysis of the relationships of the family. Amer. Mus. Novit. 2601:1–39.

Schulze, F. E., W. Kukenthal, and K. Heider, 1926–1954. Nomenclator Animalium—Generum et Subgenerum, 25 Lief., 5 vols. Berlin.

Schwaner, T. D., and R. H. Mount. 1976. Systematics and ecological relationships of the water snakes Natrix sipedon and N. fasciata in Alabama and the Florida Panhandle. Occas. Pap. Mus. Natur. Hist. Univ. Kansas 45:1–44.

Schwaner, T. D., H. C. Dessauer, and L. A. Landry, Jr. 1980. Genetic divergence of Nerodia sipedon and N. fasciata in south Louisiana. Isozyme Bull. 13.

Schwartz, J. H., I. Tattersall, and N. Eldredge. 1978. Phylogeny and classification of the primates revisited. Yearbook Phys. Anthro. 21:95–133.

Sclater, P. L. 1858. On the general geographical distribution of the members of the class Aves. J. Proc. Linnean Soc. London (Zool). 2:130–145.

Scudder, S. J. 1882. Nomenclator Zoologicus. U.S. Nat. Mus. Bull. 19. Washington, D.C.

Selander, R. K., W. G. Hunt, and S. Y. Yang. 1969. Protein polymorphism and genetic heterozygosity in two European subspecies of the house mouse. Evolution 23:379–390.

Sherborn, C. D. 1902. Index Animalium. . . , Sect. primo. Cantabrigial.

Sherborn, C. D. 1922–1933. *Index Animalium.* . . , Sect. Secunda. London.

Simpson, G. G. 1944. Tempo and mode in evolution. Columbia University Press, New York.

Simpson, G. G. 1945. The principles of classification and a classification of mammals. Bull. Amer. Mus. Natur. Hist. 85:1–350.

Simpson, G. G. 1947. Holarctic mammalian faunas and continental relationships during the Cenozoic. Bull. Geol. Soc. Amer. 58:613–688.

Simpson, G. G. 1951. The species concept. Evolution. 5:285–298.

Simpson, G. G. 1953. The major features of evolution. Columbia University Press, New York.

Simpson, G. G. 1960. The history of life. *In*: Tax (Ed.), The evolution of life. The University of Chicago Press, Chicago, Illinois. Pages 117–180.

Simpson, G. G. 1961. Principles of animal taxonomy. Columbia University Press, New York.

Simpson, G. G. 1965. The geography of evolution, collected essays. Chilton Books, Philadelphia.

Simpson, G. G. 1975. Recent advances in methods of phylogenetic inference. *In*: Luckett and Szalay (Eds.), Phylogeny of the primates, a multi-disciplinary approach. Plenum Press, New York. Pages 3–19.

Simpson, G. G., A. Roe, and R. C. Lewontin. 1960. Quantitative zoology. Harcourt, Brace and World, New York.

Slobodchikoff, C. N. (Ed.). 1976. Concepts of species. University of California Press, Berkeley.

Smith, G. R. 1973. Analysis of several hybrid cyprinid fishes from western North America. Copeia 1973:395–416.

Smith, J. D. 1972. Systematics of the chiropteran family Mormopidae. Misc. Publ. Mus. Natur. Hist. Univ. Kansas 56:1–132.

Smith, J. M. 1966. Sympatric speciation. Amer. Natur. 100:637–650.

Sneath, P. H. A. 1961. Recent developments in theoretical and quantitative taxonomy. Syst. Zool. 10:118–139.

Sneath, P. H. A., and R. R. Sokal. 1973. Numerical taxonomy. W. H. Freeman, San Francisco, California.

Sober, E. 1975. Simplicity. Clarendon Press, Oxford.

Sokal, R. R. 1962. Typology and empiricism in taxonomy. J. Theoret. Biol. 3:230–267.

Sokal, R. R. 1973. The species problem reconsidered. Syst. Zool. 22:360–374.

Sokal, R. R., and T. J. Crovello. 1970. The biological species concept: a critical evaluation. Amer. Natur. 104:127–153.

Sokal, R. R., and P. H. A. Sneath. 1963. The principles of numerical taxonomy. W. H. Freeman and Co., San Francisco, California.

Stafleu, F. A. 1967. Taxonomic literature. A selective guide to botanical publications with dates, commentaries, and types. Regnum Veg. 52. Utrecht.

Stafleu, F. A., and R. S. Cowan. 1976, 1979. Taxonomic literature, vols. I, II. Regnum Veg. 94, 98. Utrecht.

Staniland, L. N. 1953. The principles of line illustration. Burke Publishing Co., London.

Stearn, W. T. 1956. Keys; Shapes of leaves, & c. *In*: Chittenden (Ed.), Royal horticultural dictionary of gardening, Supplement:251–253, 318–322.

Stearn, W. T. 1973. Botanical Latin. Thomas Nelson and Sons, London.

Stebbins, G. L., Jr. 1950. Variation and evolution in plants. Columbia University Press, New York.

Stebbins, G. L., Jr. 1971. Chromosomal evolution in higher plants. Addison-Wesley Publishing Co., Inc. Reading, Massachusetts.

Steudel, E. G. 1821–1841. *Nomenclator Botanicus. . .* , Stuttgart.

Storer, R. W. 1971. Classification of birds. *In:* Farner and King (Eds.), Avian biology, Vol. 1. Academic Press, New York. Pages 1–18.

Strickland, H. E. 1842. Rules for zoological nomenclature. Rept. 12th meeting, British Assoc. Adv. Sci. Rpt. 1842:105–121.

Takhtajan, A. 1969. Flowering plants, origin and dispersal. Smithsonian Institution Press, Washington, D. C.

Tashian, R. E., R. J. Tanis, R. E. Ferrell, S. K. Stroup, and M. Goodman. 1972. Differential rates of evolution in the carbonic anhydrase isozymes in catarrhine primates. J. Molec. Evol. 1:545–552.

Tattersall, I., and N. Eldredge. 1977. Fact, theory, and fantasy in human paleontology. Amer. Sci. 65:204–211.

Temple, J. T. 1968. The Lower Llandovery (Silurian) brachiopods from Keisley, Westmorland. Palaeontol. Soc. Publ. No. 521.

Thoday, J. M., and T. B. Boam. 1959. Effects of disruptive selection. II. Polymorphism and divergence without isolation. Heredity 13:205–218.

Thomerson, J. E. 1969. Variation and relationships of the studfishes, *Fundulus catenatus* and *F. stellifer* (Cyprinodontidae, Pisces). Tulane Stud. Zool. Bot. 16(1):1–21.

Thompson, D. A. W. 1942. On growth and form. MacMillan Co., New York.

Throckmorton, L. H. 1978. Molecular phylogenetics. *In:* Romberger et al. (Eds.), Beltsville Symposia in agricultural research—(2) Biosystematics in agriculture. Allanheld, Osmun and Co., Montclair, New Jersey. Pages 221–239.

Tralau, H. (Ed.). 1969, 1972. *Index Holmensis,* a world phytyogeographic index. Vols 1, 2. Scientific Publishers, Ltd., Zurich.

Tuomikoski, R. 1967. Notes on some principles of phylogenetic systematics. Ann. Entomol. Fenn. 33:137–147.

Turner, B. J. 1974. Genetic divergence of Death Valley pupfish species: biochemical versus morphological evidence. Evolution 28:281–294.

Udvardy, M. D. F. 1969. Dynamic zoogeography with special reference to land animals. Van Nostrand and Reinhold Co., New York.

Van Valen, L. 1971. Adaptive zones and orders of mammals. Evolution 25:420–428.

Vari, R. P. 1978. The terapon perches (Percoidei, Teraponidae). A cladistic analysis and taxonomic revision. Bull. Amer. Mus. Natur. Hist. 159(5):175–340.

Vendler, Z. 1975. Possibility of possible worlds. Canadian J. Phil. 5:57–72.

von Baer, K. E. 1828, 1837. Ueber Entwickelungsgeschichte der Thiere: Beobachtung und Reflection. Gebrüder Kornträger, Königsberg.

Voss, E. G. 1952. The history of keys and phylogenetic trees in systematic biology. J. Sci. Lab., Denison Univ. 43:1–25.

Vuilleumeir, F. 1975. Zoogeography. *In:* Farner and Kind (Eds.), Avian biology, vol. 5. Academic Press, New York. Pages 421–496.

Waddington, C. H. 1957. The strategy of the genes. Allen and Unwin, London.

Wagner, M. 1889. Die Entstehung der Arten durch raumliche Sonderung. Gesaminelte Aufsatze, Benno Schwalbe, Basel.

Wagner, W. J., Jr. 1961. Problems in the classification of ferns. *In*: Recent advances in botany. University of Toronto Press, Montreal. Pages 841–844.

Wallace, A. R. 1855. On the law which has regulated the introduction of new species. Ann. Mag. Natur. Hist. 16, 2:184–196.

Wallace, A. R. 1876. The geographical distribution of animals. 2 vols. MacMillan and Co., London.

Wallace, D. G., M. C. King, and A. C. Wilson. 1973. Albumin differences among ranid frogs: taxonomic and phylogenetic implication. Syst. Zool. 22:1–13.

Waterhouse, C. O. 1902. *Index Zoologicus*. London.

Waterhouse, C. O. 1912. *Index Zoologicus*, No. II. London.

Webster, T. P., R. K. Selander, and S. Y. Yang. 1972. Genetic variability and similarity in the *Anolis* lizards of Bimini. Evolution 26:523–535.

Wernham, H. F. 1912. Floral evolution: with particular reference to the sympetalous dicotyledons; IX. Summary and conclusions. New Phytol. 11:272–397.

Wetmore, A. 1951. A revised classification for the birds of the world. Smithsonian Misc. Coll. 117(4):1–22.

Wetmore, A. 1960. A classification for the birds :or the world. Smithsonian Misc. Coll. 139(11):1–37.

White, M. J. D. 1973. Animal cytology and evolution (3rd Ed). Cambridge University Press.

White, M. J. D. 1978a. Modes of speciation. W. H. Freeman and Co., San Francisco, California.

White, M. J. D. 1978b. Chain processes in chromosomal speciation. Syst. Zool. 27:285–298.

Wijk, R. van der, W. D. Margadant, and P. A. Florschutz. 1959–1969. *Index muscorum*. Regnum Veg. 17, 26, 33, 48, 65. Utrecht.

Wiley, E. O. 1975. Karl R. Popper, systematics, and classification—a reply to Walter Bock and other evolutionary taxonomists. Syst. Zool. 24:233–243.

Wiley, E. O. 1976. The phylogeny and biogeography of fossil and Recent gars (Actinopterygii: Lepisosteidae). Misc. Publ. Mus. Natur. Hist. Univ. Kansas 64:1–111.

Wiley, E. O. 1977a. Are monotypic genera paraphyletic?—a response to Norman Platnick. Syst. Zool. 26:352–355.

Wiley, E. O. 1977b. The phylogeny and systematics of the *Fundulus nottii* species group. (Teleostei: Cyprinodontidae). Occ. Pap. Mus. Natur. Hist. Univ. Kansas. 66:1–31.

Wiley, E. O. 1978. The evolutionary species concept reconsidered. Syst. Zool. 27:17–26.

Wiley, E. O. 1979a. Cladograms and phylogenetic trees. Syst. Zool. 28:88–92.

Wiley, E. O. 1979b. Ancestors, species, and cladograms.—Remarks on the symposium *In*: Cracraft and Eldredge (Eds.), Phylogenetic analysis and paleontology. Columbia University Press, New York. Pages 211–225.

Wiley, E. O. 1979c. An annotated Linnean hierarchy, with comments on natural taxa and competing systems. Syst. Zool. 28:308–337.

Wiley, E. O. 1979d. Ventral gill arch muscles and the interrelationships of gnathostomes, with a new classification of the Vertebrata. J. Linnaean Soc. 67:149–180.

Wiley, E. O. 1980. Is the evolutionary species fiction?—A consideration of classes, individuals, and historical entities. Syst. Zool. 29:76–80.

Wiley, E. O., III, and D. D. Hall. 1975. *Fundulus blairae*, a new species of the *Fundulus nottii* complex (Teleostei: Cyprinodontidae). Amer. Mus. Novit. 2577:1–13.

Willis, J. C., and H. K. A. Shaw. 1973. A dictionary of the flowering plants and ferns. Cambridge University Press.

Wilson, A. 1975. Relative rates of evolution of organisms and genes. Stadler Genet. Symp. 7:117–134.

Wilson, A. C., and V. M. Sarich. 1969. A molecular time scale for human evolution. Proc. U.S. Nat. Acad. Sci. 63:1088–1093.

Wilson, E. O. 1971. The insect societies. Belknap Press of Harvard University Press, Cambridge, Massachusetts.

Wood, P. 1979. Scientific illustration. Van Nostrand Reinhold Company, New York.

Wooding, G. L., and R. F. Doolittle. 1972. Primate fibrino peptides: evolutionary significance. J. Human Evol. 1:553–563.

Wright, S. 1955. Classification of factors of evolution. Cold Spring Harbor Symp. Quant. Biol. 20:16–24.

Wulff, E. V. 1950. An introduction to historical plant biogeography. Chronica Botanica Company, Waltham, Massachusetts (trans. of the 1932 Russian text).

Yates, T. L., R. J. Baker, and R. K. Barnett. 1979. Phylogenetic analysis of karyological variation in three genera of peromyscine. Syst. Zool. 28:40–48.

Zangerl, R., and G. R. Case. 1976. *Cobelodus aculeatus* (Cope), an anacanthous shark from Pennsylvanian black shales of North America. Palaeotogr., Abt. A, Bd. 154:107–157.

Zweifel, F. W. 1961. A handbook of biological illustration. Phoenix Books, University of Chicago Press, Illinois.

Index

Italicized page numbers indicate definitions or major discussion of the term.

421